Life Cycle Management in Supply Chains:

Identifying Innovations Through the Case of the VCR

Toru Higuchi
Sakushingakuin University, Japan

Marvin Troutt
Kent State University, USA

IGI PUBLISHING
Hershey • New York

Acquisition Editor:	Kristin Klinger
Development Editor:	Kristin Roth
Senior Managing Editor:	Jennifer Neidig
Managing Editor:	Jamie Snavely
Assistant Managing Editor:	Carole Coulson
Copy Editor:	Jennifer Young
Typesetter:	Michael Brehm
Cover Design:	Lisa Tosheff
Printed at:	Yurchak Printing Inc.

Published in the United States of America by
IGI Publishing (an imprint of IGI Global)
701 E. Chocolate Avenue
Hershey PA 17033
Tel: 717-533-8845
Fax: 717-533-8661
E-mail: cust@igi-global.com
Web site: http://www.igi-global.com

and in the United Kingdom by
IGI Publishing (an imprint of IGI Global)
3 Henrietta Street
Covent Garden
London WC2E 8LU
Tel: 44 20 7240 0856
Fax: 44 20 7379 0609
Web site: http:/www.eurospanbookstore.com

Copyright © 2008 by IGI Global. All rights reserved. No part of this book may be reproduced in any form or by any means, electronic or mechanical, including photocopying, without written permission from the publisher.

Product or company names used in this book are for identification purposes only. Inclusion of the names of the products or companies does not indicate a claim of ownership by IGI Global of the trademark or registered trademark.

Library of Congress Cataloging-in-Publication Data

Higuchi, Toru.
Life cycle management in supply chains emerging technologies and techniques / by Toru Higuchi and Marvin Troutt.
p. cm.
Summary: "This book presents comprehensive, in-depth coverage of the intimate connection between the industry life cycle and supply chain management. It provides practitioners and researchers with insight into the supply chain as the basic business unit for competition, and the requisite alteration of the management of the supply chain at each stage of the life cycle"--Provided by publisher.
Includes bibliographical references.
ISBN 978-1-59904-555-9 (hardcover) -- ISBN 978-1-59904-557-3 (e-book)
1. Business logistics. I. Troutt, Marvin. II. Title.

HD38.5.H54 2008
658.5--dc22

2007040411

British Cataloguing in Publication Data
A Cataloguing in Publication record for this book is available from the British Library.

All work contributed to this book is new, previously-unpublished material. The views expressed in this book are those of the authors, but not necessarily of the publisher.

Life Cycle Management in Supply Chains:

Identifying Innovations Through the Case of the VCR

Table of Contents

Section III:
Consumer Aspects

Section IV:
Location Aspects

Foreword

A new and innovative product, if promising for commercialization, can open up an entire industry with associated supply chains. The product will evolve through various versions in response to market forces. Production processes and facility location decisions also must evolve in appropriate and predictable ways. These events have important implications for supply chain decisions. The overall history of such a product class can be called the *Industrial Life Cycle*.

In this book we discuss the various aspects of this Industrial Life Cycle. This theory subsumes the product life cycles of the various versions of the product and their overlaps, as well as their global diffusions. It also is impacted by customer life cycles and Vernon's location or product cycle theory. Thus the Industrial Life Cycle represents the interplay of many individual cycles over a longer time span.

Supply chain management and competition can benefit by knowing these Industrial Life Cycle relationships and phenomena. Supply chain managers and partners can make and anticipate appropriate decisions by being cognizant of the cues, transition points, and events in the Industrial Life Cycle.

The Video Cassette Recorder (VCR) was an excellent example product for this theory. It recently has completed its entire Industrial Life Cycle and provides many examples for illustrating the theory along with the competition, decisions, and problems faced by the supply chains operating in the industry.

We hope this work achieves our aim of a comprehensive and in-depth coverage of the intimate connections between the Industrial Life Cycle and supply chain management issues. We believe this new field will be of interest and use for both practitioners and researchers in supply chain management, strategy, and international business.

Preface

The term "life cycle" is used in various fields, including both the natural and the social sciences. Even in business alone, life cycle has a variety of meanings. One may consider the life cycle of an individual product, a category of products, or a whole industry. From the marketing perspective there also is the life cycle of consumers. This book focuses on the life cycle of an industry. This industrial life cycle concept is very close to that of the category life cycle but they have key differences. One key difference lies in the transition of sales. The target of this book is to go further into the industrial life cycle, which may be considered the central mechanism of the category life cycle. The size of the potential market is dependent on the advancement of the product and process innovations. The consumers at different stages are quite different in what they want and how much they are willing to pay. Companies should offer the right products at the right time, and at the appropriate service level. The industrial life cycle incorporates a wide variety of aspects, such as the innovation stage, the market, and facility location decisions. We integrate these into a meaningful whole that can be called the *industrial life cycle*.

In modern manufacturing, a supply chain, not a single company, is the basic business unit for competing in an industry. Although the supply chain is a relatively new concept, all products and materials have been delivered to customers from its origin. Substantially, the supply chains have sourced materials, manufactured the products, and delivered them to consumers. To offer a variety of products at competitive prices, the member companies and the structure of the supply chain play very important roles. These roles differ at each stage of the life cycle, because the mission of the supply chain changes with the stages. At the beginning of an industry, some original members of the supply chain quickly boost the R&D activity collaboratively to launch the market. After the demand grows, they must establish a system suitable for the mass production. As the competition becomes more severe in the market, supply chains next should concentrate on pursuing efficiency and cost reduction while offering a wide variety of products. At the end of the life cycle, supply chains must withdraw from the market smoothly. The life cycle stages, therefore, have a great impact on appropriate supply chain management. This book is designed to uniquely combine supply chain management and the life cycle.

This book is composed of four main parts and a concluding chapter. In the first part (Chapters I-IV), we discuss the fundamentals of supply chain management and the life cycle, along with the quantitative analysis related to life cycles. The latter is very ambitious because, although industry life cycles are pervasive in modern business reality, very few research efforts have been conducted in the field. This part summarizes the book and discusses supply chain management and life cycle theories. Supply chains have become the basic unit for competition in the market (Bradley, 1986). Effective and efficient supply chain management is very difficult because supply chains have contradictory requirements and are multi-echelon systems composed of a patchwork network of participants, stages and players (Bowersox & Closs, 1996; Bowersox, Closs, & Helferich, 1986). Supply chains also may suffer from information distortions of various kinds.

In Chapter I, SCM Models, we discuss the basis of supply chain management. A key concern for SCM is the control of the dynamic interactions among the supply chain partners (Chan & Lee, 2005; Higuchi & Troutt, 2004). Supply chains are composed of multi-echelon layers to offer the products to the customers efficiently. However, they incorporate various partners whose purposes and interests do not always harmonize. In addition, supply chain processes are so long and complex that unexpected results very often can happen. Information distortion effects can cause supply chains to be inefficient and ineffective. The bullwhip effect and the boom and bust phenomena are good examples. To highlight these characteristics of SCM, we made a comparison with the similar but less comprehensive concepts of business logistics, physical distribution, and *Keiretsu*.

In Chapter II, Review of Life Cycle Theories, we explain the major life cycle theories related to business (Abernathy, 1978; Abernathy, Clark, & Kantrow, 1983; Doyle, 1976; Kotler, 1999; Moore, 1991, 2005; Rogers, 1995; Rosenbloom & Abernathy, 1982; Utterback, 1994; Vernon, 1966, 1977, 1998). The life cycle concept has been adopted widely in marketing. The Product Life Cycle (PLC) is the most well-known one. It typically divides the life-time sales of a particular product into four stages based on the transition of the sales. PLC can be expanded by combining it with the study of the consumer types, the extreme innovators, the innovators, the early adopters, the early majority, the late majority, and the laggards. The life cycle concepts have been developed from the other views such as innovation and manufacturing facility location. Technology is a powerful driver for the diffusion of a new product. The location of manufacturing facilities has a close relationship with manufacturability and cost, which is subject to changes according to the market and technology conditions as Product Cycle Theory demonstrates.

Chapter III, Analytic Research and Quantitative Models, presents a survey of research involving life cycles and modeling. This chapter provides two kinds of background information. First, we reviewed the supply chain management, operations management, and management science literatures for those works contacting life-cycle issues and at the same time that use quantitative or modeling approaches. We also developed synoptic summaries of these publications and provide some analysis of their central topics, trends, and themes. The results will be a helpful reference guide to the related literature to date for both practicing managers and researchers. We also introduce the standard quantitative methods and models used for mathematical life-cycle models. Most of them have been developed under the label of diffusion models and most of this work has been carried out by marketing scientists. Coverage of this material was deemed necessary for completeness. Also, we hope this treatment will provide an easy reference for those supply chain students with technical and forecasting interests. Because it was developed in the marketing literature under the name of "diffusion models," it was somewhat harder to find within the supply chain management literature proper. Those readers not having related technical and mathematical interests may omit this material with no loss of continuity.

Chapter IV, Supply Chain Dynamics and Dynamic Simulation, examines the intrinsic dynamics of supply chains and the dynamic simulation modeling. In the first part of this chapter, the intrinsic dynamical interactions with supply chains are discussed. These theoretically are interesting and informative for understanding and managing supply chains. Next, the chapter explores the power of dynamic simulations for managing and understanding the workings and complex interactions in supply chains. As revealed by the literature survey in Chapter III, the short life cycle case increasingly is important in practice. The intrinsic dynamics of supply chains can especially be problematic and potentially catastrophic in the short life-cycle case. The history of the well-known *Tamagotchi*™ toy is used to provide an illustration. The *Tamagotchi* case history, simulation experiments, and analysis illustrate the disastrous consequences that can occur if information distortion is not taken into appropriate account in capacity expansion decision making.

In the second part (Chapters V-VIII), product development including performance and quality is discussed from the viewpoint of product and process innovations. Product innovation must ordinarily precede process innovation. The discussion is based on the case of the VCR industry. The industry life span of the VCR is complete and well documented and provides excellent illustrations of the various life cycle and supply chain interrelationships.

Chapter V, Prerequisite Conditions for Commercializing, discusses the road to satisfying the prerequisite conditions for commercialization and the launching of the dominant design. This chapter deals with the period before the commercialization stage occurs and uses the VCR case study for illustration. Multiple companies attempt to launch a dominant design, namely a widely adopted, initially excellent product. However, only one can succeed. A dominant design is generally created after many failed attempts. The time until the emergence of the dominant design is periodized into three, the embryo period, the fetus period, and the birth. Then, the basic requirements for commercial success are discussed. It is necessary for any new category of products to satisfy all the minimum requirements. Otherwise it could not diffuse widely. In the VCR, Sony and JVC were in a severe struggle for the de facto standard because both of them launched an excellent product in the middle of 1970's, and both of their product versions satisfied the minimum requirements for commercialization.

Chapter VI, Struggle for De Facto Standard, explains how the VHS group of companies caught up with the Betamax group in the late 1970's. In the middle of the 1970's, Sony and JVC introduced the Betamax and VHS VCRs, respectively. Both of these products had enough potential to become the de facto standard. Sony had a first mover advantage in the market. On the other hand, JVC formed the VHS group and pulled ahead by 1980. Although the capabilities for the first Betamax and VHS, SL-6300 and HR-3300, were almost equal, the consumers preferred HR-3300 because of the longer maximum recording time to videotape movies, baseball games, and football games. Most videotaped programs by any VHS machine could be played back, while Sony made a disconnection to the first Betamax when they launched the Beta II in 1977.

Chapter VII, Development of Products, illustrates the product development after the dominant design has emerged. The advancement and the price decline of products are illustrated based on the VCR case. After the dominant design emerges, the product advances incrementally or cumulatively. This is because the dominant design sets a standard design for the product and a framework for the competitors to follow. Many new generation products with new functions will have appeared in the market. Not all of them became popular. In the VCR case, typical consumers bought a monaural VHS machine first and then a HiFi VHS machine. However, most consumers did not purchase S-VHS, D-VHS and other advanced but too expensive machines. The alternation of generations of the VCR occurred only once, from the monaural to the HiFi machine.

Chapter VIII, Emergence of Destructive New Technologies, discusses the situation known as the productivity dilemma. The emergence of alternative products also is discussed. Companies launch alternative products to gain the initiative and to increase sales promotion, although the R&D and manufacturing cause dramatic cost increases. In the VCR case, there were many alternative products, such as EVR, TED, and the laser disk, especially during the 1960's. They continued to be improved many times in attempts to surpass the VHS. However, VHS (a magnetic recording system) remained the de facto standard until the DVD and HDD recorders finally replaced VHS. We also contact some recent research on destructive technologies and related phenomena from the strategy literature. This recent work also promises to be informative in better understanding cases in which a de facto standard may not completely replace one or more competitors.

In the third part (Chapter IX-XI), the development of the industry is analyzed from the perspective of the consumer roles at the differing life-cycle stages. Consumers are divided into five categories, the extreme innovators, the innovators, the early adopters, the early majority, the late majority, and the laggards. The innovators and the early adopters lead the early market and the early majority and late majority play a major role because of their size. Laggards are almost out of the picture in the model.

Chapter IX, Extreme Innovators and Innovators, discusses the two classes of customers called the extreme innovators and the innovators, respectively. Both of these groups purchased an incomplete or immature product at a high price. The characteristics and behaviors of these types of consumers are very important in the infant market. The extreme innovators dare to purchase an incomplete product at high price and contribute to its product development. The innovators adopt the new product later than extreme innovators and purchase an immature product while still at a relatively high price. Their reviews have very influential effects on the future diffusion of the product. Their second and later repeat purchases are critical factors for the alternation of product generations.

Chapter X, Early Adopters and Early Majority, discusses the characteristics and roles of these customer classes. The early adopters and early majority customer groups are reviewed from the life cycle viewpoint according to the VCR case study. Both of these classes decide to purchase when a product becomes sufficiently mature, stable in features, and the price becomes reasonable. Their adoption also signals that the product performance has attractive future possibilities. On the other hand, manufacturers should expand their manufacturing facilities quickly to cope with the growing demand. The spread is very rapid in the first phased because the early adopters (13.6 npercent) amount to almost six times as many as the innovators (2.3 percent) and the early majority (34.1 percent) account for about 2.5 times as many as early adopters.

Chapter XI, Late Majority and Laggards, explains the characteristics and the roles of the late majority and laggards classes of buyers. Both of them adopt the product sufficiently late enough that the product has become both practical and inexpensive. Their impact on the market is not very great because various types of other customers co-exist in the market by that time. Also, manufacturers will have begun to offer a wide selection of products for all customers. The late majority and the laggards therefore are only a small fraction of the total customers. In addition, these buyers will have entered the market at a time when most manufacturers will have attained adequate manufacturing ability.

In the last part (Chapters XII-XV), physical facility location aspects and decisions are discussed. Facility location decisions at the life-cycle stages can be better informed by considering the manufacturability (feasibility), sales promotion opportunities, and the cost-reduction possibilities. The appropriate allocation of facilities and the design of the supply chain network are among the most effective means for fulfilling strategy in a manufacturing supply chain.

Chapter XII, Physical Location, introduces Vernon's Product Cycle Theory. This first is used to explain the transition of facilities locations. Then, reasons are discussed to show why location factors are important at the inception of the industry life cycle. Vernon's product cycle theory is reexamined and further expanded because it was not originally able to address the recent environmental changes, the global economy, supply chain management, and the rise of developing countries. The relevance of geographic location of manufacturing facilities is demonstrated from the viewpoint of innovations. At the beginning of the VCR industry, nearly all manufacturing facilities were located in Japan. Even though a substantial global demand existed for the VCR from the beginning, Japanese companies exclusively manufactured them in Japan and exported products to the rest of the world.

Chapter XIII, Partial Dispersion, discusses the beginning of the overseas or local production, that is, production in the vicinity of the overseas customer. The strategy of locating all supply chain members in close proximity is better for most suppliers and manufacturers at the beginning of the life cycle. However, as time passes, companies start locating assembly facilities in the other advanced countries to increase the sales or to reduce costs. Overseas production is an effective and rapid way for the manufacturers to avoid trade friction and to penetrate foreign markets. Product and component standardization makes it possible to do so. The partial dispersion at the beginning of the standardized stage in the VCR Industry is demonstrated.

Chapter XIV, Total Dispersion, demonstrates the importance of global logistics or global sourcing during the standardized product stage of the industrial life cycle. In the middle of the standardized stage, companies promote total geographic dispersion of the manufacturing facilities because of the intensity of competition in the market. Leading companies, followers, and cost cutters, can all exist in the same market because the advancement of the product design and production processes invites newcomers and enables them to compete with others. The appropriate strategy differs based on the company type. The various consumer types, namely, the extreme innovators, the innovators, the early adopters, the early majority, the late majority, and the laggards, all co-exist in the market at this stage.

Chapter XV, Convergence of Facilities in Low Cost Operation Areas, discusses the convergence process of facilities under the situation that the demand for the product decline sharply. In the late standardized stage, very little room is left for the differentiation of products. The saturation in the market or the emergence of the alternative products decreases the demand for a product sharply at the end of the life cycle. For the further cost cutting under the declining demand, companies should make most of the economy of scale globally. As a result, companies start convergence of their manufacturing facilities in a low cost area or withdraw from the market.

In a concluding chapter, Chapter XVI, Application of Industrial Life Cycle Concept, we try to summarize the previous chapters from the viewpoint of the *Industrial Life Cycle* concept. The industrial life cycle concept integrates the innovation, location, and marketing aspects and is a very holistic concept in two respects, a whole category of products including their multi-generational forms, and a whole supply chain from the consumers to suppliers. We developed the industrial life cycle concept through a thorough study of the home-use VCR case history. The time frame is divided into five stages: *introduction* (before 1980); *early growth* (1980-1984); *late growth* (1985-1989); *maturity* (1990-2001); and *decline* (after 2002).

We hope this book will be of interest to a wide variety of executives; business strategists; international business managers; production, operations, and logistics experts; and marketing specialists who seek to further understand and apply supply chain management techniques. Life cycle techniques are very useful for foreseeing the future. The framework in this book, which includes the typical pattern of the innovation, the consumer behavior, the physical location, and the SCDM concept, is applicable widely in real life. The structure of the supply chain should be changed according to the mission of the supply chain which differs by stages, such as R&D oriented, the mass-production oriented, and the cost oriented. The book also should be useful for graduate students and advanced undergraduate students in business, particularly in international business and business strategy. We also hope the book will provide impetus for further research in these and related fields. A general undergraduate business education would be ideal as background for this book. However, experienced managers and advanced undergraduate students will hopefully find it quite readable.

We also hope this work achieves our aim of a comprehensive and in-depth coverage of the intimate connections between the industry life cycle and supply chain management issues. The industry life cycle of the VCR has recently come to its conclusion and provides the essential principles and illustrations for this new theory. We believe this new field will be of interest and use for both practitioners and researchers, and will contribute to both strategy and international business.

References

Abernathy, W. J. (1978). *The productivity dilemma.* Baltimore, MD: The John Hopkins University Press.

Abernathy, W. J, Clark, K.B., & Kantrow, A.M. (1983). *Industrial renaissance*. New York: Basic Books.

Bowersox, J., & Closs, D. J. (1996). *Logistical management*. New York: McGraw-Hill.

Bowersox, J. D., Closs, D. J., & Helferich, O. K. (1986). *Logistical management: A systems integration of physical distribution, manufacturing support, and materials procurement,* (3rd ed.). New York: Macmillan Publishing Company.

Bradley, P. J., Thomas, T. G., & Cooke, J. (1999). Future competition: Supply chain vs. supply chain. *Logistics Management and Distribution Report*, *39*(3), 20-21.

Chan, C. K., & Lee, H. W. (2005). *Successful Strategies in Supply Chain Management*. Hershey, PA: Idea Group Inc.

Doyle, P. (1976). The realities of the product life cycle. *Quarterly Review of Marketing*, *Summer*, 1-6.

Higuchi, T., & Troutt, M. D. (2004). A dynamic method to analyze supply chains with short product life cycle. *Computers & Operations Research, 31*(6), 1097-1114.

Kotler, P. (1999). *Marketing management*. New Jersey: Prentice Hall.

Moore, G. A. (1991). *Crossing the chasm*. New York: Harper Business.

Moore, G. A. (2005). *Dealing with Darwin*. New York: Portfolio.

Rogers, E. M. (1995). *Diffusion of innovations* (4th ed.) New York: The Free Press.

Rosenbloom, R. S. & Abernathy, W. J. (1982). The climate for innovation in industry. *Research Policy*, *11*(4), 209-225.

Utterback, J. M. (1994). *Mastering the dynamics of innovation: How companies can seize opportunities in the face of technological change*. Boston: Harvard Business School Press.

Vernon, R. (1966). International investment and international trade in the product cycle. *Quarterly Journal of Economics, 80*(1), 190-207.

Vernon, R. (1977). *Storm over the multinationals*. Boston: Harvard Business School Press.

Vernon, R. (1998). *In the hurricane's eye*. Boston: Harvard Business School Press.

Acknowledgment

We wish to thank several colleagues and associates for their help in completing this book. A special debt of gratitude is owed to Dr. Hisatoshi Suzuki for many useful comments. We are very grateful to Dr. Kousuke Takei, Dr. Toshiki Simoda, and Mr. Shin Tomita for their assistance in gathering data. We thank Dr. Eileen Weisenbach-Keller for many stimulating conversations. We also sincerely thank Laura Lin and Susan Horne for valuable help in literature reviews. This work was supported by KAKENHI (18730275), Grant-in-Aid for Young Scientists (B).

Section I

Overview

This part summarizes the book and discusses supply chain management and life cycle theories, which are the foundations of the modern manufacturing world. Supply chains have become the basic unit for competition in the market because it is much more effective and efficient for most companies to form or take part in a supply chain than to carry out all activities in isolation (Bradley, Thomas, & Cooke, 1999). The supply chains have contradictory requirements, that is, a wide variety of goods vs. lean management, high quality vs. low cost, global sourcing vs. quick response, and global compatibility vs. customization. A supply chain is a multi-echelon system composed of a patchwork network of participants, stages, and players whose preferences and intentions are not always the same. It shifts the set of players, the strategies, and structures according to the stage of products' life cycles, which in turn decides the competitive environments, such as rival and consumer behaviors. Nowadays supply chains cross the boundaries between company groups, which further complicates supply chain and life cycle management.

Chapter I

SCM Models

In this chapter, a basic concept of the SCM is discussed. A key factor for the SCM might be to control the dynamic interactions among the supply chain partners. Supply chains form a multi-echelon system to offer the products to the customers efficiently. However, they are composed of various partners whose purpose and interests do not always harmonize. In addition, supply chain processes are so long and complex that unexpected results might be happened for supply chains. The information distortion within the supply chains is the one of the major obstacles to control the supply chain efficiently and effectively. As a result, supply chains would be damaged by the bullwhip effect and the boom and bust. To highlight the character of SCM, a comparison is also made among the similar concepts, the business logistics, the physical distribution and *Keiretsu*.

Copyright © 2008, IGI Global. Copying or distributing in print or electronic forms without written permission of IGI Global is prohibited.

Fundamentals of Supply Chain Management

"Supply Chain Management encompasses the planning and management of all activities involved in sourcing and procurement, conversion, and all Logistics Management activities. Importantly, it also includes coordination and collaboration with channel partners, which can be suppliers, intermediaries, third-party service providers, and customers. In essence, Supply Chain Management integrates supply and demand management within and across companies." (Council of Supply Chain Management)

"A supply chain is dynamic and involves the constant flow of information, products and funds between different stages. Each stage of the supply chain performs different processes and interacts with other stages of the supply chain." (Chopra & Meindl, 2001, pp. 4-6)

As mentioned above, one of the most interesting aspects of supply chains concerns the dynamic interactions within them. This is due to the fact that supply chains are multi-echelon systems for delivery of products to customers from the points of origin, which consist of different departments and companies. The supply chain is consumer oriented because its goal is to sell the products to the consumer efficiently and profitably. Overall performance of the supply chain is a result of the complex and dynamic interactions among the components.

A major challenge for supply chains is to control various trade-offs (Gopal & Cahill, 1992). For instance, while sales and marketing people desire a high degree of production flexibility and rapid turnaround to catch up with recent trends for maximizing their sales, manufacturers prefer a simple product line, longer production runs, fewer set-ups, and smooth schedules for minimizing manufacturing costs or unit costs and recovering sunk costs. The trade-offs within the supply chain have a powerful impact on the overall supply chain profitability. If each player selfishly pursues its own aims, the supply chain cannot achieve maximal profitability. It perhaps is intuitively clear that the effective and efficient flow of information, products, and materials within the supply chain are required to maximize profitability.

Synchronization across the entire supply chain is one of the most critical factors for successful supply chain operation. A useful tool for improving synchronicity is the Theory of Constraints (TOC). TOC is a system

Copyright © 2008, IGI Global. Copying or distributing in print or electronic forms without written permission of IGI Global is prohibited.

improvement philosophy that emphasizes the management bottlenecks and throughput (Cox & Spencer, 1998; Dettmer, 1997; Goldratt, 1990 a, b, c; Goldratt & Cox, 1992; Goldratt & Fox, 1986; Kendall, 1998; Mabin & Balderstone, 2000; McMullen, 1998; Schragenheim, 1999). A bottleneck is a weak point, subprocess, or player in the supply chain which limits the level of throughput for the entire supply chain. The bottleneck should be identified and improved as fast as possible to maximize throughput. Throughput is an important measure related to the profitability of the entire supply chain. For example, a supply chain may have a strong R&D function and distribution channel. However, if its manufacturing ability is not able to support them, the ceiling of sales is restrained and the manufacturing capability becomes a bottleneck. The manufacturing ability is determined by the level of facilities, the efficiency of the production process, the working time, the supply of materials, and so on. For achieving the optimal performance, it is very important for supply chains to identify, reinforce, and eliminate bottlenecks to keep a good balance.

Another way to synchronize the entire supply chain is by information sharing, namely, sharing the latest market demands and inventories throughout the supply chain. The levels of inventories and backlogs are good indicators of the degree to which a supply chain can be improved by synchronization. Use of the Just in Time (JIT) system can help achieve an excellent and lean supply chain. Toyota established the JIT system through information sharing by using kanban, which means an iron signboard in Japanese. Toyota started sharing information with suppliers on production scheduling based on the latest market demand by using kanban, on which the required parts, numbers and times were indicated. As a result, Toyota and their suppliers succeeded in reducing tremendous amounts of inventories. Information sharing contributes to synchronized operations within the supply chain and increases profitability by saving inventory costs and reducing lead times.

Geographical distance, in addition to actual process times, creates unavoidable lags and costs related to global division of labor and the market. Delivery time is one of the major elements of lead time. In general, longer lead times require more safety stocks. Longer delivery distances compounded with multiple lead times between layers in the supply chain, increase costs and lead times. In addition, geographical scattering of supply chain components and players promotes independent decision-making and therefore aggravates information distortion within the system.

The modern concept of a supply chain is a wide one that incorporates both business logistics and physical distribution. A fundamental goal of the supply

Copyright © 2008, IGI Global. Copying or distributing in print or electronic forms without written permission of IGI Global is prohibited.

Table 1.

	Supply Chain	Keiretsu	Business Logistics	Physical Distribution
Base Unit	Products	Company Group	Company	Delivery Order
Goal	Customer Satisfaction	Group Profit	Company Profit	Accurate and Efficient Delivery
Scope	All Suppliers and Distributors (Very Wide and Continuous)	Closed Company Group (Related companies and long-term partners)	Company and Subsidiaries (Wide and Continuous)	Links Between Sites (Partial and Fragmentary)
Priority	Consumers	Headquarter Firm	Consumers / Company	Company (Owner)
Character	Chaotic Control (Dynamic Interactions)	Strict Control (Tightly Coupled)	Total Optimization (Static Decision Making)	Partial Optimization (Routine Work)

chain is to maximize customer satisfaction by providing desired products. This results in the requirement for supply chains to establish comprehensive, vertically and horizontally integrated systems, enabling the satisfaction of consumer product needs as efficiently and effectively as possible. Table 1 summarizes the differences among the supply chain, business logistics, and physical distribution. Keiretsu, a traditional Japanese company group, is a similar concept to the supply chain. Both emphasize the holistic approach and a sense of belonging, including the suppliers and distribution channel. However, there is an important difference between them. While Keiretsu is tightly coupled with related companies and long-term partners, the supply chain is loosely coupled and its members are generally composed of ad hoc or scratched-together players. This loose structuring promotes flexibility and in turn efficient operations. Keiretsu members are strictly controlled by a headquarter firm. On the other hand, supply chain members can decide their strategies autonomously to a larger extent.

Drivers of Supply Chain Dynamics

Since a supply chain is dynamic and chaotic, its structure must be very flexible to cope with capricious market demand and severe levels of competition. A

Copyright © 2008, IGI Global. Copying or distributing in print or electronic forms without written permission of IGI Global is prohibited.

supply chain must be able to change its partners easily to gain a competitive advantage. A great number of interactions among supply chain partners are needed to deliver the products to the consumers. These interactions may be executed based on lagged or distorted information (explained in the later sections). These distortions may lead supply chains to experience unexpected results. Figure 1 summarizes the relation among the drivers, internal factors and supply chain dynamics.

As noted, customer satisfaction and the expected consequence of higher profits are two of the major concerns of supply chain management. As the variety and availability of goods increase, customers become demanding and capricious in their wants and needs. A supply chain therefore should be an effective and efficient network for satisfying customer needs at appropriate prices and with a variety of goods and services at a given level of cost, agility, and risk. To meet such complex demands, the supply chain should improve daily operations to achieve lower costs, higher quality, and excellent service levels, as well as launch new products frequently. In daily operations, the supply chain should emphasize a smooth flow of products and materials. This helps to increase sales opportunities and decrease logistical costs. However, time lags originate from the numerous transactions required to produce and deliver goods. These lags begin with raw materials and continue throughout all stages of the SC including delivery and servicing the customer. Many such lags are unavoidable because of the nature of the production and distribution process and the multiple decision-making units within the supply chain, which lower overall SC efficiency. More recently, advertising through mass media and the Internet has speeded up the diffusion of new products and might shorten the life cycle of products. These may amplify the effects of customer behavior and can lead to the well-known problems of the bullwhip effect (Lee, Padmanabhan, & Whang, 1997a) and the boom and bust effect (Paich & Sterman 1993). Successful supply chains must be very responsive to customer needs and requirements (Ballou 1992).

Any supply chain needs intense collaboration for the development of new products with innovative functions and, in many cases, this requires entirely new partners. For example, the supply chains in the auto industry are changing their structures to accommodate hybrid engines and ITS (Intelligent Transportation Systems). This requires strategic alliances with new partners to use the latest technology immediately and without a large investment. It sometimes is inefficient or impossible for them to be in the forefront of the specific materials and parts. These advanced technologies may require too

Copyright © 2008, IGI Global. Copying or distributing in print or electronic forms without written permission of IGI Global is prohibited.

Figure 1. Drivers of supply chain dynamics

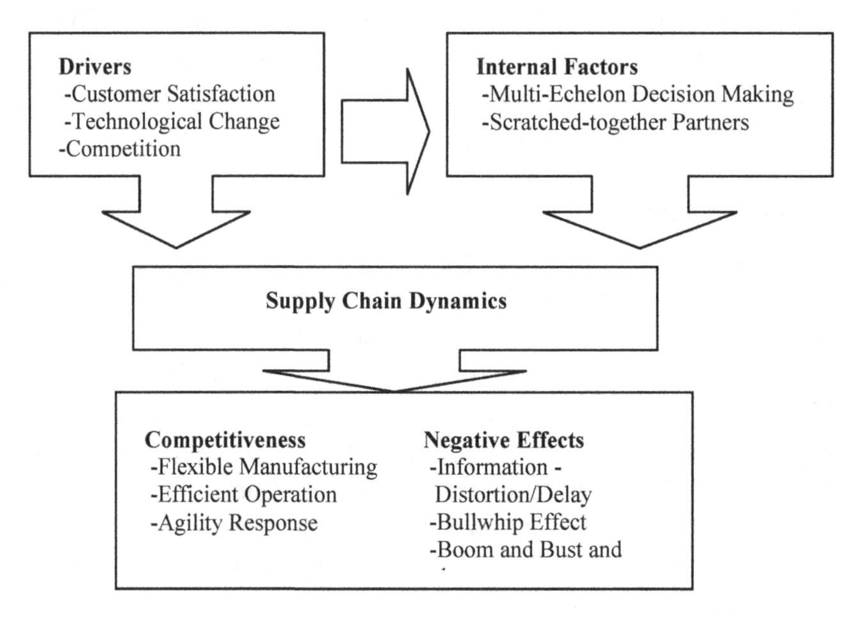

much time and cost and may be too risky for them to do R&D by themselves from the beginning. Thus, technological innovations promote strategic alliances among the different industries and former rivals.

The intensity of competition also has a significant influence on the supply chain. The supply chain (particularly manufacturers) must increase advertising and reduce total costs with the same quality and service levels (or improve quality or service levels with the same costs) as competition in the market increases. All partners in the supply chain—suppliers, manufacturers, distributors, and retailers—are under pressure to shorten cycle times and reduce costs.

Multi-Echelon Supply Chain Composed of Scratched-Together Partners

Supply chains incorporate numerous activities, such as sourcing and procurement, inventory, delivery, processing (conversion), packing, and sales activities. If an individual company were to perform these activities alone, it would be too huge, complicated, and inefficient to compete in the modern

Copyright © 2008, IGI Global. Copying or distributing in print or electronic forms without written permission of IGI Global is prohibited.

business environment. In addition to the managerial difficulties, the strategic considerations and specialties require a supply chain with suppliers, intermediaries, wholesalers, retailers, and third-party service providers. Figure 2 illustrates a simple example of the multi-echelon structure of a typical supply chain.

The goal of a supply chain is to provide products to the customers. For maximal customer convenience, retailers need to locate near their customers or demand points. This helps customers access and examine products, guides them in the choice and usage of the products, and facilitates repairs when necessary. The area a retailer covers is determined by product features, demand and purchase patterns, size, weight, condition, and the like. For example, the number of grocery stores must be much larger than that of fashion houses and appliance shops because people purchase foods and beverages much more frequently than clothing.

As the Internet society progresses, the importance of on-site retailers might wane. However, the Internet has a tendency to stimulate customer needs, which may attract customers to the on-site retailers. Retailers with a good presence on the Web give customers a chance for easy access, unrestricted by time and place, and become a powerful force in the supply chain. Therefore, through the Internet and e-commerce, retail functions are reinforced and are even more diversified.

Wholesalers and warehouses are needed to supply products to the great number of widely scattered retailers. Generally, it is inefficient to send products from a single point to all retailers mainly because of the increased transportation costs and delivery times. In a typical example involving shipments to 100 retailers in a foreign country, locating warehouses and wholesalers in-country offers substantial savings in cost and time compared to exporting products directly to individual retailers. Accordingly, wholesalers and warehouses form a substantial component of the multiple layers in the SC and are necessary for covering all areas of demand. In addition, this structure may provide certain managerial advantages as well.

In order to minimize manufacturing costs per product or unit cost, it is usually desirable that certain products be produced entirely in one factory. This is because of economies of scale and experience, or learning curve effects. Such entire or sole production in one facility is very effective for avoiding duplication of the equipment and related investments. Longer and more intensive use or full operation shortens the recovery time for equipment investments. Such comprehensive manufacture also helps concentrate learn-

Copyright © 2008, IGI Global. Copying or distributing in print or electronic forms without written permission of IGI Global is prohibited.

ing, experience, and hence, knowledge. This can have a positive impact on the reduction of safety stock and rework. However, in the real world, many companies tend to duplicate plant and equipment investments. For some products, agile response to the local market needs with little if any safety stock may be the critical factor in achieving competence. Others may not tolerate long-distance transport because of their fragileness and the loss of freshness, or may be unfavorable because of excessive transportation costs due to heavy weight and/or large size. Another case occurs when a region has an unexpectedly large volume of demand. When this occurs, it may be necessary to add another plant to produce the lines in question. For these reasons, the number of plants tends to be squeezed down in order to reduce the production costs except in the cases noted.

At the same time, and particularly so with high tech products, it often is very impractical for any plant to manufacture all parts by itself. Outsourcing enables companies to reduce managerial difficulties by eliminating inefficient activities, to allow time to catch up with the latest essential technologies in any specific area, and to spread the risks. A plant or company that requires outsourcing is called an integrator, set maker, or assembler. In particular, appliance and automobile industries have advanced the practice of outsourcing so much that the integrators now concentrate largely on the development of core parts and the design of products. But this comes at the expense of additional managerial effort in coordinating the supply system.

Suppliers constitute multiple layers of primary suppliers, secondary suppliers, tertiary suppliers, and so on. Primary suppliers deliver the modules1, parts, and materials to the integrator or plants in which products are completed. They specialize in specific modules, parts, and materials, and surpass the integrator in knowledge and efficiency of operations in these areas. Some primary suppliers function as sub-integrators by coordinating their own subsystems. For example, automakers have hundreds of primary suppliers and thousands as a whole.

Transportation and communication deal with the separation between sites that lie in different strata or levels of the SC. Raw materials often follow a very long path to the consumers. They are combined with other materials and processed many times at different places to become a finished product. Products are then distributed to consumers. The transportation functions link the sites and third party logistics carry out most transportation links. It is a critical factor for any supply chain to achieve the efficient and effective flow of materials and parts. Without communication, a supply chain cannot

Copyright © 2008, IGI Global. Copying or distributing in print or electronic forms without written permission of IGI Global is prohibited.

Figure 2. Multi-echelon structure of supply chain

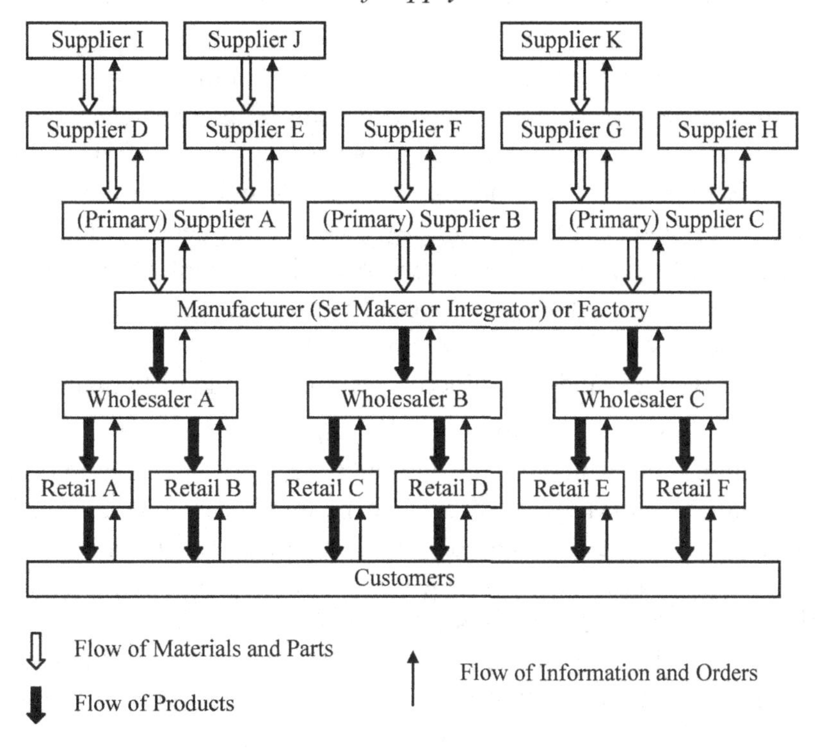

achieve the required efficiency to survive competition. The important factors of communication are speed, frequency, accuracy, and cost. It now has become the mainstream practice in modern business to outsource the liaison and information system functions, as well as to use third party logistics.

It is natural for all supply chain partners to attempt to maximize their own profits while controlling their risks. They decide and negotiate prices and quantities for sale and purchase, while considering future trends and efficiencies of production, inventory, and transportation. The overall effect of such partner behaviors makes supply-chain management very difficult. In addition to the geographical distances, what may be called mental distances create differences of viewpoints and individual self-interests, which worsen the situation. Supply chains need dynamic interactions between adjacent layers. Ironically, the openness and flexibility of supply chains may aggravate information distortion, the various lags, forecasting errors, the bullwhip effect, and tricks or deceptive practices (explained further below) of interactions within the supply chain.

Copyright © 2008, IGI Global. Copying or distributing in print or electronic forms without written permission of IGI Global is prohibited.

Information Distortion within Supply Chain

Without sharing the latest and most accurate information among all supply chain partners, information distortion can be generated and easily amplified within the supply chain. The root cause of the information distortion is independent decision-making. Partners positioned in the different layers within the supply chain do not always carry out decision-making based on the same latest and most accurate information, do not use the same inventory and order policies, and often make decisions at the same time. This causes inadequate estimates of the inventory investment, poor customer service, the loss of revenues, and misguided scheduling and facility planning.

Figure 3 illustrates how these information distortions are created in the supply chain. It is assumed that a supply chain has three echelons composed of the retailers, the wholesalers, and the manufacturer, and that all of them have enough inventories to respond quickly to weekly orders and their cycle times for orders and production are one week. The multiple decision-makers within the supply chain tend to create time lags in awareness of the latest market demands. Wholesalers have a two-month delay to place their orders to the manufacturer based on the most recent consumer demand. The manufacturer finishes refilling the products with a three-week delay.

The multiple decisions also generate and amplify variations of the orders. The retailers do not always place orders for the same amounts they sold because they determine the amount considering the level of inventory, their expectation of the future demand and their own interest. All of them need safety stock to some extent. The resulting redundant inventories increase according to the number of retailers, wholesalers, manufacturers, and suppliers. Furthermore, retailers and wholesalers want to increase the stock of good selling items but when their sales drop sharply, they abruptly stop ordering and immediately start clearing the remaining inventory. Although manufacturers make short-term production schedules with their suppliers, they typically are under long-term purchase contracts and also must deal with facility planning. They can not respond as quickly as the retailers and wholesalers. Hence, consumer purchases do not always correctly reflect the production schedule of the manufacturer.

The accuracy and timeliness of information are weakened by multiple decisions, which is a managerial problem. The manufacturer receives distorted information based on the market demand that actually occurred a few weeks ago and since has been processed many times. This sometimes causes the

Copyright © 2008, IGI Global. Copying or distributing in print or electronic forms without written permission of IGI Global is prohibited.

Figure 3. Information distortion due to delay in supply chain

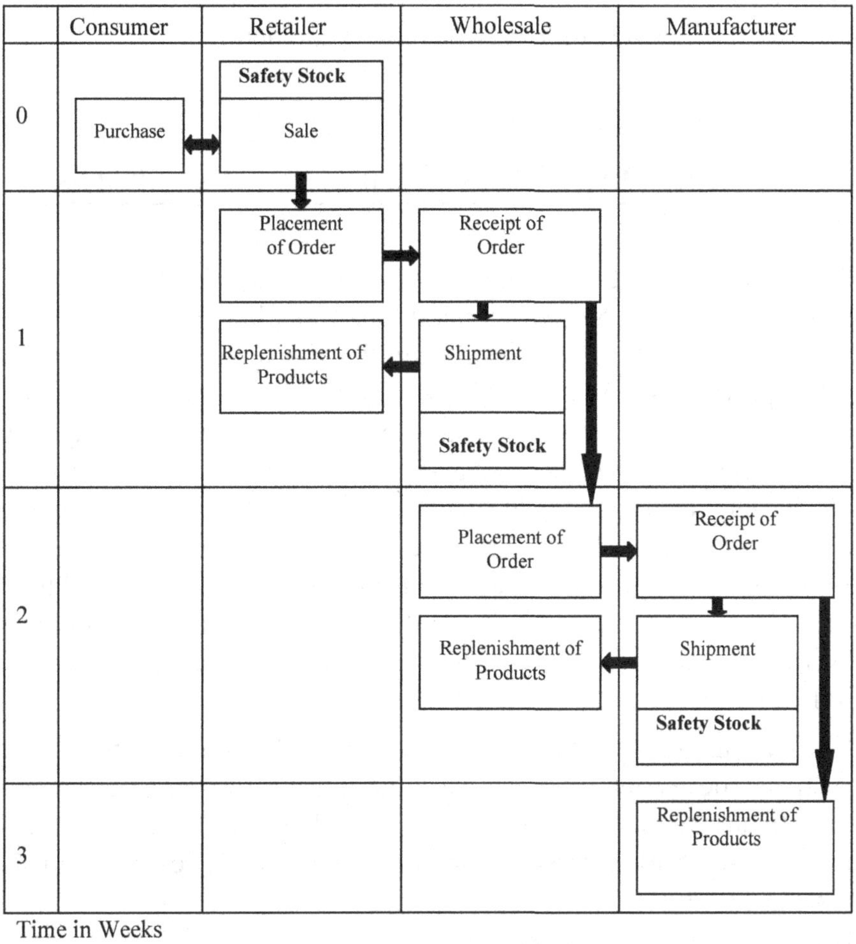

	Consumer	Retailer	Wholesale	Manufacturer
0	Purchase	**Safety Stock** Sale		
1		Placement of Order Replenishment of Products	Receipt of Order Shipment **Safety Stock**	
2			Placement of Order Replenishment of Products	Receipt of Order Shipment **Safety Stock**
3				Replenishment of Products

Time in Weeks

bullwhip effect and unnecessary phantom demand order swings, the mismatch of supply and demand, deceptive business practices, and risky capacity plans within the supply chain, which can fatally be injured, in particular, its manufacturers and suppliers.

For minimizing information distortion within the supply chain and maximizing the supply chain profitability, coordination among the levels and players through the information system is an indispensable factor (Bowersox & Closs, 1996). It is very effective for the supply chain to share information, such as the latest market demands and the amount of inventories including

Copyright © 2008, IGI Global. Copying or distributing in print or electronic forms without written permission of IGI Global is prohibited.

products, parts and materials, without redundant decision making (Gavirneni, Kapuscinski, & Tayur, 1999). The level of inventory is a good measure that reflects the efficiency of supply chain management.

Bullwhip Effect Within Supply Chains

The bullwhip effect is the exaggerated negative effects within the supply chain, which may come from small variations in customer demand. Multiple decision-making and levels amplify the order swings as mentioned in the previous section. In addition, downstream activities, such as retail and wholesale, can respond to the demand change much more quickly than the upstream activities, such as assembly and production. The retailers and wholesalers can respond to the demand changes through placement of orders and by changing the shelf and reserve spaces. On the other hand, manufacturers including suppliers need to plan for production facilities and to prepare for steady operations, such as assembling products, training employees and securing materials and parts. The following quotes concisely describe the bullwhip effect:

"The information transferred in the form of orders tends to be distorted and can misguide upstream members in their inventory and production decisions. In particular, the variance of orders may be larger than that of sales, and the distortion tends to increase as one moves upstream" (Lee, Padmanabhan, & Whang, 1997b).

"Variations in production are far more severe than variations in demand, and the more levels and stages of production there are, the more violent production level changes become" (Magee, Copacino, & Rosenfied, 1985, p. 42).

The instant noodles product provides a good example of the bullwhip effect (Figure 4). Consumers can use this product very easily. All they have to do is pour boiling water into a plastic cup containing instant noodles and wait for a few minutes. A great number of consumers, especially singles, use it daily or weekly. The special feature of instant noodles is that it will keep for a few months. People tend to purchase many instant noodles at once and store them. Many supermarkets notice this purchase pattern and sell them as loss leaders. They substantially reduce the price and clear the stock. The transaction between consumers and supermarkets creates variation and bias of the consumer demand because many consumers wait for a clearance sale.

Copyright © 2008, IGI Global. Copying or distributing in print or electronic forms without written permission of IGI Global is prohibited.

Figure 4. Bullwhip effect in the instant noodles industry

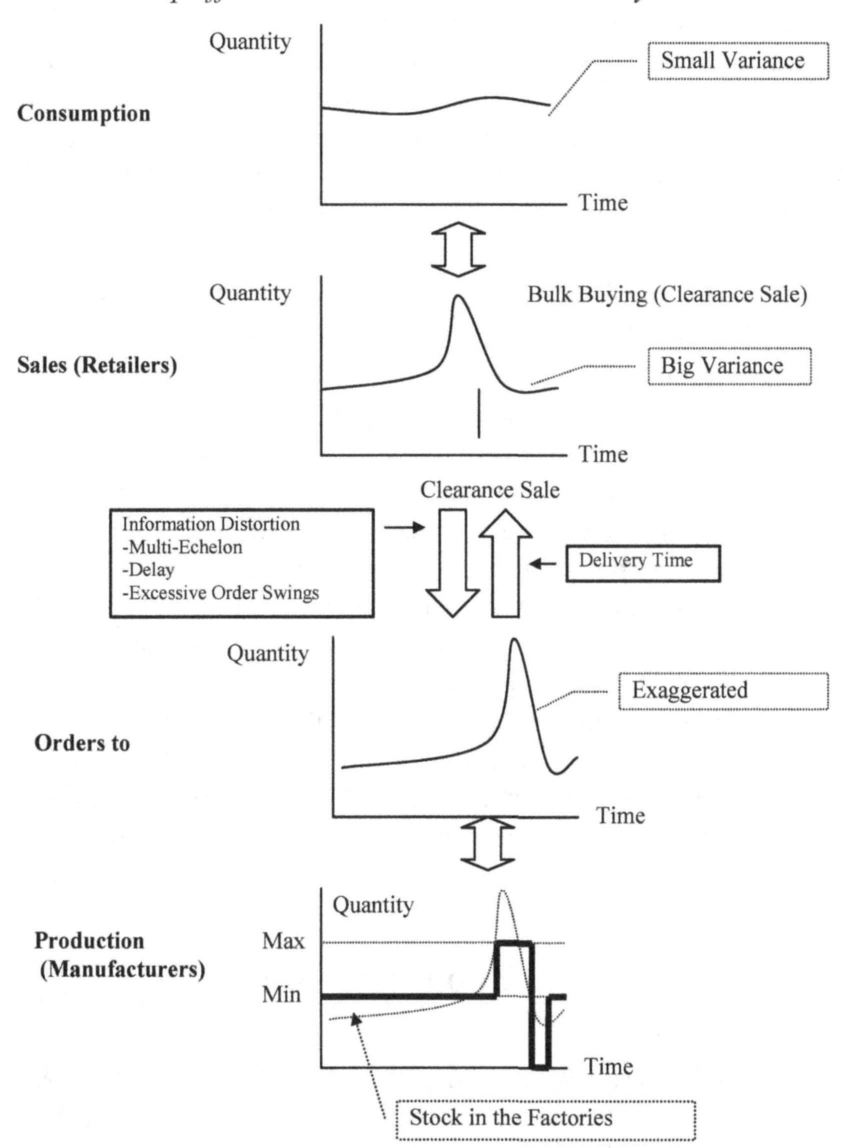

It takes a relatively short time to produce the instant noodles because the production system has been standardized, the final products having main differences only in size, package, flavor, and sauce. Tons of flour are reserved in silos adjacent to the factory. Usually, the silos and factories are linked by a pipe line. The fried noodles are added into the prepared plastic cup with

Copyright © 2008, IGI Global. Copying or distributing in print or electronic forms without written permission of IGI Global is prohibited.

flavoring and sauce and can be manufactured continuously and fluidly in the factory. When manufacturers change the production line, they can finish setting up the machines within a few hours. However, manufacturers should form a long and complicated distribution channel to deliver the products to the consumers and stock a great number of products for agile response to the variations in demand. It has been noted that the distribution channels, including wholesalers and warehouses or depots, serve as buffers. However, the diversification of products and lean management weaken their buffer function. As a result, the variation in demand has a much bigger impact on the manufacturers than on retailers and wholesalers.

Another example with the same product is as follows. In Asia, especially Japan, South Korea, Taiwan, Thailand, and others, people who live in urban areas tend to buy foods at convenience stores. Food items cost comparatively more to buy at convenience stores than from supermarkets. Convenience stores have much smaller shelf space than supermarkets but offer a sophisticated selection of products. They prefer new products because consumers are motivated to purchase by advertising and most long time, good selling items are subject to discounts and price competition. Convenience stores depend on manufacturers to launch new products to increase their sales. Otherwise, the manufacturers lose the opportunity to sell their products in the convenience stores. Various expenses related to new products, such as R&D, advertising, new packaging, and new materials, fall to the manufacturers. In addition, most new products disappear quickly, thus exhibiting a short life cycle.

Boom and Bust

Boom and Bust is an extreme case of the bullwhip effect (Paich & Sterman, 1993). It occurs when product life cycles become very short. In the boom phase, the demand for a product grows so rapidly that the supply can not keep up with it. This disparity leads to the shortage game. The shortage game is one of the major causes that makes the bullwhip effect more powerful and induces the boom and bust phenomenon. It starts when a highly popular product runs out of stock. Many customers go to multiple retailers to check their inventory. In this situation, it is typical for consumers to place multiple orders with various retailers. Retailers and wholesalers see a surge in demand and misjudge the demand because of the phantom demand or the inflated

Copyright © 2008, IGI Global. Copying or distributing in print or electronic forms without written permission of IGI Global is prohibited.

picture of the demand due to multiple orders. They then place their orders based on the inflated demand. However, they cannot receive the quantity they ordered because of a shortage in supply. In this case, they may exaggerate their orders to increase their allotments of possibly rationed supplies. The manufacturer and suppliers will have received orders far above the true demand. In turn, the manufacturer and suppliers tend to expand production capabilities to belatedly fill a huge backlog.

On the other hand, after the peak demand occurs, the product may lose its popularity and suddenly become obsolete. When the supply catches up with the real demand, backlogs will disappear and all supply chain partners suddenly suffer from the huge inventories and excess capacity. The consumers who had duplicated the orders accepted the product from the first retailer that delivered and canceled the remaining orders. Retailers and wholesalers stop receiving the products because of their excess inventories and start clearance sales. The manufacturer and suppliers experience a greater hardship than retailers and wholesalers from the bust after the boom because of the overcapacity. They expand production capacity based on an inflated picture of the real demand for the product and fall behind after the completion of the expansion (Lee et al., 1997a, b; Nehmias, 1997). As a whole, all players can be victims of boom and bust after the shortage game plays out.

The boom and bust phenomenon is amplified by the ubiquitous time lags in the supply chain. The division of labor becomes more vigorous than before, beyond the companies and the area. All supply chain processes can not finish at the same time. Materials often are processed in different places, transformed into products, and delivered to the consumers. As is generally the case, the longer the lead time, the higher the cost, and the greater the risk. In addition, the multiple decision points and times delay information processing as a whole by the supply chain and distort the information as mentioned before. In particular, the slow expansion and reduction of the manufacturing facilities may accelerate the boom and lead to a catastrophic end. The manufacturer and suppliers endure much greater risks than retailers and wholesalers because of the manufacturing facilities, which are planned based on an inflated picture of real demand, exacerbated by the natural lag and the managerial information distortion in filling orders.

A seamless information system throughout the supply chain and the control of diffusion speed on availability of the product to consumers are useful ways to mitigate or escape from the boom and bust pattern. The information system is required for timeliness, unity and traceability to estimate the latest

Copyright © 2008, IGI Global. Copying or distributing in print or electronic forms without written permission of IGI Global is prohibited.

demands correctly. Processing orders at several stages creates lags and masks phantom demands. In the case that the phantom demands are undetected for a very long time, without sharing the latest information in an on-time basis, these demands can self-amplify themselves to a certain level. Hence, prompt sharing of the correct information on the latest demands throughout the supply chain is a critical factor for effective supply chain management.

The control of the diffusion speed is a difficult problem. The typical method is to use advertising. Companies select the media in which to advertise and adjust the volume of promotion considering the impact of advertising. In the boom phase, retailers, wholesalers, manufacturers, and suppliers all tend to overestimate demand. Thus, a product can be spread widely and rapidly by word of mouth even without mass media advertising.

In the case that the demand grows far beyond the supply or manufacturability, a sizeable shortage might induce a tremendous shortage game effect and result in a catastrophic boom and bust. When the bust occurs, orders for the product are cancelled and the retailers and wholesalers tend to underestimate the demand for the product and turn their attentions to the next big hit product after the boom. However, the boom effects can be relaxed by the proper use of the information other technologies related to SCM because of the removal of the information distortion. The latest information technologies help to grasp the real demand and estimate a future demand much more exactly than before.

Figure 4. Boom and bust

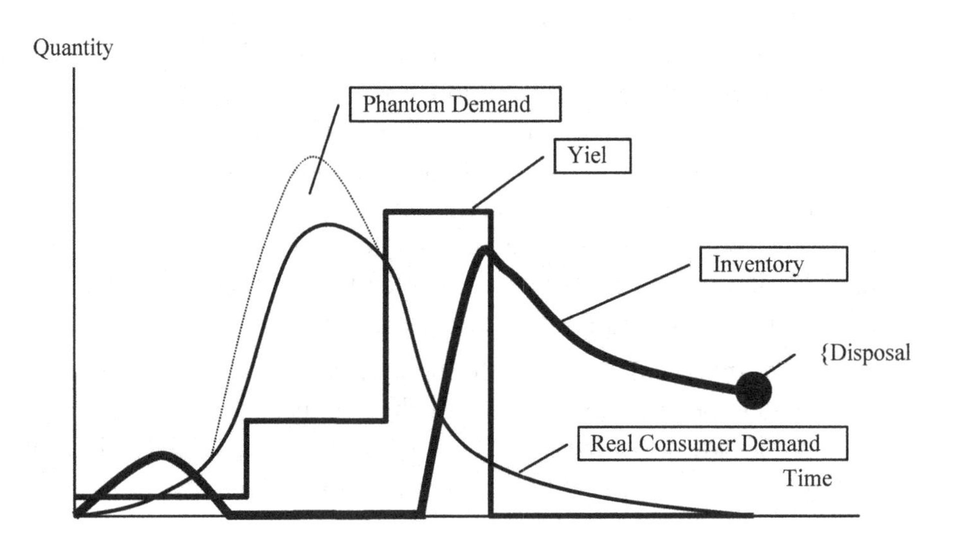

Copyright © 2008, IGI Global. Copying or distributing in print or electronic forms without written permission of IGI Global is prohibited.

The latest managerial technologies reinforce the collaboration among supply chain partners beyond the company groups and the national borders.

The boom and bust phenomenon can also be due to the time-related combined result of the shortage game and the bullwhip effect. This shows that wildly popular products are not always successful. This type of difficulty will be illustrated in a later chapter by the case of the Tamagotchi toy, a product of the BANDAI Co., Ltd. (Higuchi & Troutt, 2004) that was very popular with consumers but was catastrophically unprofitable.

Dynamic Simulation of Supply Chain Management

Computer simulations are widely used to analyze supply chain dynamics. It is too complex to manage an entire inventory by mathematical analysis because more than two echelons are involved and the inventory management is usually carried out with the aid of computer simulation (Ballou, 1992). Computer simulations can be divided into the static and dynamic models. The primary difference between them is the way in which they treat time-related events. Static simulations do not pay enough attention to time-period interplay but the dynamic simulations evaluate the system performance across time (Bowersox, Closs, & Helferich, 1986).

Static simulations set the framework for model-based analysis of the supply chain and identify the key issues, elements and relations among the players and levels. Queuing theory has also contributed to this field together with the network theory. From the logistical viewpoint, the optimum solutions for the timings and quantities of orders, the lead-times, the levels of inventory, scheduling, routing options, and the number of warehouses are all explored. Winker, Towill, and Naim (1991) proposed remedies for improving the performance of the entire supply chain by emphasizing the concept of total system stocks. Nersesian and Boyd (1996) proposed solutions for the optimum timing and quantity of orders, and the level of inventory by the use of simulation with Visual Basic. The effects of lead-time on various costs have been analyzed by stochastic models (Takeda & Kuroda, 1999; Vendemia, Patuwo, & Ming, 1995). Schwarz and Weng (1999) built a model considering the interactions between the variances of the lead-times in each link of the supply chain and system inventory holding costs. Gavirneni et al. (1999) extended existing inventory theories and simulated an overall supply chain model from the

Copyright © 2008, IGI Global. Copying or distributing in print or electronic forms without written permission of IGI Global is prohibited.

viewpoint of information distortion. Chen (1999) characterized optimal decision rules under decentralized supply chains subject to information delay in which the division managers are assumed to share a common goal to optimize overall performance of the supply chain. These analyses are performed based on routine transactions in the static framework and they treated the issues separately. In most cases, the static simulations lack appropriate feedback loops, which are essential features of supply chain dynamics.

Dynamic simulation is performed across a time interval. It is very useful to analyze and plan the supply chain options because it is a holistic approach and can be carried out interactively. The dynamic simulations are built on the framework of static simulations while incorporating feedback loops. They focus on the dynamic features and simulate the movement of the supply chain from the system perspective. Many supply chain models have been built using System Dynamics2, which is a well-elaborated methodology for deterministic simulation (Shapiro, 2001). It can simulate the impacts of causal relationships because of the existence of the feedback loops. In addition, it can consider the various delays and queues. Furthermore, it has a strong sensitivity analysis tool.

Forrester (1961) built a system dynamics model that was composed of a three-echelon production distribution system, and he used it to demonstrate how market demands were amplified through the transactions in the supply chain. He founded the Systems Dynamics Group at MIT in the early 1960s. At that time, systems dynamics was called Industrial Dynamics. It illustrated the shifting nature and behavior of the companies over the passage of time, and has been applied to various fields from the social to the natural sciences.

System dynamics models have revealed and illustrated managerial problems in the supply chain. Senge and Sterman (1992) pointed out the risk of local decision-making by demonstrating the difference between local and global optimization. Paich and Sterman (1993) analyzed the diffusion process of a new product and identified the mechanism behind the boom and bust effect mentioned previously. Cheng (1996) proposed various integrated corporate system dynamics models emphasizing the role of information technology. Higuchi and Troutt (2004) simulated the whole life cycle of a short life cycle product, the TamagotchiTM toy, which was produced by the Bandai Corporation.

Copyright © 2008, IGI Global. Copying or distributing in print or electronic forms without written permission of IGI Global is prohibited.

Globalization Issues for Supply Chains

The globalization of supply chains is a main driver of the global economy. Supply chains act beyond national borders to achieve cost reduction and sales growth. They globally source the requisite resources, quality levels and competitive prices. The pencil is a good example. The main raw materials of the pencil, which are lumber, black lead, and clay, exist in Japan. However, Japanese pencil manufacturers import the lumber from the USA, the black lead from Brazil, and the clay from the UK, and export the products all over the world. It is reasonable for manufacturers to source resources globally for the desired quality and prices. In this way, global supply chains promote production in the right regions of the world and thereby induce a global division of labor.

Geographic location issues associated with supply chains have changed dramatically in the past few decades. First, information technology has progressed and spread globally, enabling rapid transfer and processing of data, such as orders and designs, without regard to geographic distance. This contributes to a smooth global division of labor and international collaboration. Second, managerial technologies have become so sophisticated that companies located in different countries with quite different cultures can collaborate effectively. Many companies have established systematic approaches and gained ample experience with international transactions. Third, manufacturing technologies have standardized and become widely available in developing countries so the global division of labor can be performed easily. Modular manufacturing, sophisticated designs for fabrication, and digital technologies have reduced the earlier difficulties that hindered production in the developing countries. Many developing countries now have adequate technologies and techniques for the prompt production of new products. Finally, the advancement of transportation modes reduces delivery times and increases the trustworthiness of transactions. Such transportation modes guarantee the workability of global supply chains. Together with deregulation in trading and the advances in safekeeping techniques and order tracking during transportation, the amount of international trade can increase steadily. Prerequisite conditions for the global supply chain have been satisfied and management then becomes crucially important for competence in the market.

The globalization of supply chains nevertheless aggravates managerial difficulties. It lengthens the geographical and intellectual distances among the partners and leads to information distortions, lags, and deceptive business practices

Copyright © 2008, IGI Global. Copying or distributing in print or electronic forms without written permission of IGI Global is prohibited.

within the supply chain as mentioned earlier. The design of the structure of the global supply chain should be aimed at balancing global logistics, costs, and customer satisfaction. From the viewpoint of economies of scale alone, it would be most efficient to have all manufacturing processes concentrated in a single country. On the other hand, in order to increase global sales, supply chains must more typically have multiple production facilities in different countries for quick response to the customer, to lower transportation costs, and to accommodate specialized local needs. In fact, based on these criteria, the ideal situation is to have production as close to the customer as possible. The competitive global supply chain must achieve efficient and effective flow and storage of the parts and products from the point of origin to their globally dispersed consumers. Within the constraints of their structures, global supply chains must adapt according to the stages of life cycles in which products, companies, supply chains, and industries exist.

Emerging Technologies in SCM

One of the major emerging technologies in the next few decades is Information Technology (IT), which started growing in the 1990's. IT impacts SCM both directly and indirectly. The main features of IT lie in computing and telecommunications. IT makes it possible for supply chain partners to gather and analyze information easily and quickly by EDI (Electric Data Interchange), which has contributed substantially to the enhancement of SCM. On the other hand, IT poses some difficult problems for SCM because IT is almost free from the constraints of physical distance. The flow of products and parts becomes very lengthy but occurs quickly because of the ease of ordering globally.

Chopra and Meindl (2001, pp.337-338) divided the information related to SCM into the following four basic components:

1. Supplier information, which includes the list of available products, the price, the lead time, the delivery conditions, and the payment arrangement.
2. Manufacturing information, which includes the list of available products, the cost, the lead time, the batch size, the location of facilities, and so on.
3. Distribution and retailing information, which includes the list and the quantity of the products shipped and delivered, the transportation mode, the storage cost at each site, the lead time, and so on.

Copyright © 2008, IGI Global. Copying or distributing in print or electronic forms without written permission of IGI Global is prohibited.

4. Demand information, which consists of the sales information, customer information, and forecasting information.

They also provided three necessary characteristics to be useful when making supply chain decisions:

1. Accessibility: the information must be accessible in a timely manner.
2. Accuracy: the information must be accurate.
3. Necessity: the information must be of the right kind.

The Internet is an open system, on which people can exchange information quickly with others anywhere in the world. It also dramatically changes the way in which business is conducted. The Internet expands the business area so widely that the global sourcing comes to be taken as a matter of course. Materials, parts and products travel very long-distances and a price war spreads to the whole world. On the other hand, the Internet creates great pressures on time. We can order various products through the Internet in an instant without considering where the products are made. Fast and easy ordering becomes worthless if it takes a few months to receive the products and if the received products are of poor quality. The Internet accelerates the agility of businesses and, simultaneously, heightens the requirements for SCM in cost, delivery time and product quality. However, the Internet has weaknesses in the accuracy and overabundance of the information. The Internet provides too much information to check the accuracy and select the necessary information. Figure 5 shows how this occurs and which parts the Internet and ERP contribute.

At the present time, SCM has become sophisticated by the application of ERP (Enterprise Resource Planning) and other SCM software. ERP systems include the software, terminals and users function as a closed system because they deal with the secure data. ERP systems globally connect any part of the company and its supply chain, and provide real-time transactional data related to manufacturing, logistics, finance, sales (order fulfillment), human resources, and supplier management. There are a great number of the ERP software vendors who do a professional service to implement ERP. SAP and Oracle are worldwide major vendors. ERP can reduce not only direct costs such as inventory and procurement costs, but also the indirect costs by improving business processes. The office automation and the paperless

Copyright © 2008, IGI Global. Copying or distributing in print or electronic forms without written permission of IGI Global is prohibited.

Figure 5. Contributions of emerging technologies on SCM

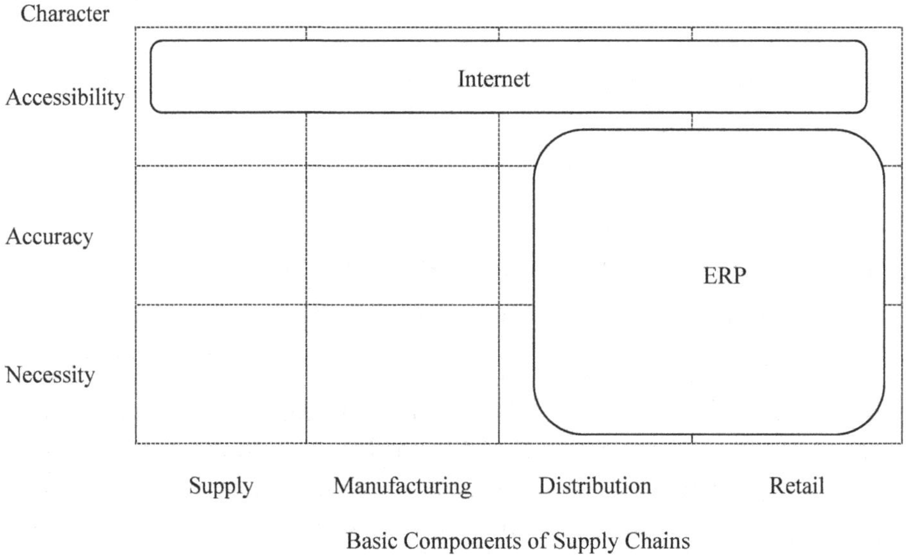

office dispense the labor and the resources. ERP contributes greatly to the efficiency and agility of the supply chain. Japanese major electrical appliance manufacturers, such as Hitachi (SAP R/3), Mitsubishi (SAP R/3), and Toshiba (ORACLE Applications), introduced ERP by 2000. However, there are some difficulties in implementing ERP (Shapiro 2001). First, ERP systems may impose too rigid requirements on data and processes for companies to operate. Second, it is impossible to employ software from the multiple vendors. Finally, all supply chain partners can not implement ERP software because it is expensive. Furthermore, although ERP systems assist the decision making by providing a wide scope of real time information, they have a weakness as an analytical tool because their main target is the distributor's efficiency in the operational and planning level, not the strategic level.

Analytical functions are reinforced by other software which can generate better forecasts and through which partners can share more accurate and timely information. Procurement and cataloging applications are analytical applications which focus on the procurement process between a manufacturer and its suppliers. Advanced Planning and Scheduling (APS) is a powerful tool, which can produce the schedules subject to the material availability and the production capacity. Manufacturing Execution System (MES) mainly assists

Copyright © 2008, IGI Global. Copying or distributing in print or electronic forms without written permission of IGI Global is prohibited.

Figure 6. Map of analytical applications (Chopra & Meindl, 2001, p. 349)

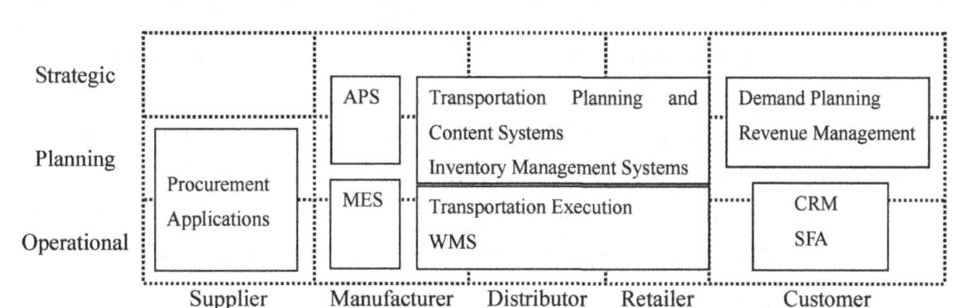

Figure 7. Relationship between emerging technologies and physical supply chain

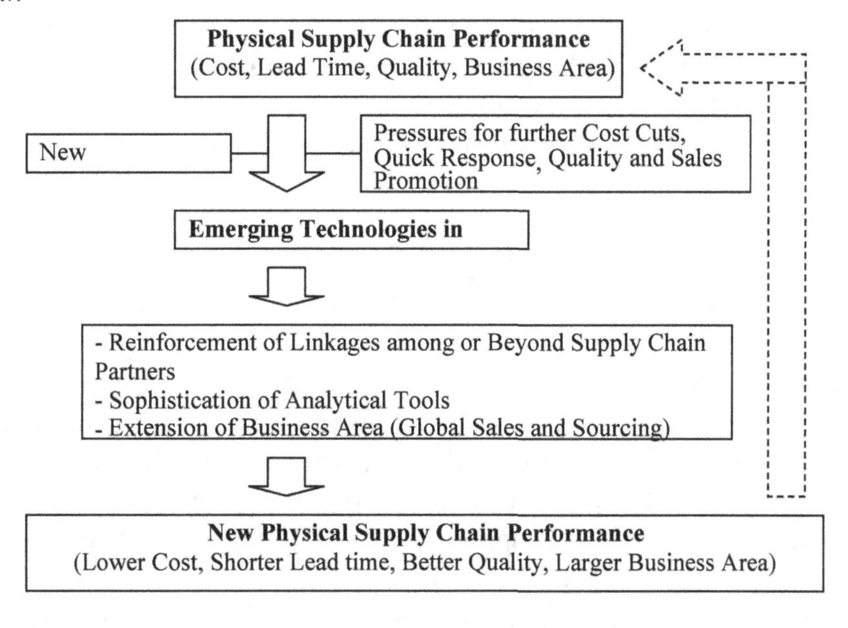

production according to short-term schedules. Transportation Planning and Content Systems propose efficient shipment plans considering the mileages and tariff requirements. Inventory systems produce the recommended inventory policy based on the demand forecasting, the costs, the margins, and the service level. Transport Execution and Warehouse Management System (WMS) make a daily plan for the transportation and inventory control.

Copyright © 2008, IGI Global. Copying or distributing in print or electronic forms without written permission of IGI Global is prohibited.

Demand Planning applications forecast the demand based on the demand trend and the seasonality and Revenue Management Applications can give recommendations on how to discriminate the price in order to maximize the customer surplus. Customer Relationship Management (CRM) and Sales Force Automation (SFA) provide the product, price and customer information in real time. Major ERP vendors and others including the consultant companies and major manufacturers who adopted ERP or developed their own systems launch various analytical applications which supplement ERP.

Figure 7 summarizes the relationship between emerging technologies in SCM and the physical supply chain. They come from technical innovations and the strong demand for the further cost declines, quick response, quality, and sales. IT such as the Internet raises the demand for the supply chain performance and, on the other hand, assists the physical supply chain to be more efficient and effective together with management technologies. Emerging technologies in SCM and the physical supply chain are two sides of the same coin. In other words, we can not enjoy the merits of emerging technologies without the advancement of physical supply chain performance.

Conclusion

We discussed the basic concepts of the SCM by using the SCM models in this chapter. The comparisons between the similar concepts of business logistics, physical distribution and Keiretsu highlight the importance of SCM. The purpose of SCM is to offer products to the customers efficiently. To do that, the headquarters of supply chains scratch together the various partners whose interests are not always identical. Supply chains are vertical systems in which supply chain partners belong to various layers between customers and materials. As a result, supply chains are multi-echelon systems whose processes are very complex.

Nowadays, global sourcing is very popular even for the products whose materials can be sourced near the big markets. Cost reduction is its main purpose. Companies can reduce costs dramatically by making the most of economies of scale in the low cost operation areas. Thus, supply chain processes become very long.

Long and complex supply chain processes can cause unexpected results because of the dynamic interactions in supply chains. One of the big tasks

Copyright © 2008, IGI Global. Copying or distributing in print or electronic forms without written permission of IGI Global is prohibited.

of SCM is to control these dynamic interactions which can have very negative influences upon supply chains. Information distortion within the supply chain is a major internal cause of the bullwhip effect and the boom and bust effects, most of which are created within supply chains.

References

Akkermans, H. (1995). Developing a logistics strategy through participative business modeling. *International Journal of Operations & Production Management*, *15*(11), 100-112.

Ballow, R. H. (1992). *Business logistics management.* New Jersey: Prentice Hall.

Barker, K. R., & Kropp. D. H. (1985). *Management science.* New York: John Wiley & Sons.

Becheck B. & Brea, C. (2001). Deciphering collaborative commerce. *Journal of Business Strategy*, *22*(2), 36-38.

Bowersox, J., & Closs, D. J. (1996). *Logistical management.* New York: McGraw-Hill.

Bowersox, J. D., Closs, D. J., & Helferich, O. K. (1986). *Logistical management: A systems integration of physical distribution, manufacturing support, and materials procurement.* (3rd ed.) New York: Macmillan Publishing Company.

Bradley, P. J., Thomas, T. G., & Cooke, J. (1999). Future competition: Supply chain vs. supply chain. *Logistics Management and Distribution Report*, *39*(3), 20-21.

Chen, F. (1999). Decentralized supply chains subject to information delay, *Management Science*, *45*(8), 1076-1090.

Cheng, H. (1996). *Enterprise integration and modeling: The meta base approach.* Boston: Kluwer Academic Publishers.

Corbett, C. J, Blackburn, J. D. & Van Wassenhove, L. N. (1999). Partnerships to improve supply chains. *Sloan Management Review*, *40*(4), 71-82.

Cox, J. & Spencer, M. S. (1998). *The constraints management handbook.* APICS Series on Constraints Management. Boca Raton, FL: St. Lucie Press.

Copyright © 2008, IGI Global. Copying or distributing in print or electronic forms without written permission of IGI Global is prohibited.

Davis, T. (1993). Effective supply chain management. *Sloan Management Review, 34*(4), 35-46.

Dettmer, H. W. (1997). *Goldratt's theory of constraints: A systems approach to continuous improvement*. Milwaukee, WI: ASQC Quality Press.

Dunning, J. H. (1988). *Explaining international production*. London: Unwin Hyman.

Fisher, M. (1997). What is the right supply chain for your product? *Harvard Business Review, 75*(2), 105-116.

Forrester, J. W. (1961). *Industrial dynamics*, Boston: MIT Press.

Gavirneni, S., Kapuscinski, R., & Tayur, S. (1999). Value of information in capacitated supply chains, *Management Science, 45*(1), 16-24.

Goldratt, E. M. (1990a). *The theory of constraints*. Croton-on-Hudson, NY: North River Press.

Goldratt, E. M. (1990b). *The haystack syndrome: Sifting information out of the data ocean*. Great Barrington, ME: North River Press Publishing Corporation.

Goldratt, E. M. (1990c). *What is this thing called the theory of constraints and how is it implemented?* Croton-on-Hudson, NY: North River Press.

Goldratt, E. M., & Cox, J. (1992). *The goal: A process of ongoing improvement* (2nd ed.) Croton-on-Hudson, NY: North River Press.

Goldratt, E. M., & Fox, R. E. (1986). *The race*. Croton-on-Hudson, NY: North River Press.

Gopal, C., & Cahill, G.. (1992). *Logistics in manufacturing*, Homewood, IL: Richard D. Irwin, Inc.

Gossain, S. (2002). Cracking the collaboration code. *Journal of Business Strategy, 23*(6), 20-25.

Gunasekaran, A. D., Macbeth, K., & Lamming, R. (2000). Modeling and analysis of supply chain management systems: An editorial overview. *Journal of the Operational Research Society, 51*, 1112-1115.

Higuchi, T., & Troutt, M.D. (2004a). Dynamic simulation of the supply chain for a short life cycle product—Lessons from the Tamagotchi case. *Computers & Operations Research, 31*(7), 1097-1114.

Higuchi, T, & Troutt, M. D. (2004b). Understanding and managing the intrinsic dynamics of supply chains. In C. K. Chan, & H. W. J. Lee (Eds.), *Successful strategies in supply chain management* (pp. 174-193). Hershey, PA: Idea Group.

Copyright © 2008, IGI Global. Copying or distributing in print or electronic forms without written permission of IGI Global is prohibited.

Kendall, G. I. (1998). *Securing the future: Strategies for exponential growth using the theory of constraints*. Boca Raton, FL: St. Lucie Press.

Lee, H. L., Padmanabhan, V., & Whang. S. (1997a). Information distortion in a supply chain: The bullwhip effect. *Management Science*, *43*(4), 546-558.

Lee, H. L., Padmanabhan, V. & Whang, S. (1997b). The bullwhip effect in supply chains. *Sloan Management Review*, *38*(3), 93-102.

Lowson, B., King, R., & Hunter, A. (1999). *Quick response*. New York: Wiley.

Luebbe, R. & Finch, B. (1992). Theory of constraints and linear programming: a comparison. *International Journal of Production Research*, 30, 1471-1478.

Mabin, V. J., & Balderstone, S. J. (2000). *The world of the theory of constraints: A review of the international literature*. APICS Series on Constraints Management. Boca Raton, FL: St. Lucie Press.

Mabin, V. J., & Gibson, J. (1998). Synergies from spreadsheet LP used with the theory of constraints: A case study. *Journal of the Operational Research Society*, *49*(9), 918-927.

Magee, J. F., Copacino, W. C., & Rosenfied. D. B. (1985). *Modern logistics management: Integrating marketing, manufacturing, and physical distribution*. New York: John Wiley & Sons.

McMullen, T., Jr. (1998). *Introduction to the theory of constraints (TOC) management system*. Boca Raton, Florida: St. Lucie Press.

Nehmias S. (1997). *Production and operations analysis* (3rd ed.). New York: McGraw-Hill.

Nersesian, R. L., & Boyd, S. G. (1996). *Computer simulation in logistics*. London: Quorum Books.

Paich, M. & Sterman, J. D. (1993). Boom, bust, and failures to learn in experimental markets. *Management Science*, *39*(12), 1439-1458.

Pidd, M. (1984). *Computer simulation in management science* (2nd ed.) West Sussex (Chichester), UK: John Wiley & Sons.

Schragenheim, E. (1999). *Management dilemmas: The theory of constraints approach to problem identification and solutions*. Boca Raton, FL: St. Lucie Press.

Copyright © 2008, IGI Global. Copying or distributing in print or electronic forms without written permission of IGI Global is prohibited.

Schwarz, L. B., & Weng, K. Z. (1999). The design of JIT supply chains: The effect of leadtime uncertainty on safety stock. *Journal of Business Logistics*, *20*(1), 141-163.

Senge, P. M., & Sterman, J. D. (1992). System thinking and organizational learning: Acting locally and thinking globally in the organization of the future. *European Journal of Operational Research*, *59*, 137-150.

Shapiro, J. F. (2001). *Modeling the supply chain*. Pacific Grove, CA: Duxbury.

Sunil, C. & Meindl, P. (2001). *Supply chain management*. New Jersey: Prentice Hall.

Takeda, K., & Kuroda, M. (1999). Optimal inventory configuration of finished products in multi-stage production/inventory system with an acceptable response time. *Computers & Industrial Engineering*, *37*, 251-255.

Tompkins, J. A. (2000). *No boundaries*. Raleigh, NC: Tompkins Press.

Troutt, M. D., White, G. P., & Tadisina, S. K. (2001). Maximal flow network modelling of production bottleneck problems. *Journal of the Operational Research Society*, *52*, 182-187.

Vendemia, W. G., Patuwo, B. E., & Ming, S. H. (1995) Evaluation of lead time in production/inventory systems with non-stationary stochastic demand. *Journal of the Operational Research Society*, *46*, 221-233.

Vennix, J. M. (1996). *Group model building*, West Sussex (Chichester), UK: John Wiley & Sons.

Winker, J., Towill, D. R., & Naim, M. (1991). Smoothing supply chain dynamics. *International Journal of Production Economics*, *22*, 231-248.

Zeleny, M. (1981). On the squandering of resources and profits via linear programming. *Interfaces*, *11*(5), 101-117.

Zeleny, M. (1986). Optimal system design with multiple criteria: De novo programming approach. *Engineering Costs and Production Economics*, *10*, 89-94.

Endnotes

[1] A modular system contributes to a decrease in the total amount of inventory because the manufacturer can offer a wide variety of goods through

Copyright © 2008, IGI Global. Copying or distributing in print or electronic forms without written permission of IGI Global is prohibited.

combinations of modules. As a result, it localizes the impact of model changes (Abernathy, 1978).

2 STELLATM is the most popular System Dynamics software (STELLA and STELLA Research software Copyright 1985, 1987, 1988, 1990, 1997, 2000, 2001 High Performance Systems, Inc. All rights reserved).

Copyright © 2008, IGI Global. Copying or distributing in print or electronic forms without written permission of IGI Global is prohibited.

Chapter II

Review of Life Cycle Theories

In this chapter, we discuss the life cycle theories related to the business. The concept of the life cycle has been widely used in marketing. The Product Life Cycle (PLC) is the most well-known one, in which the time is divided into four stages based on the change of sales. It is expanded by combining it with the study of the various consumer types. Other life cycles have been developed from the viewpoint of the innovation and manufacturing facility location. The advancement of technology is the driver for the diffusion of a new product. Sometimes it obsoletes a category of products. The location of manufacturing facilities changes according to the market and technology condition as Product Cycle Theory demonstrates. A concept of the industrial life cycle and a linkage between the life cycle and SCM also are argued in this chapter.

Copyright © 2008, IGI Global. Copying or distributing in print or electronic forms without written permission of IGI Global is prohibited.

Three Life Cycle Theories Related to Business Practice

Various life cycle theories exist in the natural and social sciences. Their similarity is that they start from the birth and come to an end or death, through growth, maturity, and decline stages. Life cycles vary based on the objects of study. In the social sciences, Product Life Cycle (PLC) is a well-known life cycle theory that focuses on the life of certain products from the viewpoint of marketing (Kotler, 1999). Products can be grouped into three classes: the individual products, a series of products, and the whole category of products. Henceforth, we call a whole category of products an industry. These life cycles are quite different from each other, in both duration and substance. Two other life cycle theories are based on innovation and location. Abernathy and Clark (1983) explained what the product and production process innovations were and how they progressed. The state of the innovation determines the function, the cost, and quality of the products and the intensity of market competition. The location of facilities is also affected by the degree of innovativeness of the product. Vernon (1966) introduced the product cycle theory, which proposes the relationship between the product and the reasonable location of production facilities.

These three theories are compatible because the state of innovation can function as their main axis. Product innovations add critical functions to products, change their basic design, and sometimes, renew them. While the basic functions and designs are fluid, manufacturers locate facilities near the customer to reflect customer needs quickly and prepare for mass production. At the first entry of the product, although it may be far from sophisticated, it is expensive because of the impracticability of early mass production. Thus, customers are largely limited to innovative adopters, which are considered to be the first 2.5 percent of potential consumers. In other words, the early products are far behind the level of the ultimately desired products and most consumers will wait for the progress (Christensen, 1997). After the basic functions and designs become stable, manufacturers standardize the manufacturing process and start mass production. Next, competitors enter the market. The price falls dramatically and the quality is improved. The range of consumers expands rapidly. Assuming that product and process innovation do not occur, foreign production becomes essentially the only way to reduce the cost. Finally, marketing activities intensify and offer consumers a wide variety of products and options.

Copyright © 2008, IGI Global. Copying or distributing in print or electronic forms without written permission of IGI Global is prohibited.

Life Cycle Patterns

Generally, the life cycle is judged by the transaction of sales. Doyle (1976) demonstrated the four-stage model, which has been widely adopted (Figure 1). During the introduction period, the sales amount does not increase very much because the product is expensive, unknown, and unrefined. In the growth stage, the amount of sales increases rapidly because of the improvement and wide acknowledgement of products. In the maturity stage, the potential for further growth diminishes because the product has been widely diffused. In the decline stage, the amount of sales decreases rapidly. The Doyle model was very practical and showed a lot of insights into the features and appropriate strategies of the stages in the life cycle. The amount of sales in money (the turnover) peaks earlier and decreases more sharply than that of units because of the effect of the price down in the long run. It is a question of which indicator, the turnover or the number of units sold, is better to measure the stage of life cycle.

The repeat purchase alters the shape of the life cycle changes. It creates a second and third peak after the first one. The intervals between peaks are regarded as the average lifetime of the product. The sales amount in units at the second or third peak might surpass the first peak when the repeat rate is very high. Repeat purchases also make the maturity stage longer because the sales amount remains at a high level. Figure 2 illustrates the life cycle including the repeat purchase activity.

Figure 1. Life cycle pattern based on sales (Based on Doyle, 1976)

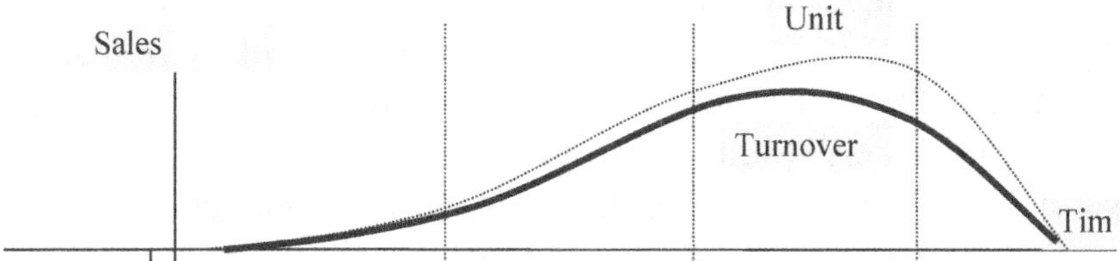

Stage	Introduction	Growth	Maturity	Decline
Sales	Slight Increase	Great Increase	Slight Increase	Decrease
Mission	Expansion	Penetration	Protection	Productivity

Copyright © 2008, IGI Global. Copying or distributing in print or electronic forms without written permission of IGI Global is prohibited.

Figure 2. Life cycle considering repeat purchases

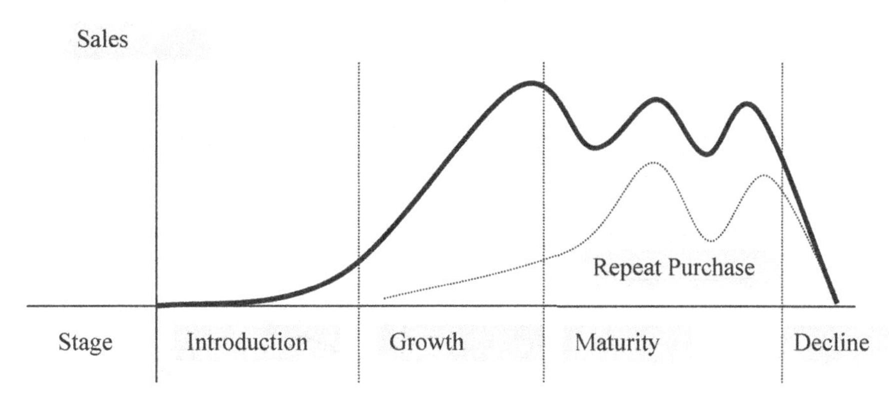

Consumer types have great impact on the life cycle. Moore (1991, 2005) categorized the various consumer types and described their characteristics. At a very scare 2.5 percent, innovators or techies are enthusiastic about new technologies and products. Early adopters (visionaries), also rare at just 13.5 percent, are very important because they motivate the majority to follow their example. The new product should display innovativeness and practicability; otherwise it will not be adopted by the visionaries. The early majority (pragmatists) amounts to 35 percent and it follows the early adopters. Pragmatists place emphasis on the practical use of the product. The late majority (conservatives) also constitutes 35 percent and waits until excellent product performance has been established. They value price over newness because the performance has been greatly improved. Laggards (skeptics) are the rest of the adopters (15 percent) who purchase the product passively.

These life cycle patterns, we believe, are the result of the combined effect of the business activity and the consumer behavior. Innovations are divided into product innovation that directly improves product performance and process innovation, including mass production and a flexible manufacturing system, that induces other companies to enter the market and which weakens restrictions on the location of factories. As a result, companies propose a variety of marketing mixes, performance and price. Although the consumers are interrelated with others to some extent as Moore (1991, 2005) mentioned, they make a final decision based on the performance and price. Figure 4 shows the relation among these components.

Copyright © 2008, IGI Global. Copying or distributing in print or electronic forms without written permission of IGI Global is prohibited.

Figure 3. Life cycle considering consumer types (Based on Moore, 1991, 2005)

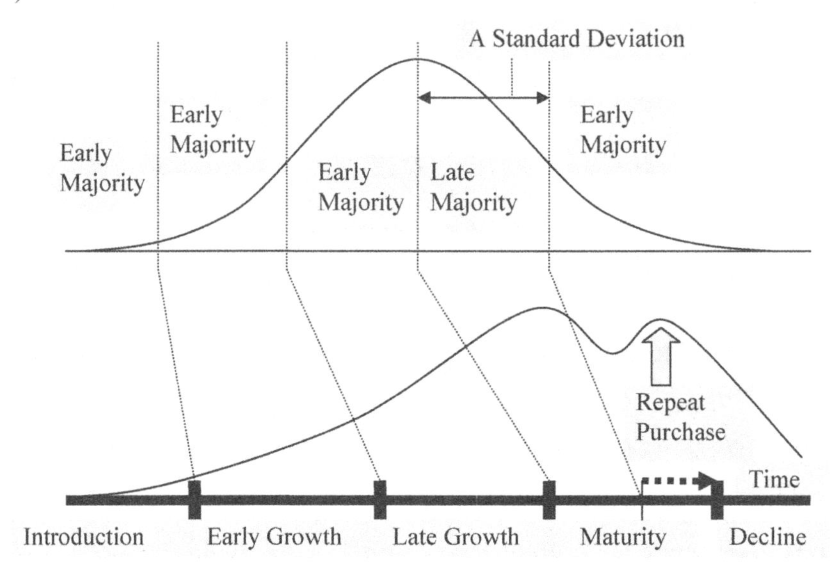

Figure 4. Relation among life cycle Components

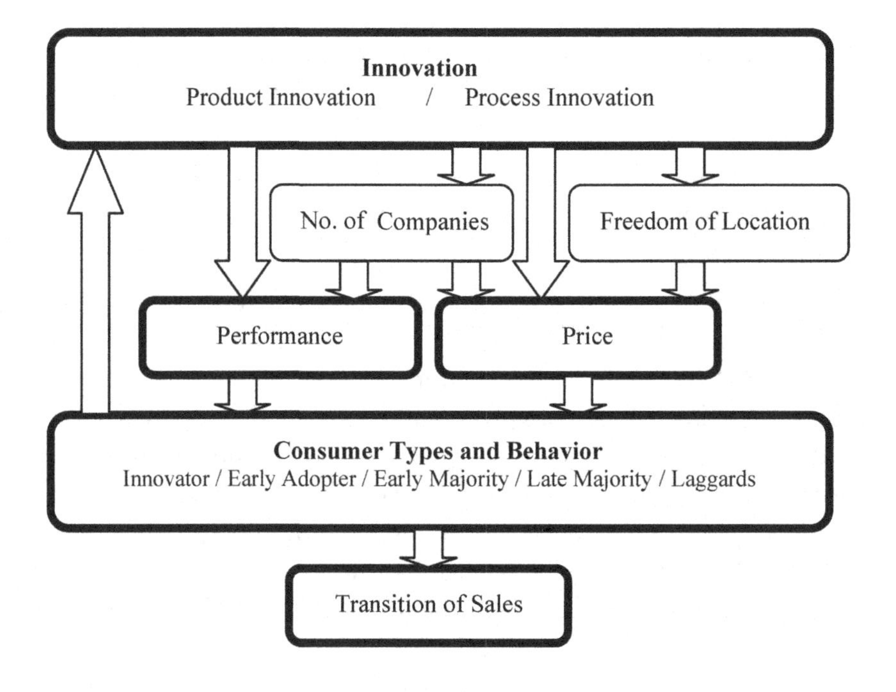

Copyright © 2008, IGI Global. Copying or distributing in print or electronic forms without written permission of IGI Global is prohibited.

Innovation and the Productivity Dilemma

Abernathy (1978) and Abernathy, Clark, and Kantrow (1983) explained the development of an industry in terms of the innovation concept. The age of an industry can be divided into two eras, namely, before and after the dominant design has been established. The former can be further subdivided into two periods: the embryonic and prenatal periods. The latter also can be subdivided into three periods: product innovation, process innovation, and the productivity dilemma. However, these processes are reversible due to new inventions and usages. Finally, a new industry that surpasses the existing one in performance related to price, obsoletes them.

The era before the dominant design can be divided into the embryonic and prenatal periods. In the embryonic period, the technical seeds for the products exist. Although some companies may have core technology capabilities for a nascent product, they are far from practical use because of the lack of peripheral technologies. Companies that have related technologies activate research and development and take various approaches in order to make the most of their technologies. Although the potential need for many emergent products clearly exists, very few such products are actually developed because companies may fail to produce even a prototype.

In the prenatal period, companies introduce an innovative product and grope for the minimum requirements from the consumers, such as the desired basic functions, quality level, price, and size. Some companies succeed in producing prototypes. However, most of them cannot satisfy the minimum requirements. In addition, companies that do succeed in producing a prototype fulfilling the minimum requirements in function and size, will face the next difficult problem, that of commercialization. Commercialization requires the companies to reproduce high quality prototypes at a low cost. In addition, the potential for mass production is necessary. This period ends with the emergence of a dominant design that satisfies the minimum requirements and considerations of manufacturability.

The dominant design is the first product in the class to show significant potential to satisfy latent customer needs in the near future and to be highly valued by the innovative adopters. The dominant design is called the de facto standard because it establishes the standards of the product, such as specifications and usages. For example, the Model T, introduced at a price of less than $1,000 by Ford in 1908, became a dominant design in the auto industry. It synthesized the existing technologies and established the framework of the

Copyright © 2008, IGI Global. Copying or distributing in print or electronic forms without written permission of IGI Global is prohibited.

automobile, and it continues to be the original model of modern automobiles (Abernathy, 1978). The company that establishes the de facto standard has a big advantage because it owns the patent rights and it makes the most of its knowledge and facilities.

After the emergence of the dominant design, the product can be developed incrementally. Product innovations continue to have a great effect on the product's performance because the initial products have a great potential or room to develop. Newcomers who own the technologies that contribute to the product's development, enter the industry and compete with others based on the dominant design by adding new functions and by significantly improving the existing functions. Consumer electronics companies have especially utilized this strategy by continually repeating the process of adding new functions or enlarging the number of products, and by greatly downsizing or miniaturizing the products. When the product design is unstable, companies hesitate to automate the whole manufacturing process because they have substantial risk that their customized machines may need improvement in the very near future.

Process innovations become the main focus in the industry for expanding the range of customers after the product innovations have become stagnant. Process innovations have greater effects on the cost and quality of the product than do product innovations. Standardization and modularization play a very important role in process innovation (Abernathy, 1978). Standardization isolates and localizes the impact of design changes. In the event that some parts change, supply chains can accommodate them without any changes or with only minor changes. Modularization guarantees compatibility with other suppliers in the chain. It enables the supply chain to produce a wider range of products while maintaining low levels of inventories. Process innovations promote other newcomers, or industry competitors, to enter the market. Often such entrants have strength in cost cutting or they serve a niche market.

Gradually, the impact of process innovations becomes dilutive. Companies advance the standardization and modularization of the production process in order to more efficiently handle a wide variety of products. In other words, most companies have difficulty differentiating their products from others. At this stage the industry is said to be mature. Although it achieves excellent productivity and high quality production at a low cost, the opportunities for improvements and innovations of the product diminish. Abernathy (1978) called this phenomenon the productivity dilemma.

Copyright © 2008, IGI Global. Copying or distributing in print or electronic forms without written permission of IGI Global is prohibited.

Under the productivity dilemma, the industry becomes stifled. In order to improve this situation, some companies search for a brand-new technology, usage, material, or design. If the industry succeeds in purging the old design, sales grow again and the status of competition in the industry once again becomes fluid. This process or phase is called de-maturity. For example, the de-maturity process is evident in the TV industry. Tube or cathode ray TVs started switching to LCD (Liquid Crystal Display) TVs, Plasma TVs, and Rear Projection TVs at the beginning of the 21st century. These types of televisions utilize distinct techniques and core parts. Once the de-maturity starts, some existing technologies and production facilities may become obsolete. Hence, most de-maturity processes are destructive.

Regardless of the emergence of the de-maturity process, an existing industry may disappear if it is surpassed and taken over by a new industry. This industry takeover assumes that the cost to the consumer for switching products is relatively low. The VCR industry faced this crisis at the beginning of the current century. At that time, consumers began to prefer DVD (Digital Video Disk) and HDD (Hard Disk) recorders to VCRs (Video Cassette Recorders). VCRs previously dominated in recording TV programs and the playing and reproduction of movies in the home. Other devices such as LDs (Laser Disk) and DVDs challenged the VCR but ended in failure in the 1980s and 1990s because the switching cost was too high. Since the mid-1990s, some PCs and home video game machines have provided a function for recording DVDs; in addition, some PCs can now record TV programs. The VCR thus lost its supremacy little by little. In the mid-2000s, most household appliance manufacturers withdrew production from the VCR and shifted to the DVD and HDD recorders.

Figure 1 shows the shift in type of innovation from product innovation to process innovation and summarizes the development of products. The vertical axis indicates the impact of product and process innovation on product performance and cost. The first step is to identify the innovations and count the number of both types of innovations. The second step is to classify these into major and minor innovations. The third step is to estimate the impact of the product and process innovations on product performance and cost. Generally, the impact of product innovations on product performance achieves its peak when a dominant design has been established. After that, it steadily weakens. The impact of process innovations remains at a low level while the impact of product innovations is strong. After product innovations subside, process innovations become active. This also weakens after its peak as Figure 5 shows.

Copyright © 2008, IGI Global. Copying or distributing in print or electronic forms without written permission of IGI Global is prohibited.

Figure 5. Transition of impacts of product and process innovation (Based on Abernathy, 1978, p. 72)

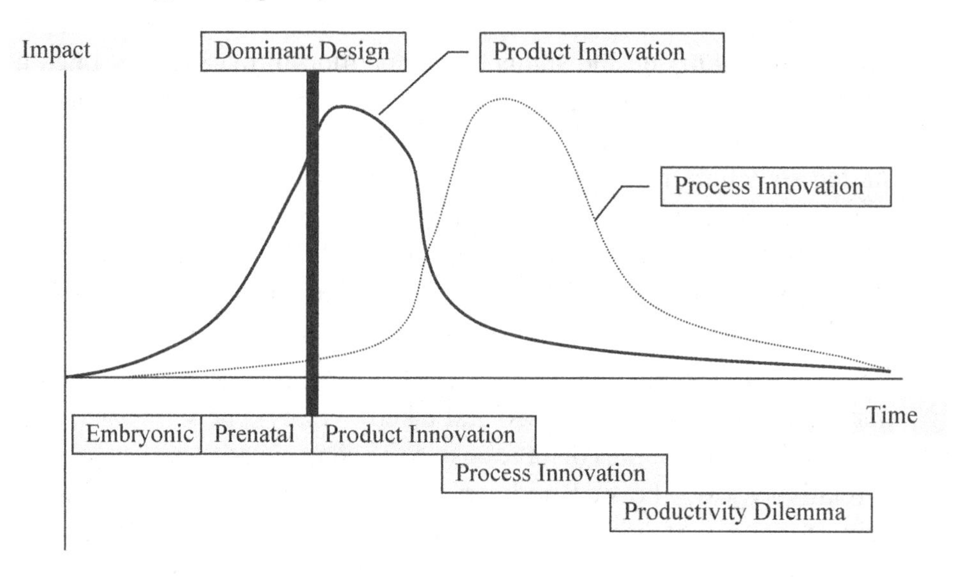

Figure 6. De-maturity process (Based on Shintaku, 1994, p. 16)

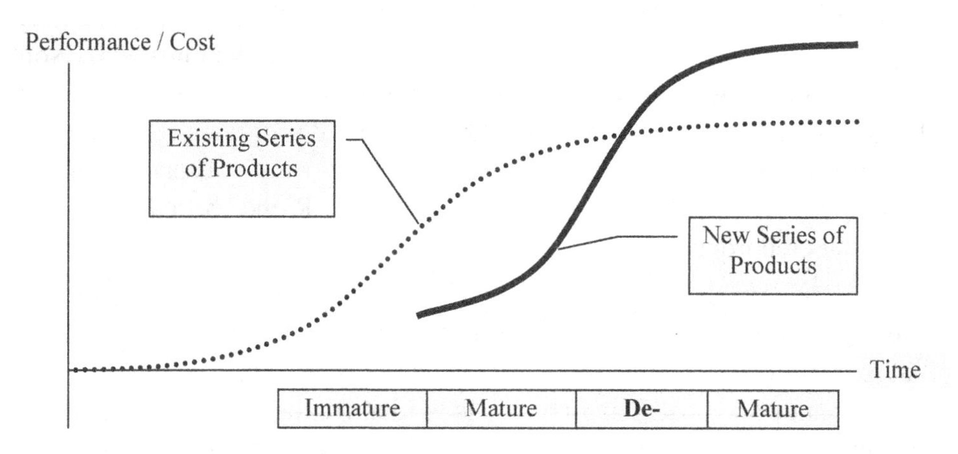

Figure 6 illustrates the process of de-maturity. The vertical axis indicates the performance/cost ratio. At the beginning, it is very low because the product performance is very low and the cost is huge. After the product innovations occur, the ratio rises rapidly. It continues to grow while process innovations are in effect. It plateaus in the productivity dilemma phase. Sometimes a new technology and/or usage enter here and demonstrate a bigger potential than the existing one.

Copyright © 2008, IGI Global. Copying or distributing in print or electronic forms without written permission of IGI Global is prohibited.

SCM Considering Innovation and the Productivity Dilemma

Recently, products have become so complicated and specialized that supply chains should unite to launch a brand-new product. Supply chains are continually seeking the seeds of future success, such as new technologies, materials, and usages, for a new market. It has become much more difficult than before to make a new product based on a simple idea or technology. After they identify the related technologies, they coordinate operations therein or construct new partnerships to commercialize it quickly. SCM has a considerable effect on innovation and has different roles according to the periods classified by the condition or stage of the innovation.

Before the dominant design appears, some companies seek a new market. In the embryonic period, although these companies have the ideas and the technical seeds for the new industry, its actual realization is still quite distant. They must take a great risk in research and development and concentrate on the core technologies to establish the new product's basic functions and specifications. This requires understanding the future feasibility related to the peripheral parts that suppliers would manufacture. Supply chains are not formed at this time, but future supply chain partners can have a say in establishing the basic concept and specifications to some extent.

In the prenatal period, supply chains are formed to produce the prototype. Based on an initial design, supply chain partners can cooperate to complete the first prototype. As a general rule, the weakest parts determine the performance of products. Latent customer needs also must be articulated appropriately to succeed in launching a new market. After the first prototype, more sophisticated designs appear for improving product performance and in preparing for mass production. Supply chains face a time competition as to how fast a product that satisfies the minimum requirements in the market can be developed. The range of supply chain partners determines the feasibility and the collaboration in R&D among them and affects the speed of developing a new category of products.

Just before the dominant design emerges, some supply chains may succeed in developing the products that satisfy the minimum requirements and are able to be mass-produced. In this case, supply chains compete with each other to achieve the dominant design by being popular, and thus recognized as the de-facto standard in the market. Success depends not only on the excellence of performance and the price of the product, but also on the strength of the

Copyright © 2008, IGI Global. Copying or distributing in print or electronic forms without written permission of IGI Global is prohibited.

distribution channel. Even if a new product exhibits technical excellence, it sometimes can fail to be popular because of a weak distribution channel[1]. Hence, the core company plays a very important role as director and coordinator of the whole supply chain during incubation of the new market.

Just after the dominant design is determined, product innovations sophisticate the product and the basic design. It can be said that the industry is still immature at this time because supply chains emphasize the development of product performance rather than the establishment of a mass production system. Many supply chain partners need to adjust to changes and to contribute to the development of the product. In this period, consumer demand for the new product often grows so rapidly that supply chains should start preparing for mass production. Gradually, supply chains enable modular manufacturing. As a result, the impact of changes can be localized.

Maturity proceeds as process innovations occur frequently. After the basic design and the specifications are set, supply chains standardize the manufacturing process and establish a mass production system to reduce costs and prices. Compatibility of parts makes it easy for supply chain partners to collaborate together. This also increases the quality of the products and decreases manufacturing costs. In addition, supply chains review their logistical costs to provide the product more efficiently. As a result, the performance/cost ratio continues to increase steadily as in Figure 6.

The advancement of the maturity phase leads to the productivity dilemma. Although the performance/cost ratio reaches a high level, little room is left for product and process innovations. By then, supply chains have established and sophisticated the supply system. They lose the effect of the technological advantages. For further cost cutting, they concentrate the product line according to their core competence, enhance the compatibility of the parts and pursue global sourcing opportunities.

Some supply chains with strength in relevant technologies try to initiate the de-maturity process by making major innovations such as introductions of new technologies and usages. However, the impact of most of these innovations is confined to a limited area due to modular manufacturing. Supply chains can absorb the impact of a major change through smaller changes in some peripheral parts. Otherwise, supply chains take a great risk. If maturity advances too much without the maturity process, the possibility that an alternative industry may obsolete the existing industry increases.

Copyright © 2008, IGI Global. Copying or distributing in print or electronic forms without written permission of IGI Global is prohibited.

Location of Manufacturing Facilities

Multinationals that have developed a global network commonly see the world as a chessboard on which they are conducting a wide-range of campaigns. The chessboard's squares are nation-states, and an enterprise can consider entering any one of them by a number of different means—by trading with dependent firms in the country, by developing alliances with enterprises already operating in the country, or establishing a subsidiary of its own in the country (Vernon, 1998, p. 22).

Rational location or geographic positioning of manufacturing facilities is a major business decision. It can have a great effect on the price, quality, and service level of products, and should be very flexible according to the specifics of a firm's situation. Vernon (1966) developed a theory of manufacturing facility location decisions, which is called the product-cycle model. It explained the location decisions of such facilities by taking into account the state of innovation or age of the product, international trade, and economies

Table 1. The main activities of supply chains based on the innovation

Period of	Activities
Embryo	Search for "seeds" and "needs" Formation of the initial supply chain
Fetus (Prenatal)	Production of prototype • Articulation of customer needs • Matching seeds and needs • Cooperation among the supply chain
Just Before DD (Dominant Design)	Establishment of DD • Mass-manufacturability • Diffusion of the product
Product Innovation	Development of the product • Improvement of product performance • Addition of new functions • Localization of the impact of product innovations
Process Innovation	Development of the efficient manufacturing system • Reduction of the cost • Improvement of product quality • Automation of the manufacturing process • Creation of varieties of the product Reduction of the logistics cost • Sharing the information on time • Shortening the cycle time • Reduction of inventories
Productivity Dilemma	Global sourcing for further cost cutting Search for new materials and usages for de-maturity

Copyright © 2008, IGI Global. Copying or distributing in print or electronic forms without written permission of IGI Global is prohibited.

of scale. The age of a product is divided into three stages depending on the state of innovation.

The propensity of multinational enterprises to use their home markets to develop and introduce new products stems from a series of powerful forces. It has been confirmed again and again in empirical studies of various sorts that successful innovations tend to be those that respond to the market conditions surrounding the innovation. The original idea may be developed almost anywhere, but successful innovation depends strongly on the compelling character of the demand (Vernon, 1977, pp. 40-41).

In the first stage the product is new or first introduced. It includes the period before a product is born and the period in which the product is considered technically new. In this stage, it is reasonable for companies to locate facilities in the most advanced countries. These countries not only have the latest technologies, they also have sophisticated consumers. Equal access to the latest technology does not always mean equal probability of the application of it in the generation of new products. To bring excellent new products to the market, companies need to adjust by interchanging usage information with sophisticated consumers on a close and frequent basis.

The second stage is product maturity. This stage is recognized by significant progress in process innovations. The demand for the product grows domestically as well as globally because of mass production using capital-intensive facilities that contribute to the simultaneous enhancement of quality and cost reduction. At this stage, many companies achieve manufacturability of the product and enter the market. They increase the variety of products and features, as well as pressure for price competition. For the growth of global sales, companies start locating manufacturing facilities near foreign consumers.

In the final stage, the product and processes become very standardized. It becomes difficult for any company to differentiate their products from others because the product innovations have little impact on product performance. In addition, most companies have nearly the same production efficiency. Very few process innovations can create competitive advantages and little room is left for differentiation in quality and manufacturing cost, except for labor and rent costs. Finally, manufacturing facilities tend to migrate to certain developing countries to further reduce labor cost.

However, the worldwide economic environment has changed dramatically in the past few decades. The product life-cycle theory should take into account changes in production and consumption in the global economy. First,

Copyright © 2008, IGI Global. Copying or distributing in print or electronic forms without written permission of IGI Global is prohibited.

companies have more options in global production than before. National borders are becoming a smaller problem because of the advancement of accessibility and tele-collaboration. The international flow of goods, materials, and information has become very efficient. In addition, many developing countries have accumulated the manufacturing know-how and increased ability to quickly respond to innovations. The advancement of modularity of product design makes collaboration in production very easy. Furthermore, the international capital market has also advanced. As a result, the feasible areas of global collaboration have expanded more rapidly than ever before. Second, the uniqueness of the market becomes a very important factor in creating new products. It has been said that the gap between rich and poor in the world is widening. On the other hand, many developing countries have strengthened their purchasing power. In addition, the Internet accelerates the global diffusion of products. People can order the products through the homepage and receive them allover the world which is supported by the advancement of the international courier service and account settlement system. Companies cannot neglect developing countries and need to plan for them at the beginning. Companies must also struggle to find the best timing and ways to customize their products for many local markets.

Location of Manufacturing Facilities from SCM View

Before the emergence of a product, the supply chain partners require effective and efficient communication among themselves. In response to the existence of innovative or sophisticated customers, many excellent suppliers locate in the most advanced countries. To launch a successful product to the market first, supply chains should reflect their customer needs correctly and embody them in the product quickly. Supply chain partners make the prototypes by trial and error. Generally, it is very effective for the supply chain to locate in the same market country because the speed of feedback determines the time required to complete the prototypes.

Immediately following the successful production of a prototype that satisfies minimum market requirements, supply chains start preparing for commercialization. To do this, they need to establish a mass production system that guarantees the quality of products and response to customer needs at a low cost. The limit of the supply of products may be a supplied part whose manufacturing capacity is lowest among all parts, and the estimates of the products may likewise be decided by the lowest quality among all parts. Therefore,

Copyright © 2008, IGI Global. Copying or distributing in print or electronic forms without written permission of IGI Global is prohibited.

supply chains continue to need close communication to find the bottleneck in their mass production system and improve it quickly as a whole.

Where a supply chain succeeds in commercializing a product, it should respond to market growth. Many opportunities exist for a single company and a supply chain to make improvements in mass production. Before a supply chain can localize the most change in design of a product, all partners should locate closely. Hence, it is reasonable that most partners locate their expanded manufacturing facilities in advanced countries.

As a product matures, demand in foreign markets grows. By this time, the production process is so advanced that the ratio of standardized parts has increased, and many labor-intensive processes, such as assembly, can be done in other countries. It often is easy to source cheap standardized in developing countries. Sometimes, local governments pass regulations to encourage relocation to their countries. The assembly process shifts to market countries or to low cost operation countries. It becomes feasible and effective for supply chains to transfer some processes to other countries to some extent.

A product becomes standardized according to the advancement of the product design and the manufacturing process. All supply chains experience difficulty in differentiating their products from others by performance. For further cost cutting, most supply chain partners relocate and concentrate their manufacturing processes in several developing countries. Supply chains can gain a competitive advantage through economies of scale in a low cost operation area. Table 2 summarizes the location implication from the SCM view.

Table 2. The location of manufacturing facilities from SCM view

Stage	Source of Advantage	Activities
New	Linkage among the Supply Chain Partners and the Innovative Customers	• Seeking for a Seed • Close Communication among Concerns • Advancement of Product
Maturing	Overseas Investment and Strategic Alliances	• Rationalization of Product Process • Sales Promotion in Foreign Markets • Global Sourcing for Cost Reduction
Standardized	Global Supply Chain Coordination	• Transfer of Manufacturing facilities to a few developing countries • Economies of scale by Concentrating the facilities • Reduction of Global Logistics Costs

Copyright © 2008, IGI Global. Copying or distributing in print or electronic forms without written permission of IGI Global is prohibited.

Product Life Cycle (PLC)

The last life cycle is the well-known Product Life Cycle (PLC). PLC views the life of certain products from the marketing perspective. Product life cycles are analyzed by the different base units, an industry (the whole category of products), a series of products, and an individual product. PLC is composed of four stages: introduction, growth, maturity, and decline (Kotler, 1999, pp. 354-373).

The life cycle of an industry incorporates the life cycles of series and individual products, and therefore, has the longest life among them. That of series also incorporates some individual products. That of an individual product has the shortest life among the life cycles. In the home-use VCR, a category of products including all home-use VCR products is regarded as an industry. The Monaural (standard) products, HiFi (High-Fidelity) products and SVHS (Super Home Video System) products are the series of products. Certain products, such as HR-3300, HR-F6, and others are individual products. All of them experience the introduction, growth, maturity and decline stages. We will explain these stages only at the industry level.

In the first or introduction stage a brand-new product is launched in the market. Sales grow very slowly because of the high price and low awareness of the product in the market. In addition, most customers are reluctant to change their behaviors and the distribution channels are still under formation in the beginning. Companies spend huge amounts of money on the manufacturing facilities and promotion of the new product. They combine the pricing and sales promotion approaches (Kotler, 1999, p. 361). The price strategy is divided into the skimming and penetration strategies. In the skimming strategy, companies set the price at a high level to capitalize on the first mover advantage. In the penetration strategy, they emphasize the great potential of the markets to attain commercial success in the long run by using the learning curve effect. In general, profit may be very low or negative in the introduction stage.

The second stage is the growth stage. The sales volume increases rapidly because the consumers acknowledge and start buying the product. The expansion of the market attracts new competitors who have various market-penetration strategies such as adding new functions, creating new market segments, and capitalizing on distribution channels, the price down, and so on. Profits may increase in this stage in accordance with the expansion of sales volume.

Copyright © 2008, IGI Global. Copying or distributing in print or electronic forms without written permission of IGI Global is prohibited.

The third stage is the maturity stage, which is divided into three phases: the growth maturity in which the growth rate starts declining, the stable maturity in which the sales volume plateaus, and the decaying maturity in which the sales volume starts declining. The slowdown of growth makes the competition severe. A few volume leaders pursue maximum profit through economies of scale. Others, including cost cutters, focus on cost-cutting and niches to obtain a certain sales volume and a high margin in a small target market. Many companies may have the maximum profit by concentrating their resources on profitable products while abandoning others.

The last stage is the declining stage, in which the sales volume declines. In this stage, the numbers of both the new and repeat purchases decrease because of the spread to most of potential customers or the emergence of alternative products. The decline in sales volume leads to overcapacity and price competition. As a result, the profits shrink rapidly. Most competitors withdraw from the market and those who remain abandon the unprofitable segments.

PLC from SCM Perspective

PLC has a great impact on SCM. PLC directs the supply chain to the appropriate market strategy in each stage. Appropriate strategies, such as pricing, promotions, model changes, distribution channels, service level, and others, are different according to the stages. On the other hand, SCM has an important role in accelerating product development and price down in modern business. In other words, SCM contributes greatly to shortening the durations of each stage in the PLC.

In the introduction stage, it is uncertain whether the new product will succeed or not. Supply chains compete to launch new a high-quality product first and to distribute it quickly. They need to simultaneously establish a supplier system and a distribution channel. However, a supplier system and a distribution channel seem to behave separately in this stage. A supplier system is composed of upstream partners such as a set-maker and numerous suppliers, whose main purpose is to improve product quality and process. A distribution channel is composed of downstream partners, such as retailers and wholesalers, who promote and deliver the new products.

Copyright © 2008, IGI Global. Copying or distributing in print or electronic forms without written permission of IGI Global is prohibited.

In the growth stage, demand for the product grows rapidly. Supply chains sometimes may face bottlenecks in the production process and incur opportunity losses. They start showing their real ability in the coordination between a supplier system and a distribution channel. The supplier system needs a very long time and takes a huge risk to expand their capacity. On the other hand, a distribution channel can respond much easier and more quickly to the expansion of customer needs than a supplier system subject to the capacity of a supplier system. Supply chains smooth the flow of materials, parts and products for market expansion.

In the maturity stage, the competition among supply chains becomes severe and the diversification of customer needs proceeds. Supply chains widen the variety of products to satisfy customer needs in order to increase or maintain the sales volume. In this stage, the mission of the supply chain is to respond to customer needs quickly and efficiently. To do this, the close communication between the supplier system and the distribution channel is an essential factor. In addition, supply chains advance the standardization of parts and global sourcing. They should guarantee a wide selection of products to customers while reducing the logistical costs.

In the decline stage, supply chains need to decide whether to persevere as a survivor or to withdraw from the market. To be a survivor, supply chains need to attain further cost cutting by squeezing the product selection and restructuring the supplier system. Some supply chains offer and receive OEM (Original Equipment Manufacturing) products for the economies of scale in the manufacturing and maintenance of product selection. Supply chains strengthen open transaction and global sourcing. Global sourcing is targeted for low cost operations based on cheap resources and economies of scale.

Industrial Life Cycle

The Industrial Life Cycle concept is developed in this chapter. Numerous versions of life cycles have been applied to many industries (Debresson & Lampel, 1985). The three major life cycle theories are mentioned above. Although these theories appear to be closely connected, they have not yet been sufficiently correlated. In this section, we attempt to integrate them through creation of the Industrial Life Cycle.

The above three life cycles are coaxial and complementary to each other. All of them use a time frame as a horizontal axis and their research object is a

Copyright © 2008, IGI Global. Copying or distributing in print or electronic forms without written permission of IGI Global is prohibited.

Table 3. Industrial life cycle

Stage	Introduction	Early Growth	Late Growth	Maturity	Decline
Innovation (Era of ---)	Dominant Design	Product Innovation	Process Innovation	Productivity Dilemma	(De-maturity)
Location	Close	Local Production	Global Production	Convergence	Withdrawal
Consumer	Innovator	Early Adopter	Early Majority	Late Majority Laggard	Laggards

category of products. The time frames of these life cycles can be integrated and divided into five stages as shown in Table 3. Separately, they explain the nature of innovation, rational location, and market trend. The growth stage is divided into two parts, the early and late growth, because the type of target consumer and the number of competitors are different between these stages. In the early growth, early adopters decide to purchase a new product with foresight after evaluating its feasibility and potential. Followers get a chance to enter the market because of market expansion. In the late growth, the early majority that is guided by the early adopters plays a very important role placing emphasis on newness and practical use. The further market expansion gives a chance to cost-cutters to enter into the market. The increase in sales amount (turnover) becomes lower than that of sales units because of severe competition. This type of integrated model is very helpful in understanding the movement of an industry. Hereafter, in this book the time frame listed in Table 3 is used to indicate the stages. Our standpoint is that industry prepares a production system for a targeted type of consumer and proposes product performance and price to them, and that consumers decide whether or not to purchase. In other words, the transition of sales is a combined result of the industrial activity and consumer behavior.

Emerging Technologies and Life Cycles

In this section, we consider durable goods that offer essential services and whose demands continue as objects of study. The list of these essential services could go on and on and include the washing, refrigeration, mobility, calculation, word processing, videotaping, telecommunications, and so on. The products offering these essential services demonstrate the typical pattern of the life cycle. Multiple generations of the products and the purchase patterns are the key concepts around which to arrange the life cycle.

Copyright © 2008, IGI Global. Copying or distributing in print or electronic forms without written permission of IGI Global is prohibited.

It is true that the necessity is the mother of invention. Companies or inventors succeed in launching a brand new product which satisfies a necessity after several attempts. Products are developed one by one with the new ideas and technologies so that these come to be put to practical use. It can be said that this is the beginning of the first stage of the life cycle.

Generally, the products in the first generation are so immature that they have much room for improvement in both product performance and cost. It follows that only a limited number of consumers, the innovators, adopt the products at this stage. Emerging technologies have important effects on other first generation products by the two routes of product development and mass production. In the first generation, product development has a much greater effect on the expansion of the market to new potential customers than on mass production because a critical problem for the first generation is achieving adequate product performance. The succeeding purchasers, the early adopters and the early majorities, are unwilling to pay for immature or incomplete products.

The second generations tend to appear when the diffusion speed has slowed down. Companies launch new generation products to spur the market and increase sales. The adopters such as innovators and the early adopters have a willingness to buy new generation products because they offer better performance and have a better performance to cost ratio than those of the first generation. This occurs because of the advancement of the product development and the mass production will have reduced costs.

New generation products are launched one after another because of the emerging technologies and the market needs. However, the impact of the new generation products becomes weaker and weaker. It is not possible for each new generation product to lead the alternation of generations. Sometimes, new generation products can not satisfy those buyers who put a high value on the innovativeness, particularly when there is only a slight difference in performance between new and former generation products. This situation is the so-called the productivity dilemma.

Sometimes alternative products de-mature the market by way of basic technologies that are different from previous ones. While previous products reach the margin of their product performance, alternative products must show a big leap. Otherwise, innovative consumers such as the innovators and the early adopters might not change their purchase behavior.

The life cycle can be also arranged according to purchase pattern. The purchase pattern is divided into four parts, the first adoption of a new product,

Copyright © 2008, IGI Global. Copying or distributing in print or electronic forms without written permission of IGI Global is prohibited.

the repeat purchase of an adopted product, and two kinds of replacements, to new generation products and to an alternative product, respectively. The choice among these decides the occurrence of the alternation of generations and the shape of the life cycle.

The life cycle starts with the innovator's first adoption. Then, the early adopters follow the innovators. At that time, only one generation of the product exists in the market. The demand expands rapidly because the earlier the consumers purchase a product, the less their number is in the first half of the diffusion process. As a result, a product reaches a critical level that it is recognized as a promising product by a wide variety of people. Then, the early majority follows the early adopters as the price goes down.

After the second generation products appear in the market, multiple generation products can exist in the market simultaneously. Companies do not want to launch alternative products for a while because they should recover the sunk costs first. Adopters such as innovators and the early adopters have a choice to repeat purchasing the first generation products or to replace to the second generation products. They would decide to replace to latest generation if there is enough difference in the product performance / cost ratio because they highly value innovativeness. Even under the same situation, the early and late majority and laggards may choose the former generation products because these are typically simpler and cheaper than the latest generation products.

Table 4. Product generations and life cycle

	First Generation	Second Generation	Nth Generation	Alternative Product
Introduction	Adoption by Innovators			
Early Growth	Adoption by Early Adopters			
Late Growth	Adoption by Early Majority or Repeat Purchase by Innovators	Replacement by previous Adopters or Adoption by Early Majority		
Maturity	Adoption by Late Majority or Repeat Purchase by Previous Adopters	Replacement and Repeat Purchase by Previous Adopters and Adoption by Late Majority	Replacement by Previous Adopters and Adoption by Late Majority	
Decline	Adoption by Laggards or Repeat Purchase by Previous Adopters	Replacement and Repeat Purchase by Late Majority and Adoption by Laggards	Replacement and Repeat Purchase by Late Majority and Adoption by Laggards	Replacement by Adopters or Adoption

Copyright © 2008, IGI Global. Copying or distributing in print or electronic forms without written permission of IGI Global is prohibited.

In the situation that a product has been diffused widely, consumers have three choices, a repeat purchase, replacement by the new generation products, and replacement by the alternative products. It depends on the difference of the product performance to cost ratio and the consumer behavior type. The innovators and the early adopters are critical for the alternation of generations and the end or beginning of the life cycle. Table 4 demonstrates the typical pattern of the product generations and the life cycle from the viewpoint of the consumer behavior.

Conclusion

We discussed the existing life cycle theories related to business. The concept of the life cycle has widely been used and various life cycles have been developed in the various business fields. The Product Life Cycle (PLC) is the most well known of these. In the PLC, time is divided into four stages based on the change of the sales rate. Other major life cycles are related to product or product class innovation and manufacturing facility location. The innovation point of view highlights the ratio of product performance to cost. This ratio provides an indicator for the size of the potential consumer pool or limits to the types of consumers. Product Cycle Theory demonstrates the importance of location of manufacturing facilities. Physical location of facilities should change according to the states of the market and technology.

The basic concept of the industrial life cycle was introduced in this chapter to integrate the various individual life cycles. The advancement of technology is the driver for the launch of the new product. The performance/cost ratio which is improved by the process innovation limits the potential types of customers and, therefore, has a close relationship with the diffusion of a new product. Finally, for further cost cutting, most products are eventually produced in a low cost operation area exclusively. Critical factors change according to the life cycle stages.

References

Abernathy, W. J. (1978). The productivity dilemma. Baltimore, MD: The John Hopkins University Press.

Copyright © 2008, IGI Global. Copying or distributing in print or electronic forms without written permission of IGI Global is prohibited.

Abernathy, W. J, Clark, K.B., & Kantrow, A.M. (1983). Industrial renaissance. New York: Basic Books.

Barry, L. B. (1994). Are product life cycles really getting shorter? Journal of Product Innovation Management, 11(4), 300-308.

Bass, F. M. (1969). A new product growth model for consumer durables. Management Science, 15, 215-227.

Bowersox, J. & Closs, D. J. (1996). Logistical management. New York: McGraw-Hill.

Buzzell, R. D. (1966). Competitive behavior and product life cycles. In J. S. Wright & J. L. Goldstucker (Eds.), New ideas for successful marketing (pp. 46-68). Chicago: American Marketing Association.

Christensen, C. M. (1997). The innovator's dilemma. Boston: Harvard Business School Press.

Cooper, R. G. (2005). Product leadership (2nd ed.) New York: Basic Books.

Davis, T. (1993). Effective supply chain management. Sloan Management Review, 34(4), 35-46.

Debresson, C. & Lampel, J. (1985). Beyond the life cycle: Organizational and technological design. I. An alternative perspective. Journal of Product Innovation Management, 2(3), 170-187.

Dhalla, N. K. & Yuspeh, S. (1967). Forget the product life cycle concept! Harvard Business Review, 45, 102-112.

Doyle, P. (1976). The realities of the product life cycle. Quarterly Review of Marketing, Summer, 1-6.

Dummer, G. W. (1983). Electronics inventions and discoveries: Electronics from its earliest beginning to the present. New York: Pergamon Press.

Dunning, J. H. (1988). Explaining international production. London: Unwin Hyman.

Higuchi, T. (2006). Industrial life cycle in the VCR industry. Sakushin Policy Studies, 6, 19-34.

Higuchi, T., & Troutt, M. D. (2004). A dynamic method to analyze supply chains with short product life cycle. Computers & Operations Research, 31(6), 1097-1114.

Higuchi, T, Troutt, M. D., & Polin, B. A. (2004). Life cycle considerations for supply chain strategy. In Chan, C. K. & Lee, H. W. J. (Eds.), Success-

Copyright © 2008, IGI Global. Copying or distributing in print or electronic forms without written permission of IGI Global is prohibited.

ful strategies in supply chain management (pp. 67-89). Hershey, PA: Idea Group.

Itami, H. (1989). Nihon no VTR sanngyou naze sekai wo seiha dekitanoka. Tokyo: NTT Publishing.

JEITA (Japan Electronics and Information Technology Industries Association). (2005). Minnseiyou Dennsi kiki Deta Shyu. Tokyo: JEITA.

Kohli, R., Lehman, D. R, & Pae, J. (1999). Extent and impact of incubation time in new product diffusion. Journal of Product Innovation Management, 16(2), 134-144.

Kotler, P. (1999). Marketing management. Englewood Cliffs, NJ: Prentice Hall.

Lambe, C. J. & Spekman, R. E. (1997). Alliances, external technology acquisition, and discontinuous technological change. Journal of Product Innovation Management, 14(2), 102-116.

Levitt, T. (1965). Exploit the product life cycle. Harvard Business Review, 43, 81-94.

Lilien, G., Kotler, P., & Moorthy, K. (1992), Marketing models. Englewood Cliffs, NJ: Prentice-Hall.

Loch, C., Stein, L., & Terwiesch, C. (1996). Measuring development performance in the electronics industry. Journal of Product Innovation Management, 13(1), 3-20.

McIntyre, S. H. (1988). Market adaptation as a process in the product life cycle of radical innovations and high technology products. Journal of Product Innovation Management, 5(2), 140-149.

Millson, M. R., Raj, S. P., & Wilemon, D. (1992). A survey of major approaches for accelerating new product development. Journal of Innovation Management, 9(1), 53-69.

Moore G. A. (1991). Crossing the chasm. New York: Harper Business.

Moore G. A. (2005). Dealing with Darwin. New York: Portfolio.

NHK (Japan Broadcasting Corporation.). (2000). Project X: Challengers 1. Japan Broadcast Publishing, Tokyo: JPN.

Norman, D. A. (1998). The invisible computer. Cambridge, MA: MIT Press.

Rogers, E. M. (1995). Diffusion of innovations (4th ed.) New York: The Free Press.

Copyright © 2008, IGI Global. Copying or distributing in print or electronic forms without written permission of IGI Global is prohibited.

Rosenbloom, R. S. & Abernathy, W. J. (1982). The climate for innovation in industry. Research Policy, 11(4), 209-225.

Sato, M. (1999). The story of a media industry. Tokyo: Nikkei Business Publications.

Sheth, J. N. (1971). Word of mouth in low-risk innovations. Journal of Advertising Research, 11(3), 15-18.

Shintaku, J. (1990). Nihon kigyou no kyousou sennryaku (Competitive strategies of Japanese firms). Tokyo: Yuhikaku.

Utterback, J. M. (1994). Mastering the dynamics of innovation: How companies can seize opportunities in the face of technological change. Boston: Harvard Business School Press.

Vernon, R. (1966). International investment and international trade in the product cycle. Quarterly Journal of Economics, 80(1), 190-207.

Vernon, R. (1977). Storm over the multinationals. Boston: Harvard University Press.

Vernon, R. (1998). In the hurricane's eye. Boston: Harvard University Press.

Endnote

[1] In these decades, global sales become very important in determining the de-facto standard because various *de jure* standards compete with each other, such as in the case of the cell phone. Hence, the importance of global strategic alliances is increasing for achieving the de-facto standard.

Copyright © 2008, IGI Global. Copying or distributing in print or electronic forms without written permission of IGI Global is prohibited.

Chapter III

Analytic Research and Quantitative Models

This chapter provides two kinds of background information that we consider important to the subject area. First, we surveyed the supply chain management, operations management, and management science literatures for those works contacting life-cycle issues and at the same time that use quantitative or modeling approaches. We then developed synoptic summaries of these publications and provide some analysis of their central topics, trends, and themes. Hopefully the results will be a helpful reference guide to the related literature to date for both practicing managers and researchers. In the second part of the chapter, we introduce the standard quantitative methods and models used for mathematical life-cycle models. These have been developed under the label of diffusion models and most of the work has been carried out by marketing scientists. This topic should be useful to practitioners in making forecasts, constructing estimates related to capacity, and other supply chain management forecast and planning issues. We also note that some research needs in this area.

Copyright © 2008, IGI Global. Copying or distributing in print or electronic forms without written permission of IGI Global is prohibited.

Introduction

In this chapter, we first review a number of articles related to life cycles in the supply chain management and operations management literatures. Next we discuss some theory behind the mathematical analysis of product life-cycle curve models. Much of this work has been developed by marketing scientists under the label of product diffusion modeling. Two classes of such models can be distinguished. Several models are described in detail along with the problem of parameter estimation. We also suggest some topic areas for further research.

Background

We next present a survey of research involving life cycles and modeling. The review of these articles is presented in chronological order based on year of publication and alphabetical on first author name within year of publication. For authors with multiple publication years the more recent year was used for ordering. Under the term modeling, we broadly include quantitative modeling as well as tabular, diagram-based, and miscellaneous typologies. We also review some articles that deal with short product life cycles as we provide a more thorough dynamic modeling example in the next chapter. Some additional related articles are referred to later in the chapter. These articles were obtained by searching a number of databases. Every effort was been made to include relevant works as of the date of writing. While completeness can not be assured, we believe this collection of works will be representative of the general trends and types of results that make contact with life-cycle considerations.

Main Focus of the Chapter

Review of Life-Cycle Related Research

Parlar and Weng (1997) consider joint coordination between a firm's manufacturing and supply departments in the short life-cycle case. Demand is mod-

Copyright © 2008, IGI Global. Copying or distributing in print or electronic forms without written permission of IGI Global is prohibited.

eled by a probability distribution assumption as in the Newsvendor Model. They analyze the cases of independent and joint decisions on the number to produce and the amount of raw materials to be supplied for manufacturing.

Beamon (1998) provided an interesting and useful survey of the supply chain modeling literature up to that time. Shorter product life cycles was mentioned as one of the forces leading to increased attention to the supply chain as a whole rather than the focus on individual operations in earlier times. In addition to shorter life cycles, the two other forces were rising costs and shrinking resource bases. This work also provides a thorough coverage of supply chain performance measures, along with articles that have concentrated on models focused on those measures. A useful table is also provided that categorizes the principle contributions according to model type (deterministic, stochastic, analytical, economic, and simulation), performance measures (cost, customer responsiveness/backorders, activity time, and flexibility), and decision variable types used (production/distribution scheduling, inventory levels/ordering or batch size, number of stages, distribution center to customer assignment, plant-product assignment, buyer-supplier relationships, product differentiation, and number of product types to be held in inventory). An agenda for future research was also proposed.

Bollen (1999) considers real options theory for valuation of capacity change decisions for a product over its demand life cycle. Dynamic programming is the technique used. It is demonstrated that significant differences exist in such optimal capacity expansion decisions as compared with simplistic assumptions such as constant or increasing demands.

A now virtually classic reference is the comprehensive supply-chain modeling exposition given by Tayur, Ganeshan, and Magazine (1999). However, interest in the life-cycle connections and influences on the supply chain had not yet been significantly examined. Nonetheless, the chapter by Raman (1999) was an early treatment of the important special case of short life-cycle products. Such products were often associated in the past with such industries as fashion products or style goods. However, with increasing customization and competition more and more supply chains must deal with stockouts and markdowns. This area continues to be of importance. Raman called for more management science studies to address these problems. Many of the following works have attempted to address this need particularly in connection with various inventories.

Cohen, Ho, and Matsuo (2000) make the important observation that most strategic operations-planning decisions use the assumption of stationary and

Copyright © 2008, IGI Global. Copying or distributing in print or electronic forms without written permission of IGI Global is prohibited.

time-invariant demand, and therefore neglect the innovation-diffusion or life-cycle dynamics. They examine the effects of life-cycle dynamics on the three important areas of (1) production and inventory planning, (2) technology planning, and (3) capacity planning. In addition, they also address the other direction of how those functions affect the diffusion process itself. They therefore obtain insights on the marketing-operations interface and the benefits that may be achieved by simultaneous decision making across those functions. They use probability analysis with the normal distribution assumption. They also contact the short life-cycle case and forecasting issues. They provide models for both process choice and capacity expansion.

Taylor (2001) studies channel coordination. Although not expressly defined, channel coordination appears to be a term arising from marketing but essentially equivalent to cooperation among the players in a supply chain. In this study, dynamic programming models are applied to analyze the three policies of price protection, midlife returns, and end-of-life returns. These are used in declining price environments and therefore become potentially important in the maturity and decline phases of the product life cycle; however, they are most often used with short life-cycle products. In price protection contracts the manufacturer pays the retailer a credit applying to the retailer's unsold inventory when the wholesale price drops during the life cycle. Conditions are determined under which all or a combination of these features achieve coordination and provide mutual advantages to the parties.

Angelis and Porteus (2002) derive results on the optimal simultaneous capacity and production plan for short-life-cycle goods in a produce-to-stock supply chain under stochastic demand. Capacity expansion or contraction unit costs are assumed to be given. Optimal decisions take the form of target intervals. When initial capacity for a period is below the lower target limit then the capacity is raised to the limit. If the initial capacity for the period is above the upper target limit then capacity is reduced to that limit. Otherwise no change of capacity is made. Both the cases of carry-over or no carry-over of product to the next period are considered. The approach is analytic based on optimality functions similar to those occurring in dynamic programming.

Aitken, Childerhouse, and Towill (2003) do not give a quantitative analysis per se but provide some important considerations about the relationship of life-cycle stages with supply-chain strategy. Using what they call generic modeling within a case study, they discuss the matching of products to pipelines for maximizing competitiveness with respect to order-winner and market-qualifier product characteristics.

Copyright © 2008, IGI Global. Copying or distributing in print or electronic forms without written permission of IGI Global is prohibited.

Crandall and Crandall (2003) present a diagram-based model aimed at conceptualizing and better informing the management of excess inventories. The choice of the term life cycle seems to derive from considering the life cycles of goods, materials, and parts inventories within the production system or even the whole supply chain. In addition, they argue that even excess inventories have a life cycle of their own. They identify four stages of this cycle as: Create/Prevent, Identify/Classify, Analyze/Value, and Dispose/Recycle. They also note that shorter product life cycles are among the forces contributing to excess inventories. Reverse logistics is also discussed as an extension to the supply chain that can help deal with excess inventories.

Weng and McClurg (2003) consider a single-buyer single-supplier supply chain characterized by a short life-cycle product with uncertainty in both delivery time and demand. Using stochastic analysis similar to that of the "newsboy" or "newsvendor" problem, they study the effects of coordination. Here, coordination means buyer and supplier share information regarding their relevant costs and the buyer's demand and that they jointly select the amount to be ordered by the buyer and sold by the supplier. Their models seek to maximize total (i.e., system) expected costs. An illustrative case and numerical experiments demonstrate the usefulness of the approach. They show parameter values for which the coordination results in higher expected profits that when independent decisions are made.

Damodaran and Wilhelm (2004) propose a stochastic programming model to maximize profit and scenario analysis to support decisions and "what-if" questions on upgrading features for a family of products over its life cycle. The method is motivated by high-technology products such as notebook computers that have short life cycles. The model intends to integrate the various functions of design, marketing, manufacturing, production planning, and supply chain management. Random customer demand is assumed. Computational performance of the method is favorably compared to existing commercial software. As noted above, the method was motivated by products with short life cycles. A useful extension of this model would be that to longer life-cycle cases, so that more insight might be obtained on how data and decisions are affected by other stages of the common life-cycle diffusion pattern.

Fandel and Stammen (2004) developed a complex linear mixed-integer programming model to plan operations and optimize after-tax profit of a company principally taking into account recycling and multi-country considerations. More attention has been given to the stages of the supply chain than to the product life cycle per se. However, the model does appear to permit the

Copyright © 2008, IGI Global. Copying or distributing in print or electronic forms without written permission of IGI Global is prohibited.

flexibility to follow a particular forecasted life-cycle pattern by defining the appropriate minimum throughputs for various future macro and micro time periods in the forecast horizon.

Klimenko (2004) studies the early stages of the industry or production life cycle. In the context of the quality of the match between the inputs offered by suppliers and sought by suppliers, the relative strengths of the centripetal force leading to clustering of firms and the centrifugal force preventing it are considered. The industry spatial distribution is found to depend on the relative strengths of these forces. An economic model is used to study the equilibrium profits of the assembler and supplier under the influences of competition and congestion.

Wang, Huang, and Dismukes (2004) consider the supplier selection problem using a combination of the Analytic Hierarchy Process (AHP) and preemptive goal programming (PGP). Of special interest, this work provides a table that assigns the type of supply chain (Agile, Lean, or Hybrid) most suited to each stage of the product life cycle. The AHP is first applied using criteria in the SCOR model, which provides a set of metrics for supply chain operation. The SCOR model was constructed by the Supply Chain Council (1999). SCOR has four main categories. The first is Delivery Reliability, with the metrics of delivery performance, fill rate, order fulfillment lead time, and perfect order fulfillment. Category two is Flexibility and Responsiveness with the metrics of supply chain response time and production flexibility. The third category is Cost having the metrics of total logistics management cost, value-added productivity, and warranty cost or returns-processing cost. The fourth category is Assets with the metrics of cash-to-cash cycle time, inventory days of supply, and asset turns. The AHP is used to obtain priorities of the metrics for each type of supply chain. Then the results are used in the PGP model as follows. At Priority 1 the total value of the purchases are maximized and at Priority 2, the total costs are minimized. This paper also provides a good review of other quantitative models for the supplier-selection problem. One strength of the approach is the ability of the AHP to combine qualitative and quantitative criteria. Another strength is that PGP permits constraints on supplier capacities to be factored in, so that a lower priority supplier might have to be selected if a top priority one has inadequate capacity.

Fixson (2005) observes that shorter product life cycles in combination with increasingly heterogeneous markets force companies to compete simultaneously in the three domains of products, processes and supply chains. Three-dimensional concurrent engineering (3D-CE) may be applied to address the

Copyright © 2008, IGI Global. Copying or distributing in print or electronic forms without written permission of IGI Global is prohibited.

attendant complexity. A framework is developed based on existing product characteristics of component commonality, product platforms, and product modularity. Two examples are given that show how individual product architecture dimensions link decisions across different domains. The framework can be further used to obtain advice for product design with respect to operational strategy and to develop dynamic capabilities in planning effective product-operational combinations. The main technique used is Function-Component Allocation (FCA) and in particular, the function-component matrix.

Higuchi and Troutt (2004, 2005) discuss supply chain dynamics in depth. They give an overview of dynamic simulation and demonstrate its use in the case of the supply chain for a short life-cycle product, the well-known *Tamagotchi*™ toy. We mention this work briefly here for completeness and devote the next chapter to an intensive coverage and integration of those results.

Slikker, Fransoo, and Wouters (2005) consider the single-supplier n-retailers supply chain in the case of a short life-cycle product. This work extends that of Müller, Scarsini, and Shaked, (2002); Slikker and van den Nouweland (2001); and Slikker, Fransoo, and Wouters (2001). By short life cycle is meant a small finite demand period as in the single-period inventory situation known as the News-vendor problem, and in fact the article takes the point of view that each retailer faces that kind of problem. The primary focus is on whether the retailers can improve their joint (i.e. total) expected profits by cooperation and sharing of information, resulting essentially in combined ordering. However, an interesting additional aspect is considered in that transshipments are permitted between retailers. Thus retailers must order in advance of knowing actual demand. Then, shortages might be made up from other retailers or overages sold similarly. N-person cooperative game theory is used as the methodology. It was noted that transshipments may be more costly in case rework of non-identical products is required. They also provide a good survey of similar treatments including those that focus on Nash equilibria and Shapley value (Shapley,1967, 1953) solutions. Evidently, the primary issue is the division of cooperative gains and Shapley value appears to be promising for this purpose. The authors note that further extensions would be useful along the lines of work by Suijs (1999) and Klaus (1998).

Tan and Khoo (2005) is an application of chemical engineering relationships to environmental supply chain considerations. Various scenarios of operations of a three-stage supply chain for aluminum production are examined for minimization of undesirable by-products. This work is also noteworthy as an instance of the field of Life Cycle Assessment (LCA) and green production

Copyright © 2008, IGI Global. Copying or distributing in print or electronic forms without written permission of IGI Global is prohibited.

chain considerations. *The International Journal of Life Cycle Assessment* considers the societal life cycles of products, components and materials from the point of view of environmental impacts along with recycling issues.

Weng and Parlar (2005), (see also Gunasekaran, 2005), are among a number of studies that focus on short life-cycle products. They consider price incentives for firms that can produce either standardized or customized build-to-order (BOP) products and employ a stochastic dynamic programming model. This work obtains conditions associated with the optimality of offering price incentives for customer commitment to the standard version.

Chang, Wang, and Wang (2006) observe that supply-chain product-development strategy should depend on the particular phase of the product life cycle. This in turn affects core competencies and outsourcing synergies. They consider a multi-attribute quantitative approach for decision support in the supplier-selection context.

Kainuma and Tawara (2006) extend the range of the supply chain to re-use and recycling with reference to the life cycle. They propose the use of multiple attribute utility for assessing the supply chain along these lines and thus add to the dimensions of the metrics associated with lean and green considerations.

Krishnan and Jain (2006) use the Generalized Bass diffusion model (Bass, Krihnan, & Jain, 1994) (GBM) to obtain results on the optimal dynamic advertising policy for a new product introduction. The GBM model is discussed further in this chapter in the section on Mathematical Modeling of Life-Cycle and Diffusion Curves. They found that the optimal policy depended on advertising effectiveness, discount rate, and the ratio of advertising to profits. Depending on the relative values of these, the optimal advertising pattern can take one of four forms, namely, first decreasing—then increasing, first increasing—then decreasing, monotone increasing, or monotone decreasing. Dynamic programming was the technique used for optimization.

Narasimhan, Talluri, and Mahapatra (2006) propose that in industrial purchasing contexts firms often procure a set of products from the same suppliers to benefit from economies of scale and scope. These products often are at different stages of their respective product life cycles (PLCs). Moreover, firms consider multiple criteria in purchasing products, and the relative importance of these criteria varies depending on the PLC stage of a given product. Therefore, a firm should select suppliers and choose sourcing arrangements such that product requirements across multiple criteria are satisfied over time. Their work proposes a mathematical model that addresses this issue

Copyright © 2008, IGI Global. Copying or distributing in print or electronic forms without written permission of IGI Global is prohibited.

and contributes to the sourcing literature by demonstrating an approach for optimally selecting suppliers and supplier bids given the relative importance of multiple criteria across multiple products over their PLC. Application of the model was demonstrated on a hypothetical dataset to illustrate the strategic and tactical significance of such considerations.

While Punakivi and Hinkka (2006) do not discuss modeling, they provide a logistics case study on transportation mode selection from the industry point of view. Thus, this work also contacts the industry life cycle and the development of product markets. Their study hopes to better inform current and future logistics service needs. They identify the shortening of product and service life cycles as one of the four major influential trends, along with globalization, concentration on core competencies, and the growth and expansion of e-business in supply chain networks. A useful table ranks the importance of the four criteria of convenience, price, quality, and speed in the industries of construction, electronics, machinery, and pharmaceuticals, respectively. Also useful is a figure showing the relationship of price and speed in the same industries.

Vonderembse, Uppal, Huang, and Dismukes (2006) develop a typology for supply chain design based on product characteristics, customer expectations, and stage of the product life cycle. Importantly, they observe that the key success factors for a product are expected to change as it moves through its life cycle. Through case studies and literature review, the typology is developed and research questions are proposed. A very useful structured literature review is provided on areas researched with respect to supply chains in general, lean supply chains, and agile supply chains. Especially useful and relevant for the present work is a table that associates the stage of the product life cycle with product type (standard, innovative, or hybrid) and with supply chain classification (lean, agile, or hybrid). The case studies included are those for Black and Decker (a lean supply chain), IBM (an agile supply chain), and Daimler Chrysler (a hybrid supply chain).

Geyer, Van Wassenhove, and Atasu (2007) focus on remanufacturing in the finite product life-cycle case. Applying economic analysis this work quantifies the cost savings potential of production systems related to collection, remanufacture, and end-of-use products. For example, end-of-life computers still may have components useful for remanufacture into lower specification applications such as toys. Their models suggest considerations for optimal coordination of production cost structure, collection rate, product life cycle, and component durability.

Copyright © 2008, IGI Global. Copying or distributing in print or electronic forms without written permission of IGI Global is prohibited.

Kamath and Roy (2005, 2007) also have considered the systems dynamics modeling of a supply chain for short life-cycle products as has been considered in Higuchi & Troutt (2004, 2005). Their work focuses on loop-dominance analysis in feedback loops and its usefulness for extracting key relationships between demand uncertainty and capacity-augmentation decisions.

An Analysis of the Literature Sample

The foregoing literature sample consists of 30 articles and book chapters, where those in which authors had more than one article or chapter are counted just once. An analysis can be done in terms of topics contacted and research methodologies used. Here we present topic areas and methodologies along with their frequencies of occurrence across the sample.

Topic Area

Topic Area	Frequency of Occurrence
Capacity changes and decisions	7
Convergent markets and combining technologies	1
Coordination of buying, decisions, and planning	9
Costs	1
Diffusion curves and optimal advertising policy	1
E-business	1
Forecasting	3
Function-Component Allocation (FCA)	1
Generalized Bass diffusion model (GBM)	1
Globalization	2
Inventory issues	8
Life-Cycle Assessment (LCA), recycling, and green considerations	6

Copyright © 2008, IGI Global. Copying or distributing in print or electronic forms without written permission of IGI Global is prohibited.

Life-cycle stages	5
New-product introduction	3
Performance measures or metrics	5
Price protection and declining price environments	4
Process choice	1
Product-cycle model and spatial distribution	3
Product attributes	8
Product families	1
Production planning	4
Service life cycles	1
Sharing of Information	2
Short or finite life cycle	19
Customization, modularity and standardization	4
Reverse logistics	1
Strategy and core competencies	9
Supplier selection and outsourcing	3
Supply chain characteristics, design, and types	9
Transshipments between retailers	1

Methodology

Methodology	**Frequency of Occurrence**
Case study method	2
Dynamic Programming	5
Game theory and Shapley value	1
Literature review	3
Loop-dominace analysis	1
Mathematical Programming	3

Copyright © 2008, IGI Global. Copying or distributing in print or electronic forms without written permission of IGI Global is prohibited.

Miscellaneous methods (Real Options Analysis, economic models, dynamic optimization)	3
Multiattribute analysis and the Analytic Hierarchy Process (AHP)	4
Probability analysis, Bayesian techniques, or the newsvendor model	5
Systems Dynamics and Dynamic Simulation modeling	2

Some Promising Directions for Further Research

Here we mention three research contributions that we anticipate could benefit by extensions to include additional life-cycle considerations. We also revisit two of the previously mentioned above contributions in connection with what we consider to be needed additional research.

Lee and Whang (2002) consider the relationship between supply chains and secondary markets. The latter may be regarded as existing to dispose of or acquire excess inventories. An important rationale for this research is the difficulty of forecasting demand for short life-cycle products, resulting in frequent stockouts or excess inventories. They developed a two-period model with a single manufacturer and many resellers. Optimal decisions are derived for the resellers and the equilibrium price for the secondary market. While it is found that the secondary market always improves allocative efficiency, the welfare of the supply chain may or may not improve due to the secondary market. This work was motivated in part by short life-cycle products. However, it would be of interest to extend the results to product families in which, while individual versions have short life cycles, the family itself may have a longer marketable life time.

As noted above, Narasimhan et al. (2006) studied supplier selection in connection with the life-cycle stage of products. Li and O'Brien (2001) have obtained quantitative relationships between the performance of supply chain strategies and product attributes. The product attributes considered might reasonably be enlarged or subcategorized in terms of the life-cycle stage to obtain an extension to both those works.

Copyright © 2008, IGI Global. Copying or distributing in print or electronic forms without written permission of IGI Global is prohibited.

Srinivasan, Haunschild, and Grewal (2007) recently have addressed a topic that may be considered a micro-analysis of the early stage of the life cycle. Specifically, they consider learning issues in new-product introductions. Given the importance of short life-cycle product issues as revealed by the above survey, such micro-analyses should be illuminating and perhaps also extended to other phases of the product life cycle. Examples of convergent markets are those for products such as digital cameras, multi-media players, and video-on-demand. These products can be developed by combining technologies across different industries and may have the further interesting feature of being approached by different supply chains. Their analysis uses count models and Bayesian techniques.

The paper of Damodaran andWilhelm (2004) was reviewed above. This paper also raises issues about the life cycle of families of products. We believe that topic is of both academic and practical interest for further study. Repeat purchases can have a profound effect as will be seen in the next chapter. Some repeat purchases might be replaced by a different choice or upgrade within one and the same product family. Thus, a clearer understanding of diffusion of product families should be helpful in understanding and forecasting future demands.

Similarly, the work of Klimenko (2004) was reviewed above. Along with Punakivi and Hinkka (2006), this work was one of only two that our survey found dealing with the subject area of the product-cycle model and industry spatial distributions. Thus, more work should be expected in that general subject area. In addition, we believe that this subject area can benefit from additional quantitative modeling techniques, especially game theory considerations in the case of competition between entire supply chains.

Mathematical Modeling of Life-Cycle and Diffusion Curves

In this section, we concentrate on the life-cycle or diffusion model itself and what is known about its mathematical properties. We also consider some related issues such as estimation and applications. We may distinguish two general approaches to constructing models for product life cycles. One, which appears to have originated with the work of Bass (1969), can be called the *fundamental-structural* approach and develops models based on hazard-rate considerations. The other approach we call the *empirical-technical* approach. This is based on the observation that the sales rate curve rises from zero to some maximum value and then declines (Kotler, 1999; Kotler & Keller,

Copyright © 2008, IGI Global. Copying or distributing in print or electronic forms without written permission of IGI Global is prohibited.

2006). This unimodal pattern is commonly occurring in the class of probability density function (pdf) models. We consider the Pearson system of such curves as one very flexible class of such pdf models. We also consider the problem of fitting such models to observed data. Finally we point out some research needs in this area.

The Fundamental-Structural Approach

As far as we could determine, most of the research on the topic of mathematical modeling of the life-cycle curve pattern has been carried out by marketing scientists under the name of diffusion or new-product diffusion modeling. This approach has focused on target-market sales patterns. A thorough and rigorous mathematical treatment of the topic is given by Bass, Jain, and Krishnan (2000), which we follow extensively in this discussion. We first consider the Bass model.

The Bass Model

This model proposed by Bass (1969) describes the cumulative sales over time for a product in a given target market. Following Bass et al. (2000), the cumulative sales form of the model is given by,

$$\begin{aligned} dN(t)/dt = N'(t) &= p(m - N(t)) + qm^{-1}N(t)(m - N(t)) \\ &= (p + qm^{-1}N(t))(m - N(t)) \qquad (1) \end{aligned}$$

where

t = time

$N(t)$ = Cumulative number of sales at time t

m = the total market potential sales (maximum of $N(t)$)

p = coefficient of innovation or external influence

q = coefficient of imitation or internal influence

Copyright © 2008, IGI Global. Copying or distributing in print or electronic forms without written permission of IGI Global is prohibited.

If we further define $F(t)$ as the cumulative fraction of potential sales, then $N(t) = mF(t)$ and (1) can be put into the more elegant normalized cumulative market fraction form,

$$f(t) = dF(t)/dt = F'(t) = p(1 - F(t)) + qF(t)(1 - F(t)) = (p + qF(t))(1 - F(t)). \quad (2)$$

Assuming $F(0) = 0$, the solution of this differential equation is given by,

$$F(t) = (1 - \exp\{-(p+q)t\})/(1 + qp^{-1}\exp\{-(p+q)t\}) \quad (3)$$

From which it follows that,

$$f(t) = [(p+q)^2 p^{-1}] \exp\{-(p+q)t\}/[1 + qp^{-1}\exp\{-(p+q)t\}] \quad (4)$$

The function $f(t)$ given by (4) may be regarded as the resulting model of the sales rate, or in short, the product life-cycle curve.

Equation (2) can be put into Hazard-Rate form, or developed from that perspective, as follows. Let f(t) be the pdf and let h(t) be the hazard-rate so that,

$$h(t) = f(t)/(1 - F(t)) \quad (5)$$

From (2) we have,

$$f(t) = F'(t) = (p + qF(t))(1 - F(t)) \quad (6)$$

from which it follows that,

$$h(t) = f(t)/(1 - F(t)) = p + qF(t). \quad (7)$$

Thus, the Bass (1969) model assumes $h(t) = p + qm^{-1}S(t)$ where $S(t)$ is $N(t)$ or $mF(t)$ and $h(t)$ is the hazard rate. The hazard rate in this context is sometimes described as the likelihood of adoption at time t given no prior adoption. It

Copyright © 2008, IGI Global. Copying or distributing in print or electronic forms without written permission of IGI Global is prohibited.

might also be described as the instantaneous force of adoption by its analogy to the force of mortality in actuarial mathematics.

We may also define $n(t)$ as the noncumulative number of adopters at time t and $N(t)$ as the cumulative number. Then in the discrete case $N(t) = \sum_{-\infty}^{t} n(t)$ and in the continuous case $N(t) = \int_{-\infty}^{t} n(s)ds$. Thus, in the continuous case $n(t) = dN(t)/dt$. In either case, we may call $n(t)$ the product life-cycle or sales rate curve.

Proportional hazards models (Cox, 1972) are based on assuming the hazard rate s_i given by $h(t|\mathbf{x}) = h_0(t)\varphi(\mathbf{x},\beta)$, where $\mathbf{x}$ is a vector of covariates, β **is a** parameter vector that might be estimated by the maximum likelihood estimation (MLE) method. In effect they make the hazard rate conditionally dependent on covariates. The Bass model may be regarded as a special case in which $\varphi(\mathbf{x},\beta)$ is a constant. Proportional hazards models are considered in Helsen and Schmittlein (1993) and their estimation is addressed in Heckman and Singer (1985). Some additional modified hazard-rate based models have been proposed for special applications. Sinha and Chandrashekaran (1992) propose a *split-hazard* model for the adoption timing of ATM machines. Chandrashekaran and Sinha (1995) considered a *Split-population Tobit model* to analyze factors influencing volume of PC purchases. Weerahandi and Dalal (1992) combined diffusion and binary choice modeling for forecasting fax machine market segment sales. Haldar and Rao (1998) study the relationship of adoption timing and various covariates.

Let $S(t)$ and $Y(t)$ be the sales and cumulative sales, respectively, regarded as a continuous quantities. A few simple models are those given by the logistic and negative exponential pdfs.

The logistic model is as follows. One can directly model the cumulative sales by the logistic function by way of,

$$Y(t) = a_1/(1 + \exp\{-(a_2 + a_3 t)\}) \tag{8}$$

This has been used by Mansfield (1961). Then the sales rate is given by,

$$S(t) = dY/dt = (a_3/a_1)Y(t)[\,a_1 - Y(t)]$$
$$= a_3 Y(t) - (a_3/a_1)[Y(t)]^2. \tag{9}$$

Copyright © 2008, IGI Global. Copying or distributing in print or electronic forms without written permission of IGI Global is prohibited.

The Negative Exponential model results if we let $q = 0$. Then the solution of (2) is:

$$F(t) = 1 - \exp\{-pt\}. \tag{10}$$

Therefore $f(t) = p\exp\{-pt\}$ with resulting sales-rate or life-cycle curve given by,

$$S(t) = mp\exp\{-pt\}. \tag{11}$$

A number of models that incorporate price and/or advertising are reviewed in Bass, Jain, and Krishnan (2000). See also the edited volume by Mahajan, Muller, and Wind (2000). Here we emphasize a few of the most prominent ones.

The Bass (1980) Model

Bass (1980) assumed the constant elasticity of demand function $Q(t) = f(t)\,c[Pr(t)]^{\eta}$, where $Q(t)$ is the demand at time t, $Pr(t)$ is the price at time t, and $f(t)$ is given by the Bass (1969) diffusion rate specified in (2). Under assumption of maximization of profits the following solution is obtained.

$$Q(t) = [m/(1 - \lambda\eta)]f(t)F(t)^{\lambda\eta/(1-\lambda\eta)}. \tag{12}$$

Bass et al. (2000) note that $f(t)$ must be specified for estimation with this model. They further reported that based on data for six durable products the model gave good fit to empirical sales and prices series.

The Kalish (1985) Model

Kalish (1985) proposed a simultaneous pair of differential equations based on the interplay between consumer awareness level $I(t)$ and adoption or purchase. These equations are:

Copyright © 2008, IGI Global. Copying or distributing in print or electronic forms without written permission of IGI Global is prohibited.

$$I'(t) = [1 - I][f(A(t)) + bI + b'(m^{-1}Y(t))] \text{ (Awareness)} \tag{13}$$

$$Y'(t) = k[g(Pr(t)/u(Y(t)/m))I - Y(t)] \text{ (Adoption)} \tag{14}$$

where in addition to the previously defined terms we have $A(t)$ is the advertising at time t, g, f, and u are specified functions, and b, b', and k are constants. If $I = 1$ is assumed in the Awareness equation it becomes redundant and the system reduces only to the second or Adoption equation. An explicit solution for $S(t)$ was obtained by assuming g as exponential and u as quadratic functions, respectively. Bass et al. (2000) provide more detailed discussion of this model and note some comparisons to an earlier approach of Robinson and Lakhani (1975).

The GBM Model

Krishnan and Jain (2006) used the Generalized Bass (Bass et al., 1994) diffusion or life-cycle model (GBM) to obtain results on the optimal dynamic advertising policy for a new product introduction. The GBM model assumes that the probability of adoption at time t, and not earlier, is given by the product of the diffusion force and the marketing impact force. Thus,

$$S(t)[M - CS(t)]^{-1} = [p + qCS(t)/M]x(t) \tag{15}$$

where

$S(t)$ = sakes at time t
$CS(t)$ = cumulative sales at time t
M = market potential
p and q are the diffusion parameters
$x(t)$ = current market effort

The solution of this differential equation was obtained by Bass (1994) and is given by,

Copyright © 2008, IGI Global. Copying or distributing in print or electronic forms without written permission of IGI Global is prohibited.

$$S(t) = M(p + q)^2 p^{-1} x(t)[\exp\{-(p + q)X(t)\}][\ 1 + (q\ p^{-1}\exp\{-(p + q)X(t)\}]^{-2} \quad (16)$$

where $X(t)$ is defined as the cumulative marketing effort up to time t.

Both $x(t)$ and $X(t)$ can be further related to price and advertising expenditures. Let $v(t)$ be the price and let $a(t)$ be the advertising expenditure at time t. Then,

$$x(t) = 1 + \beta_v v'(t)/v(t) + \beta_a a'(t)/a(t) \quad (17)$$

and

$$X(t) = \int_0^t x(\tau)\, d\tau = t + \beta_v\, ln\ (v(t)/v(0)) + \beta_a\, ln\ (a(t)/a(0)) \quad (18)$$

The foregoing relations can be combined by substituting the right-hand sides of (17) and (18) into (16) to give a final overall analytic expression for $S(t)$.

The Empirical-Technical Approach

The logistic and negative exponential models discussed above are examples of what we call empirical technical models. In short, these simply make a suitable assumption about the shape of the diffusion curve. As noted above, the typical life-cycle or sales-rate curve rises from zero to some maximum value and then declines. This is a unimodal pattern similar to common probability density functions like the normal and gamma pdfs. Thus, if a parametric family of such curves is available then the analyst can consider a member of the class as a potential model for real data. Bass et al. (2000) also note that other specifications for $f(t)$ such as the gamma, generalized gamma, and Gompertz are possible models for the sales rate. Thus, one might start directly with any such pdf model. It is that implied process we call as the empirical technical approach to modeling the life-cycle curve. The Pearson system of frequency curves contains a wide variety of potentially suitable shapes for these purposes and is discussed next.

The Pearson System (Pearson, 1895) is generated by solutions to the following differential equation for the pdf $p = p(x)$:

Copyright © 2008, IGI Global. Copying or distributing in print or electronic forms without written permission of IGI Global is prohibited.

$$p^{-1}p' = -(a + x)/(c_0 + c_1x + c_2x^2) \tag{19}$$

Depending of the values of and relationships between the four parameters a, c_0, c_1, and c_2, a wide variety of curve shapes result. These include U, inverted-U, J, and reverse-J shapes. Importantly, the special cases of the Beta, normal and gamma distributions are included. Depending on the roots of the equation $c_0 + c_1x + c_2x^2 = 0$ a number of different types of Pearson models result and Types $I - IX$ are discussed in Johnson, Kotz, and Balakrishnan (1994). We note that in the context of pdf models there will be another parameter, the density constant K that insures that the resulting pdf model is indeed a pdf. However, in the present context K may be regarded as a scale constant. Thus in all, we have five parameters.

Estimation of Parameters and Applications

Parameter estimation with the fundamental-structural approach has been addressed in Bass et al. (2000). See also Putsis and Srinivasan (2000) and Lilien, Rangaswamy, and Van Den Bulte (2000). Software for the Generalized Bass Model and a tutorial are provided in Mahajan et al. (2000).

For the empirical-technical approach, Elderton and Johnson (1969) provide a study on fitting Pearson curves to data. It is known that the parameters can be expressed as functions of the first four moments of the pdf $p(x)$ so that the method of moments (Mood, Graybill, & Boes, 1974) estimation method might be applied in principle. However, in practice the analyst may not have a random sample from which to compute sample moments if the goal is to estimate the rest of the life cycle from early sales experience. A nonlinear regression or spreadsheet genetic algorithm approach might also be very convenient for modern users. Applications of diffusion modeling are discussed in Lilien, Rangaswamy, and Van Den Bulte (2000).

Future Trends

The existing research has been aimed primarily at modeling the sales rate of a product in a particular target market. These therefore may be regarded as having a macro or whole-industry focus. There exists a need for a more

Copyright © 2008, IGI Global. Copying or distributing in print or electronic forms without written permission of IGI Global is prohibited.

micro-focused approach to modeling individual firm-level sales rates or life-cycle curves. Perhaps the most likely common practice would be to combine an estimate of market share with an industry life-cycle projection. However, the market share is at least partially under the policy control of the firm, or more generally the firm has a degree of control over its product's sales rate. We believe a combination of the Bass (1980) and Krishnan and Jain (2006) approaches may be fruitful as one research direction. We note for completeness that some results have already been obtained on another micro-focused viewpoint, described as disaggregate-level diffusion models (Roberts & Lattin, 2000).

Firm-level, or even regional and firm-level sales rates and ultimate market capture should be dependent on the intensity and number of competitors and the number and types of channels available to the buyer. Intuitively, as time from introduction increases, the numbers of competitors and channels should increase and exert a downward pressure on both prices and the sales of any individual firm. Systems dynamics models should prove helpful in simulating possible scenarios. We devote most of the next chapter to that important technique.

The foregoing considerations also suggest the possibility of developing profit-based life-cycle models. It is doubtful that realistic assumptions can be accommodated while still leading to analytical solutions at that level of complexity. However, again we should expect that dynamic simulations perhaps along with optimal control solution techniques should produce new and useful planning and what-if analysis tools. In particular, some of the more restrictive assumptions of the above models need not be imposed for the sake of analytical tractability. A model of this type might be used to evaluate various sophisticated what-if scenarios by first making assumptions about possible competitor supply-chain behaviors and then observing apparent optimal policies for one's own chain.

Conclusion

In this chapter, we have attempted to give both breadth and depth to the mathematical modeling aspects related to life cycles within the supply-chain and operations management literatures. We first reviewed the literature for works in this area and gave a discussion of a substantial number of publica-

Copyright © 2008, IGI Global. Copying or distributing in print or electronic forms without written permission of IGI Global is prohibited.

tions along those lines. Then we concentrated on models for the new product diffusion or life-cycle curve itself. Together with other quantitative articles contacted in Chapter IV, a reasonable coverage has hopefully been achieved. We believe that dynamic modeling may give more promise than the analytical modeling approach considered in this chapter. We discuss this kind of modeling in the next the chapter.

References

Aitken, J., Childerhouse, P., & Towill, D. (2003). The impact of product life cycle on supply chain strategy. *International Journal of Production Economics*, *85*, 127-140.

Angelis, A., & Porteus, E. L. (2002). Simultaneous capacity and production management of short-life-cycle, produce-to-stock goods under stochastic demand. *Management Science, 48*(3), 399-413.

Bass, F. M. (1969). A new product growth model for consumer durables. *Management Science, 15*, 215-227.

Bass, F. M. (1980). The relationship between diffusion rates, experience curves, and demand elasticities for consumer durables technical innovations. *Journal of Business, 53*, 51-67.

Bass, F. M., Jain, D. C., & Krishnan, T. (2000). Modeling the marketing-mix influence in new-product diffusion. In V. Mahajan, E. Muller, & Y. Wind (Eds.), *New-product diffusion models*. (pp. 99-122). Norwell, MA: Kluwer Academic Publishers.

Bass, F. M., Krishnan, T. V., & Jain, D. C. (1994). Why the Bass model fits without decision variables. *Marketing Science, 13*, 203-223.

Beamon, B. M. (1998). Supply chain design and analysis: Models and methods. *International Journal of Production Economics, 55,* 281-294.

Bollen, N. P. B. (1999). Real options and product life cycles. *Management Science, 45*(5), 670-684.

Chandrashekaran, M., & Sinha, R. K. (1995). Isolating the determinants of innovativeness: A split-population Tobit (SPOT) duration model of timing and volume of first and repeat purchase. *Journal of Marketing Research, 32*, 444-456.

Copyright © 2008, IGI Global. Copying or distributing in print or electronic forms without written permission of IGI Global is prohibited.

Chang, S. L., Wang, R. C., & Wang, S. Y. (2006). Applying fuzzy linguistic quantifier to select supply chain partners at different phases if product life cycle. *International Journal of Production Economics, 100*, 348-359.

Cohen, M. A., Ho, T. H., & Matsuo, H. (2000). Operations planning in the presence of innovation-diffusion dynamics. In V. Mahajan, E. Muller, & Y. Wind (Eds.), *New-product diffusion models*. (pp. 237-259). Norwell, MA: Kluwer Academic Publishers.

Cox, D. R. (1972). Regression models and life tables (with discussion). *Journal of the Royal Statistical Society, Series B, 34*, 187-220.

Crandall, R. E., & Crandall, W. (2003). Managing excess inventories: A life-cycle approach. *Academy of Management Executive, 17*(3), 99-113.

Damodaran, P., & Wilhelm, W. E. (2004). Branch-and-price methods for prescribing profitable upgrades of high-technology products with stochastic demands. *Decision Sciences, 35*(1), 55-82.

Elderton, W. P., & Johnson, N. L. (1969). *Systems of frequency curves*. London: Cambridge University Press.

Fandel, G.., & Stammen, M. (2004). A general model for extended strategic supply chain management with emphasis on product life cycles including development and recycling. *International Journal of Production Economics, 89*, 293-308.

Fixson, S. K. (2005). Product architecture assessment: A tool to link product, process, and supply chain design decisions. *Journal of Operations Management, 23*, 345-369.

Geyer, R., Van Wassenhove, L., & Atasu, A. (2007). The economics of remanufacturing under limited component durability and finite product life cycles. *Management Science 53*(1), 88-100.

Gunasekaran, A. D., Macbeth, K., & Lamming, R. (2000). Modeling and analysis of supply chain management systems: An editorial overview. *Journal of the Operational Research Society, 51*, 1112-1115.

Haldar, S., & Rao, V. R. (1998). A micro-analytic threshold model for the timing of first purchases of durable goods. *Applied Economics, 30*(7), 959-974.

Heckman, J. J., & Singer, B. (1985). Longitudinal analysis of labor market data. *Econometric Society Monographs, No. 10*, Cambridge, MA: Cambridge University Press.

Copyright © 2008, IGI Global. Copying or distributing in print or electronic forms without written permission of IGI Global is prohibited.

Helsen, K., & Schmittlein, D. C. (1993). Analyzing duration times in marketing: Evidence for the effectiveness of hazard-rate models. *Marketing Science, 11*(4), 395-414.

Higuchi, T, & Troutt, M. D. (2004a). Understanding and managing the intrinsic dynamics of supply chains. In C. K. Chan & H. W. J. Lee (Eds.), *Successful strategies in supply chain management* (pp. 174-193). Hershey, PA: Idea Group.

Higuchi, T., & Troutt, M.D. (2004b). Dynamic Simulation of the Supply Chain for a Short Life Cycle Product - Lessons from the *Tamagotchi* Case. *Computers & Operations Research, 31*(7), 1097-1114.

Johnson, N. L., Kotz, S., & Balakrishnan, N. (1994). *Continuous univariate distributions - 1*. New York: John Wiley & Sons.

Kainuma, Y., & Tawara, N. A. (2006). A multiple attribute utility approach to lean and green supply chain management. *International Journal of Production Economics, 101*, 99-108.

Kalish, S. (1985). A new-product adoption model with price, advertising , and uncertainty. *Management Science, 31*, 1569-1585.

Kamath, N. B., & Roy, R. (2005). Supply chain structure design for a short lifecycle product: A loop dominance based analysis. Proceedings from *38th Hawaii International Conference on System Sciences (HICSS-38 2005), CD-ROM / Abstracts , January 3-6, 2005, Big Island, HI, USA.*

Kamath, N. B., & Roy, R. (2007). Capacity augmentation of a supply chain for a short lifecycle product: A system dynamics framework. *European Journal of Operational Research, 179*(2), 334-351.

Klaus, B. (1998). *Fair allocation and reallocation: An axiomatic study*. Unpublished doctoral dissertation. Maastricht University, Maastricht, The Netherlands.

Klimenko, M. M. (2004). Competition, matching, and geographical clustering at early stages of the industry life cycle. *Journal of Economics and Business, 56*, 177-195.

Kotler, P. (1999). *Marketing management, the millennium edition*. Englewood Cliffs, NJ: Prentice Hall.

Kotler, P., & Keller, K. L. (2006). *Marketing management*, 12th ed. Upper Saddle River, NJ: Pearson Prentice Hall.

Copyright © 2008, IGI Global. Copying or distributing in print or electronic forms without written permission of IGI Global is prohibited.

Krishnan, T. V., & Jain, D. C. (2006). Optimal dynamic advertising policy for new products. *Management Science, 52*(12), 1957-1969.

Lee, H., & Whang, S. (2002). The impact of the secondary market on the supply chain. *Management Science, 48(6)*, 719-731.

Li, D., & O'Brien, C. (2001). A quantitative analysis of relationships between product types and supply chain strategies. *International Journal of Production Research, 73,* 29-39.

Lilien, G. L., Rangaswamy, A., & Van Den Bulte, C. (2000). Diffusion models: Managerial applications and software. In V. Mahajan, E. Muller, & Y. Wind (Eds.), *New-product diffusion models*. (pp. 295-311). Norwell, MA: Kluwer Academic Publishers.

Mahajan, V., Muller, E., & Wind, Y. (Eds.). (2000). *New-product diffusion models*. Norwell, MA: Kluwer Academic Publishers.

Mansfield, E. (1961). Technological change and the rate of imitation. *Econometrica, 29*(4), 71-766.

Mood, A. M., Graybill, F. A., & Boes, D. C. (1974). *Introduction to the theory of statistics*. New York: McGraw-Hill Book Company.

Müller, A., Scarsini, M., & Shaked, M. (2002). The newsvendor game has a non-empty core. *Games and Economic Behavior, 38,* 118-126.

Narasimhan, R., Talluri, S., & Mahapatra, S.K. (2006). Multi-product, multi-criteria model for supplier selection with product life cycle considerations. *Decision Sciences, 37*(4), 577-603.

Parlar, M., & Weng, Z. K. (1997). Designing a firm's coordinated manufacturing and supply decisions with short product life cycles. *Management Science, 43*(10), 1329-1344.

Pearson, K. (1895). Contributions to the mathematical theory of evolution. II. Skew variations in homogeneous material. *Philosophical Transactions of the Royal Statistical Society of London, Series A, 186*, 343-414.

Punakivi, M., & Hinkka, V. (2006). Selection criteria of transportation mode: A case study in four Finnish industry sectors. *Transport Reviews, 26*(2), 207-219.

Putsis, W. P., Jr., & Srinivasan, V. (2000). Estimation techniques for macro diffusion models. In Mahajan, V., Muller, E., & Wind, Y. (Eds.). (2000). *New-product diffusion models*. (pp. 264-291). Norwell, MA: Kluwer Academic Publishers.

Copyright © 2008, IGI Global. Copying or distributing in print or electronic forms without written permission of IGI Global is prohibited.

Raman, A. (1999). Managing inventory for fashion products. In A. Tayur, R. Ganeshan, & M. Magazine (Eds.), *Quantitative models for supply chain management* (pp. 789-806). Boston: Kluwer Academic Publishers.

Roberts, J. H., & Lattin, J. M. (2000). Dissaggregate-level diffusion models. In V. Mahajan, E. Muller, & Y. Wind (Eds.), *New-product diffusion models*. (pp. 207-236). Norwell, MA: Kluwer Academic Publishers.

Robinson, B. & Lakhani, C. (1975). Dynamic price models for new product planning. *Management Science, 21*, 1113-1122.

Shapley, L. (1953). A value for n-person games. In: Tucker, A, & Kuhn, H. (Eds.), *Contributions to the theory of games II* (pp. 307-317). Princeton, NJ: Princeton University Press.

Shapley, L. (1967). On balanced sets and cores. *Naval Research Logistics Quarterly, 14*, 453-460.

Sinha, R, K, & Chandrashekaran, M. (1992). A split-hazard model for analyzing the diffusion of innovations. *Journal of Marketing Research, 29*, 116-127.

Slikker, M., & van den Nouweland, A. (2001). Social and economic networks in cooperative game theory. Boston: Kluwer Academic Publishing.

Slikker, M., Fransoo, J., & Wouters, M. (2001). *Joint ordering in multiple news-vendor situations: A game-theoretical approach.* Working paper. Eindhoven University of Technology. Eindhoven, The Netherlands:.

Slikker, M., Fransoo, J., & Wouters, M. (2005). Cooperation between multiple news-vendors with transshipments. *European Journal of Operational Research, 167*, 370-380.

Srinivasan, R., Haunschild, P., & Grewal, R. (2007). Vicarious learning in new product introductions in the early years of a convergent market. *Management Science 53*(1), 16-28.

Suijs, J. (1999). *Cooperative decision-making under risk.* Boston: Kluwer Academic Publishers.

Supply-Chain Council. (2007). Overview of the SCOR Model 2.0. Supply Chain Council, Inc. Retrieved February 7, 2007, from http://www.supplychain.org/

Tan, R. B. H., & Khoo, H. H. (2005). An LCA study of a primary aluminum supply chain. *Journal of Cleaner Production, 13*, 607-618.

Copyright © 2008, IGI Global. Copying or distributing in print or electronic forms without written permission of IGI Global is prohibited.

Taylor, T. A. (2001). Channel coordination under price protection, midlife returns, and end-of-life returns in dynamic markets. *Management Science, 47*(9), 1220-1234.

Tayur, S., Ganeshan, R., & Magazine, M. (Eds.). (1999). *Quantitative models for supply chain management*. Boston: Kluwer Academic Publishers.

Vonderembse, M. A, Uppal, M., Huang, S. H., & Dismukes, J. P. (2006). Designing supply chains: Towards theory development. *International Journal of Production Economics, 100*, 223-238.

Wang, G.., Huang, S. H, & Dismukes, J. P. (2004). Product-driven supply chain selection using integrated multi-criteria decision-making methodology. *International Journal of Production Economics, 91*, 1-15.

Weerahandi, S., & Dalal, S. R. (1992). A choice-based approach to the diffusion of a service: Forecasting fax penetration by market segment. *Marketing Science, 11*(1), 39-53.

Weng, Z. K., & McClurg, T. (2003). Coordinated ordering decisions for short life cycle products with uncertainty in delivery time and demand. *European Journal of Operational Research, 151*, 12-24.

Weng, Z. K., & Parlar, M. (2005). Managing build-to-order short life-cycle products: Benefits of pre-season price incentives with standardization. *Journal of Operations Management, 23*, 482-493.

Copyright © 2008, IGI Global. Copying or distributing in print or electronic forms without written permission of IGI Global is prohibited.

Chapter IV

Supply Chain Dynamics and Dynamic Simulation

This chapter first focuses on the intrinsic dynamic interactions within supply chains. These are theoretically interesting and informative for understanding and managing supply chains. Next, the chapter explores the power of dynamic simulations for managing and understanding the workings and complex interactions in supply chains. Dynamic simulation grew out of the work of Forrester on Systems Dynamics. While general influences of the differing players and levels can be qualitatively described, the use of dynamic simulations permits exploration of the processes by which small changes and interactions of variables and parameters can have surprisingly large impacts on supply chain performance. As revealed by the literature survey in Chapter III, the short life-cycle case is increasingly important in practice. At the same time, the intrinsic dynamics of supply chains are especially problematical and potentially catastrophic in the short life-cycle case. The history of the well-known *Tamagotchi*™ toy is used to provide an illustration both for the technique of dynamic simulation and for the difficulties that can occur in such supply chains.

Copyright © 2008, IGI Global. Copying or distributing in print or electronic forms without written permission of IGI Global is prohibited.

Introduction

This chapter is devoted to examining the intrinsic dynamics of supply chains and to dynamic simulation modeling. We emphasize the short life-cycle product case. An example case is based on the experience of virtual pet toy called *Tamagotchi*™. We follow the results in Higuchi and Troutt (2004, 2005).

From the viewpoint of both theory and practice, one of the most interesting aspects of supply chains is their intrinsic-dynamic behavior. Dynamic interactions can cause unexpected and undesirable results. There are both external and internal reasons for this. Externally, severely competitive environments, capricious consumer behavior and technological innovations are major concerns for supply-chain management. These affect both the structure and behavior of the supply chain and force it to be more flexible and agile. Internally, the supply chain is a multi-echelon system composed of a scratched-together and shifting set of players whose preferences and intentions often differ. This may create and amplify the information distortions, lags, and tricks of business within the supply chain. The combined effects of these external and internal factors make Supply Chain Dynamics (SCD) active and complex. In this chapter, we discuss the influences, mechanisms, and effects of SCD. The following quote due to Chopra and Meindl (2001, pp. 4-6) captures the essence of SCD. "A supply chain is dynamic and involves the constant flow of information, products and funds between different stages. Each stage of the supply chain performs different processes and interacts with other stages of the supply chain."

Background

A supply chain may be regarded as an interaction of processes among participants from different companies and levels. The purpose of supply chain management is to maximize supply chain profitability, which is total profit to be shared appropriately across all supply chain stages and players. It can be concluded that the essence of the supply chain is its dynamics and that how to manage it is one of the most important tasks to maximize supply chain profitability.

SCD features arise from external and internal factors. External factors are drivers for the supply chain to change its structure and therefore have a close

Copyright © 2008, IGI Global. Copying or distributing in print or electronic forms without written permission of IGI Global is prohibited.

relationship with the planning and designing process. Major external factors are the intensity of competition, consumer behavior and technological innovations. Severe competition in the market puts pressure on manufacturers, distributors and retailers to cut costs while maintaining the same quality or service level (or increase the quality or service level under the same costs). The supply chain becomes the major arena for competition in modern business because it is often inefficient for any single company to produce a whole product and maintain the entire, often global, distribution channel (Bradley, Thomas, & Cook, 1999). As supply chains become worldwide, they must cross the boundaries between company groups. Hence, severe competition in the market makes the supply chain very important and forces it to change continuously.

We believe the following analogy may be helpful in connection with these concepts. Consider the first order vector linear differential equation system given by $\mathbf{x}'(t) = \mathbf{A}\mathbf{x}(t) + \mathbf{u}(t)$. Here, $\mathbf{u}(t)$ is an exogenous vector input series that may be regarded as analogous to the external influences on the system. The autonomous term $\mathbf{A}\mathbf{x}(t)$ may be regarded as representing the internal dynamics of the system. Although the supply chain is more complex, nonlinear, and involves stochastic as well as deterministic aspects, this model may help to clarify the distinction between internal and external influences, as well as symbolically suggest the potential for their interactions.

Customers tend to become more and more demanding and capricious in their wants and needs as the variety and availabilities of goods increase. Supply chains must deal with this routinely. It takes time and many transactions to produce and deliver goods from the raw materials because of the nature of the production and distribution processes. This process frequently and usually involves long distances between the retailers and the manufacturer and a multi-echelon system. This may amplify the effects of customer behavior and can cause the well-known problems called the bullwhip effect (Lee, Padmanabhan, & Whang, 1997a) and the boom and bust (Paich & Sterman, 1993). On the other hand, increasing customer needs require the performance of the supply chain to continuously improve. Successful supply chains must be ever more responsive to customers according to their needs and requirements (Ballou, 1992). To keep abreast of the growing customer needs, the supply chain should enhance its collaborations or change its structure by making the most of strategic alliances.

Advertising through mass media and the development of the Internet has speeded up the diffusion of new products because consumers who are informed

Copyright © 2008, IGI Global. Copying or distributing in print or electronic forms without written permission of IGI Global is prohibited.

about a product can check the review. The distance and the national borders become meaningless. At the same time, technical innovation and severe competition in the market promote rapid obsolescence of existing products and technologies. When a company succeeds in developing a new product category, other competitors may soon emerge. The market originator must endure not only the substantial risk of whether the market will materialize or not, but also the difficulty of recovering major costs, such as research, development and advertising. Increasingly, the supply chain becomes the mechanism for coping with these problems because it is often inefficient for any single company to produce a whole product. Hence, modern business is essentially the competition of one supply chain with another (Bradley et al., 1999). Supply chain dynamics is the interaction processes of the participants from different departments and companies. A positive aspect of supply chain dynamics is effective collaboration, which may lead to higher performance. A negative aspect is independent decision making, which may create various delays and aggravate the forecasting error. Tompkins (2000) introduced the concept *Supply Chain Synthesis*. It is a holistic, continuous improvement process of ensuring customer satisfaction and is all about using partnerships and communication to integrate the supply chain. Bowersox and Closs (1996, p. 37) observed that "coordination is the backbone of overall information system architecture among value chain participants." Therefore, it is necessary to coordinate the activities appropriately within a supply chain to achieve better overall, that is, system, performance.

Technological innovations, especially, radical innovations, are very disruptive and change the nature and balance of the market. They can make obsolete the existing technologies and product categories (Abernathy, Clerk, & Kantrow, 1983). In order to avert the risk that another company succeeds in a radical innovation, it is better not to cover all the related processes. The auto industry is a good example. A vast number of materials and parts are required to produce a car. Principal firms like GM, Toyota, Daimler-Chrysler, and other automakers are called the set makers or integrators because they have key suppliers or sub-integrators. It is impossible for them to be in the forefront of the all materials and parts. In addition, they currently are competing with others to commercialize the hybrid car, alternative fuel cars, and Intelligent Transportation Systems, among others. These advanced technologies cost too much and are too risky for most of them to do the relevant research and development solely by themselves. Therefore, technological innovations encourage the strategic alliances among the different industries and former rivals and make the players of the supply chain fluid. Thus, due to external

Copyright © 2008, IGI Global. Copying or distributing in print or electronic forms without written permission of IGI Global is prohibited.

factors, the first essence of the supply chain is its dynamic nature and the concomitant requirement of flexibility.

Internally, the supply chain is a multi-echelon system composed of scratched-together or ad hoc players whose intentions may and often do differ. To fulfill a customer's needs, a lot of stages and people, retailers, wholesalers, warehouses, transporters, manufactures, and suppliers are needed. It is natural for everyone in a supply chain to attempt to maximize their own profits by deciding combinations of the selling and buying prices, the production, inventory, and transportation costs, and so on. The accumulation of these individual behaviors makes supply-chain management very difficult. The geographical and mental distances and the differences of the viewpoints and individual self-interests worsen the situation because the supply chain is a loosely coupled organization. This is in sharp contrast to *Keiretsu*, a traditional Japanese company group, whose members are related companies and are long-term partners, and in which members are strictly and centrally controlled by a headquarters firm.

The supply chain can also globalize easily and dramatically. Globalization enhances the production at the right place for the right cost with sales promotion, tax minimization, and so on. On the other hand, globalization lengthens the geographical and mental distance among the players. This amplifies in-

Figure 1. Outline of supply chain dynamics

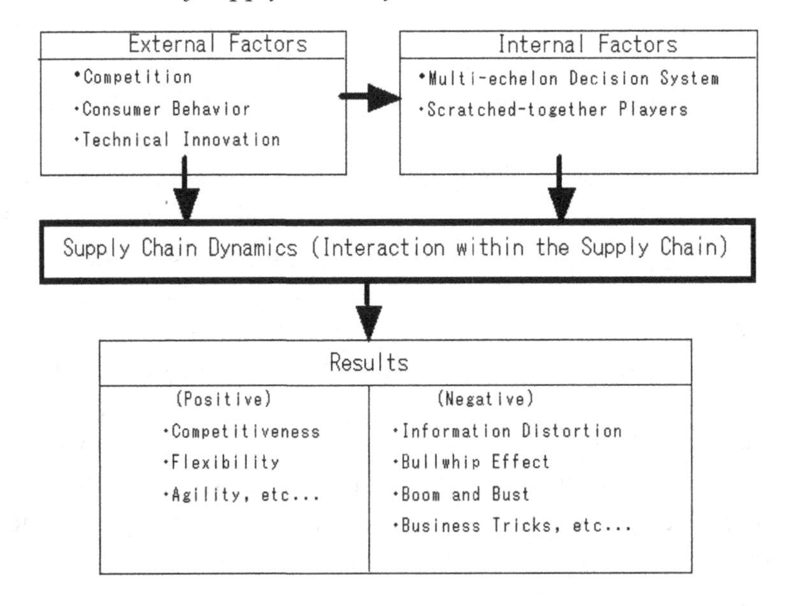

Copyright © 2008, IGI Global. Copying or distributing in print or electronic forms without written permission of IGI Global is prohibited.

formation distortions (Lee, Padmanabhan, & Whang, 1997b), lags, and tricks of business within the supply chain, and makes its management much more difficult than before. From the internal factors, the second main influence on SCD is the autonomous decision making of the players in the system.

SCD therefore results from the combined effects of these external and internal factors as Figure 1 illustrates. External factors, such as competition, consumer behavior, and technological advances, have two aspects, stimulating the supply chain to change its structure and creating the internal supply chain dynamics. Some parts of the former are strategic matters. The purpose of this chapter is how to manage SCD, not design or classify it. Hence, in the rest of the chapter, we limit the discussion to the management of internal supply dynamics set by the external and internal factors and continue the explanation of SCD from the interactions within the supply chain.

In the following sections, we discuss a number of related issues including the role of collaboration within supply chains, negative phenomena in SCD, the Theory of Constraints (TOC), and dynamic computer simulation. In the next section, we examine problems that confront supply chains and which cannot be handled adequately without collaboration within supply chains. On the other hand, supply chain management is so difficult that it may lead to unpredictable and undesirable results. TOC gives a good framework and dynamic computer simulation is an effective aid in the management of SCD.

Main Focus of the Chapter

Collaboration within Supply Chains

The supply chain can be an effective and efficient network for satisfying customer needs at appropriate prices and variety of goods and services at a

Figure 2. Examples of contradictory requirements for supply chain

Wide Variety of Goods	vs.	Lean Management
High Quality Products	vs.	Low Cost
Global Manufacturing	vs.	Quick Response
Global Compatibility (Standardization)	vs.	Customization

Copyright © 2008, IGI Global. Copying or distributing in print or electronic forms without written permission of IGI Global is prohibited.

given level of cost, agility, and risk. It is the modularity and the commitment to enhance collaborations that can guarantee such effectiveness and efficiency. Collaborative commerce is a good milestone, which makes product development and the daily operations more effective and efficient by using a website (Bechek & Brea, 2001). It integrates the company's core business processes within the supply chain including the customers (Gossain, 2002). It is the accumulation of effective and flexible collaborations with appropriate partners that leads to higher competitiveness in the market.

The supply chain may be said to face contradictory requirements, such as providing a wide variety of goods versus lean management, high quality versus low cost, and so on as in Figure 2. To achieve these tasks simultaneously, modularity and standardization play very important roles. A modular system ensures that the total amount of inventory decreases, that the manufacturer offers a wide variety of goods with relatively few parts by combinations of modules, and that it localizes the impact of model changes (Abernathy, 1978, pp. 86-113). Technically, modularity guarantees the feasibility of collaborations within the supply chain.

A supply chain component, partner, or player's commitment is a precondition for inducing positive interactions within the supply chain. Sharing the goals, strategy, and profits is necessary to enhance such a commitment. The goal is to maximize profitability not only for the supply chain as a whole, but also for the individual players, because the vehicle for competition is the entire supply chain in modern business as has been noted earlier. Strategy is an indispensable component of any supply chain in order to achieve higher competitiveness in the market. It also functions as a guide for each player to understand and carry out its role. The policy and rules for the profit sharing have a great effect on the incentive and behavior of each player. Sharing these appropriately and fairly will strengthen the unity of the supply chain.

From the point of view of operations, synchronization is the key word in the supply chain. In addition to the above contradictory requirements, and although the supply chain can become geographically complex and global, it nevertheless is required to become increasingly agile. There are two ways to synchronize the supply chain. First, bottlenecks must be identified and eliminated (or reduced) for maximizing throughput (Goldratt & Cox, 1992). Throughput is an important measure of synchronization of the supply chain. The bottleneck, the weakest point in the chain, limits the level of throughput for the entire supply chain. For example, even though a supply chain has a strong distribution channel, it may not achieve optimal performance if its

Copyright © 2008, IGI Global. Copying or distributing in print or electronic forms without written permission of IGI Global is prohibited.

manufacturing and R&D departments are not agile enough. On the other hand, a supply chain that has strong such departments may not be able to make the most of them without a strong distribution channel. In this sense, it is useful to find the bottleneck and rectify it as quickly as possible. Keeping the network well balanced contributes to higher throughput and increases the ability to synchronize the entire system.

The second way to synchronize is to be a lean supply chain by information sharing on the latest market demands and inventories among the players. The levels of inventories and backlogs are also good measures for the degree to which synchronization is being achieved in the supply chain. JIT is a well-known approach to achieving a lean supply chain. It was originally called the "kanban" system. "Kanban" means a signboard in Japanese. To share information on the latest market demands and the production scheduling with suppliers, Toyota used and sent to suppliers an iron signboard on which the required parts, numbers and times were indicated. Toyota and their suppliers reduced the amounts of inventories dramatically by sharing information through this system[1]. Sharing the information saves costs and time, and makes it possible to synchronize operations within the supply chain.

The stages in Figure 3 might appear to be very different at first glance, but they are similar. They may be regarded as a pair of dual problems. The first one, the commitment, is to minimize inventories by information sharing. On the other hand, the second one, the synchronization, is to maximize throughput by eliminating bottlenecks and keeping the supply chain well balanced. Their goal is to maximize the supply chain's profitability. Both of them incorporate continuous improvement processes, *supply chain synthesis* or *kaizen*, and pursue synchronization throughout the supply chain, whose basis is the customers' wants and needs.

Figure 3. Collaboration within supply chain

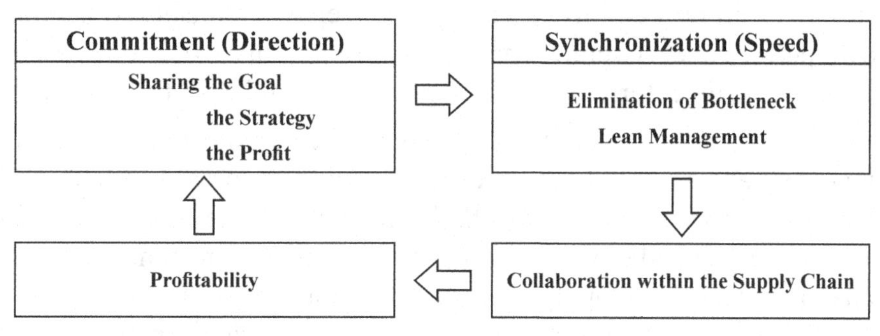

Copyright © 2008, IGI Global. Copying or distributing in print or electronic forms without written permission of IGI Global is prohibited.

The business process standardization is the collaboration within the supply chain. It enables supply chain partners, even newcomers, to behave efficiently and effectively for the supply chain by sharing the information and the strategy because the business process standardization makes the necessary information and its format clear. As a result, supply chains can recruit new companies beyond the national borders and the company groups in order to improve the bottleneck in the supply chain.

Undesirable Phenomena in Supply Chain Dynamics

By this term, we mean those obstacles, characteristics and features that work against the goals of shared information, bottleneck reductions, and synchronicity. Most of these come from the independent decision-making at the different stages of the chain because of its multi-echelon nature and the often fluid set scratched-together players. A *keiretsu* organization or chain is composed of affiliated companies with long-term established partnerships, and which has exclusionary business practices. In contrast, a supply chain is a much more open and flexible network. Ironically, this may aggravate information distortions, various lags, forecasting errors, and tricks of business within the supply chain.

There are numerous trade-offs in the supply chain (Gopal & Cahill, 1992). For instance, the sales and marketing departments desire a high degree of production flexibility and rapid turnaround to catch up with recent trends and to maximize their sales. From the logistical way of thinking, their behavior is based on the short-term. They would try to dramatically increase the stock of goods that are rising in popularity, but reduce the stock of goods decreasing in popularity. On the other hand, to reduce unit costs or recover sunk costs, manufacturers favor longer production runs, fewer set-ups, smooth schedules, and a balanced line. These types of trade-offs have a great influence on supply chain profitability. If each player behaves selfishly to maximize its own performance, overall supply chain profitability would decrease rapidly because the inventories and backlogs increase and cash flows shrink.

Various lags come from the geographical factors, the actual process times, inventory and order policies, decision timing, and so on. The causes of these lags can be divided into an essentially unavoidable one and a managerial one. The delivery times between stages and the manufacturing times at factories are unavoidable. As the supply chain becomes longer and the pressure stronger, the reduction of these times and more efficient management are required.

Copyright © 2008, IGI Global. Copying or distributing in print or electronic forms without written permission of IGI Global is prohibited.

Independent decision-making of players might enlarge the unavoidable lags and, in the worst case, create other lags. The order system is a good example. Both batch and periodic order systems create lags at each echelon. If a supply chain consisted of four stages (retailers, wholesalers, manufacturers, and suppliers) and they order weekly, suppliers receive the market demand that actually occurred three weeks before. This leads not only to the mismatch of supply and demand, but also to misguided capacity plans and decisions.

Without sharing the most current market demands, information distortion can be generated and enlarged at each stage in the supply chain. The lack of such information causes tremendous inefficiencies, such as the bullwhip effect, excessive or inadequate estimates of the inventory investment, poor customer service, lost revenues, misguided capacity plans, and missed production schedules. From the viewpoint of information distortion, Gavirneni, Kapuscinski, and Tayur (1999) simulated an overall supply chain model emphasizing the value of information and extended existing inventory theory to supply chain level. The level of inventory is a further measure reflecting the efficiency of supply chain management.

The following quotes have been offered by various experts to explain the bullwhip effect, which is the exaggerated order swings phenomena caused by information distortion in the supply chain.

The information transferred in the form of orders tends to be distorted and can misguide upstream members in their inventory and production decisions. In particular, the variance of orders may be larger than that of sales, and the distortion tends to increase as one moves upstream (Lee et al., 1997b, p. 546).

Magee, Copacino, and Rosenfeld (1985, p. 42) suggest that "variations in production are far more severe than variations in demand, and the more levels and stages of production there are, the more violent production level changes become."

The phenomenon called the "shortage game" or "shortage gaming" may be regarded as starting the phantom demand phenomenon and the boom and bust effects, and these can also be escalated or amplified by it (Lee et al., 1997a). Figure 4 summarizes the movement of the supply chain from the view of the flow of orders and products. The shortage game starts when product demand greatly exceeds supply and some customers begin to make duplicate orders with multiple retailers. Retailers and wholesalers misjudge the circumstances, seeing a surge in demand and a shortage of the supply.

Copyright © 2008, IGI Global. Copying or distributing in print or electronic forms without written permission of IGI Global is prohibited.

They then may place excessive orders to the manufacturer. This process often is not very risky for the customer who can accept the first order filled and cancel any outstanding duplicates. Later, when the supply catches up with the real demand, backlogs will suddenly disappear and large excess inventories emerge. Both the retailer and the manufacturer can be victims of the shortage game. Retailers may be stuck with hard-to-sell excess inventories. While the same also is true of manufacturers, their risk can be even greater if capacity changes have been made without a real basis. Based on an inflated picture of real demand, exacerbated by the natural lag in filling orders, the manufacturer might disastrously place larger amounts of capital into capacity expansion. Thus, these overstated demands called "phantom" demands result from the shortage game (Nehmias, 1997). The boom and bust phenomenon (Paich & Sterman, 1993) can be the result of the shortage game exaggerated by the bullwhip effect from the viewpoint of the life cycle of the product. These difficulties are well illustrated by the case of the Bandai Co., Ltd. and the Tamagotchi™ toy (Higuchi & Troutt, 2004). A seamless information system possessing timeliness, unity, and traceability is very useful to estimate phantom demands quickly. Processing orders at several stages masks phantom demands and worsen the situation. Without sharing the latest information in an on-time basis, phantom demands are masked so long that they can incur self-reproduction to a certain level. Hence, sharing the raw information (orders from the customers) throughout the supply chain promptly is a key. If possible, it is very helpful to identify the customers who placed multiple orders at different retailers and grasp their purchase records.

We believe a possible aid to early identification of phantom demand might be based on the sharing of information about cancelled orders. It is known that sharing information about real demand at the customer or point of sale level can be useful in reducing the bullwhip effect (Chopra & Meindl, 2001). Large increases in orders coupled with increases in cancelled orders could be indicative of phantom demand as follows. If demands are inflated then as these are filled, even partially in the case of rationing, then we can expect a higher level of activity in cancelled orders as well.

Another possibility for further research would be to compare ratios of returns to orders and look for outlier patterns in the series of ratios. Also, regression models may similarly be useful. Simulation models such as described in this chapter may be helpful in this regard by expanding them to include returns assumptions. Perhaps some benchmark ratios might then be associated with periods of higher or lower over-ordering and returns activity.

Copyright © 2008, IGI Global. Copying or distributing in print or electronic forms without written permission of IGI Global is prohibited.

Figure 4. A model of the supply chain process (three-echelon: customer-retail- manufacturer)

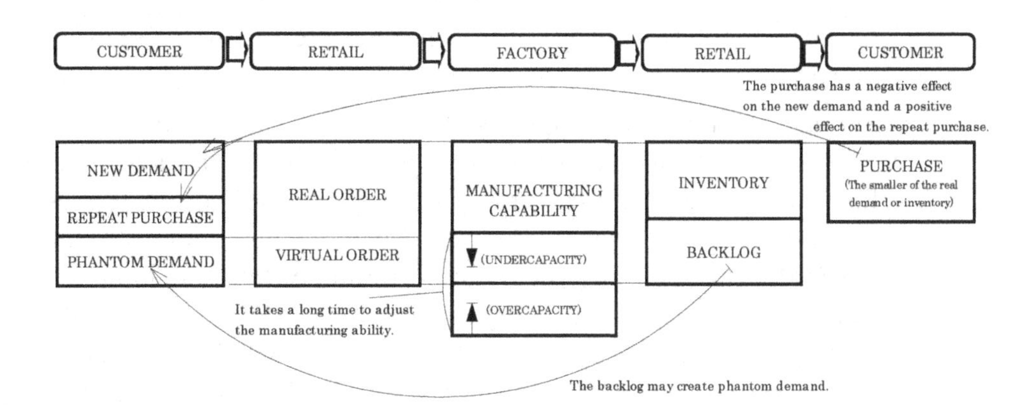

The Theory of Constraints

Given the potential importance of bottleneck management in supply chains, we discuss in this section some approaches to their management. The Theory of Constraints (TOC) is a system improvement philosophy that focuses heavily on bottleneck management. While TOC is generally well known in the operations management literature, it does not appear to be discussed frequently in the supply chain literature. Also in this section, we discuss a more recent mathematical programming tool of use in bottleneck identification and management.

TOC was developed by Goldratt and others. Some key contributions are: Cox and Spencer (1998), Dettmer (1997), Goldratt (1990a, b, c), Goldratt and Cox (1992), Goldratt and Fox (1986), Kendall (1998), Mabin and Balderstone (2000), McMullen (1998), and Schragenheim (1999). TOC has produced a number of principles and methods for improving the flow of constrained systems. TOC is a kind of common sense management philosophy. It has continuous improvement as a major feature. It also uses various diagrams for thinking through business problems. Identification and management of bottlenecks is a major feature of TOC. Some general tenets of TOC are the following:

Copyright © 2008, IGI Global. Copying or distributing in print or electronic forms without written permission of IGI Global is prohibited.

1. Managers should verbalize their intuitions about business problems and solutions. They probably know the true problems and good solutions. If they don't then they may do the opposite of what they believe in.
2. Improvements are changes and changes are always perceived as threats.
3. Surface the hidden assumptions. TOC Logic Trees, Compromise, or "Evaporating Cloud" diagrams and other diagrams are used in TOC for this purpose. The dice game is a good exercise for this. It goes as follows: Set up a series of "stations" where the output of each station is based on the roll of a die. Output from one station is input to the next. Everyone assumes that the system output will be 3.5 per period (the average of the values on a die), but it's actually much less because each station can only output as much as it received from its predecessor. This exercise produces interesting discussion about having equal capacity at all processes, and about making assumptions, more generally.
4. It is important to understand relationships (correlations) and especially the "why" of relationships (effect-cause-effect).
5. Concentrating on optimizing the present system may be a dead end. Improvement of the system should be the goal. For example, using EOQ may prevent the search for improvement in setup times.

The major process steps in TOC are:

i. Identify the major constraint(s) to the system. These are called bottlenecks and limit the system in one way or another. Bottlenecks might be limiting production, information, sales, and so on.
ii. Decide how to exploit the constraint.

Subordinate all other activities to the bottleneck.

iii. "Elevate" the constraint, that is, remove it, reduce it, and so on until it is broken and something else becomes the constraint.
iv. Iterate this process continually.

Copyright © 2008, IGI Global. Copying or distributing in print or electronic forms without written permission of IGI Global is prohibited.

In addition, TOC suggests scheduling according to the "Drum, Buffer, and Rope." The "Drum" is the bottleneck process. It sets the pace for the whole system. The "Buffer" is a time element for establishing realistic commitment dates for shipments. We want to continually reduce the buffer. The "Rope" is a time element for starting new jobs, like a setup time. The rope pulls new jobs into the process. We want to shorten it as much as possible. TOC addresses other time elements as well.

TOC applications are numerous and are considered in more detail in the above-cited references. Here we mention a few examples of how TOC thinking can be insightful and beneficial. What one may think is a bottleneck often is not. For example, a process may have sufficient capacity for average demand but be unable to handle peak load demand. That process is not necessarily a bottleneck but may be a resource that does not have enough "protective capacity." There may be other ways to solve the problem than by treating it as a bottleneck. Policies and procedures often create situations in which the bottleneck is not fully exploited. Before examining ways to increase capacity of the bottleneck it is important to examine these policies and procedures to see whether they should be changed to fully exploit the bottleneck. For example, transferring work to the bottleneck in smaller lot sizes could improve the flow to the bottleneck thus insuring that the bottleneck does not lose time waiting for work. Also, changing set-up procedures could reduce lost time due to set-ups at the bottleneck (Goldratt, 1990c; Goldratt & Cox, 1992; Goldratt & Fox, 1986). Subordinating other processes to the bottleneck can often increase system flow. This may mean changing the scheduling procedures to ensure that the bottleneck is never idle (Goldratt & Fox, 1986).

As noted in point 5 above, use of optimization models comes with cautions. In fact, early work by Zeleny (1981, 1986) noted the difference between optimizing a poorly designed system and designing an optimal system. Nevertheless, several useful connections between Linear Programming (LP) and TOC are potentially important tools for the management of bottlenecks in supply chains (Luebbe & Finch, 1992; Mabin & Gibson, 1998). Luebbe and Finch compare the steps of LP to those of TOC. Mabin and Gibson stress how LP and TOC were used together effectively in a system with product mix features. Troutt, White, and Tadisina (2001) show how LP-based maximal flow network modeling can help both with bottleneck identification and with capacity expansion decisions for improving system throughput.

Copyright © 2008, IGI Global. Copying or distributing in print or electronic forms without written permission of IGI Global is prohibited.

Dynamic Computer Simulation

The supply chain, especially in the multi-echelon inventory system case, is too complex to control effectively as a whole without the aid of computer simulation (Ballou, 1992). Computer simulation is very useful, if not actually essential, for analyzing SCD. Computer simulations can be divided into two types, static and dynamic. Queuing and inventory theories are the bases of both. Static simulations provide the foundation for the dynamic ones. Dynamic simulations incorporate feedback mechanisms that change the structure or behavior of the supply chain. The following statement highlights a difference between them.

The primary difference between them is the manner in which time-related events are treated. Whereas dynamic simulation evaluates system performance across time, in static simulation no attempt is made to structure time-period interplay. Dynamic simulation is performed across time so that operating dynamics may impact the planning solution. (Bowersox, Closs, & Helferich, 1986, pp. 412-421).

Systems dynamics is a suitable tool to analyze dynamic systems from the viewpoint of the whole system. It was originally called Industrial Dynamics because it focused on the shifting nature and behavior of the companies over the passage of time. Forrester (1961) built a system dynamics model of the three-echelon production distribution system and demonstrated how market demands are amplified through the transactions in the supply chain. It has since been applied to wide ranging areas from the social to the natural sciences. In fact, many supply chain models have been built by using System Dynamics. Some System Dynamics models are integrated models that simulate the effects of interactions within the supply chain. Paich and Sterman (1993) provide a typical one. They analyzed the "boom and bust" process by simulating the diffusion process of a new product. This is caused by the fact that the product life cycles become very short and diffusion and obsolescence occur and are amplified by SCD. There are time lags everywhere for catching the latest market demands, expanding or reducing the manufacturing facilities according to revealed demands, and so on. Due to the lack of stability of diffusion speed and obsolescence, the players, retailers, wholesalers, and manufacturers tend to overestimate the demand when the demand is growing rapidly, and underestimate it when it is decreasing.

Copyright © 2008, IGI Global. Copying or distributing in print or electronic forms without written permission of IGI Global is prohibited.

However, their levels of agility are quite different. Retailers are most agile and, on the other hand, manufacturers are least agile in the supply chain. System Dynamics is a powerful tool to simulate SCD, particularly the combined effects of these lags and differences.

A Dynamic Simulation Illustration

In this section we describe in detail and illustrate the dynamic simulation technique. The example case is a special type of supply chain associated with a product of very short life cycle. This case is exemplified by the experience of Bandai Company, a Japanese toy manufacturer, which introduced the first virtual pet toy called *Tamagotchi*™. We develop a simplified dynamic simulation model to analyze this type of supply chain and use the *Tamagotchi*™ case for illustration. The model uses STELLA™ (2001), a software package for dynamic simulation and focuses on logistics and marketing. The key elements are inventory, backlog, and delay. The purpose is to aid decision makers by simplifying a complex and interactive supply chain. Supply chain phenomena such as the bullwhip effect and boom and bust have been widely studied and discussed. However, their interaction with other factors has not been extensively elaborated. We use scenario-based dynamic simulations to study the short product life-cycle case, exemplified by *Tamagotchi*™. The model has three parts: a market, a retail, and a factory components. To simulate the supply chain dynamics, all parts consist of scenarios based on the *Tamagotchi*™ case and are integrated into a dynamic model. Such models should be helpful to decision makers and planners faced with similar short life-cycle product introductions as well as useful experimental laboratories for supply chain research.

Simulation Background

Gopal and Cahill (1992) discussed trade-offs within the supply chain. For instance, sales and marketing wish for a high degree of production flexibility and rapid turnaround. They want to catch up with recent trends. From the short-term view, they would increase the stock of goods rising in popularity, but reduce the stock of goods decreasing in popularity. On the other hand, manufacturing favors longer production runs, fewer set-ups, smooth schedules, and a balanced line. These types of trade-offs have a great influ-

Copyright © 2008, IGI Global. Copying or distributing in print or electronic forms without written permission of IGI Global is prohibited.

ence on the supply chain. Magee et al. (1985, pp. 42) argued that "variations in production are far more severe than variations in demand, and the more levels and stages of production there are, the more violent production level changes become." From the viewpoint of distribution, the main character of the supply chain is the multiple echelon(s) including suppliers, manufacturers, wholesalers, and retailers. Overall, performance is a result of the complex interactions among them.

Static simulations set the framework for dynamic simulations. They help identify and model the key issues, elements, and relations among them in the supply chain. However, it is very difficult for static simulations to analyze supply chain dynamics because of the lack of appropriate feedback loops. Under specific conditions, they also propose the optimal solutions, such as EOQ, the timing of orders, the level of inventory, the number of warehouses, and so on, and the effect of the key factors, such as the lead-time and the cost structure (Pidd, 1984; Schwarz & Weng, 1999; Takeda & Kuroda, 1999; Vendemia, Patuwo, & Hung, 1995). Nersesian and Swartz (1996) systemized the use of simulation in logistics. With Visual Basic, they introduced the ways to decide the issues separately, such as the timing and quantity of orders, the level of inventory, the number of warehouses and so on. Much research has been conducted on the effect of the lead-time on performance (Takeda & Kuroda, 1999; Vendemia et al., 1995). Schwarz and Weng (1999) have built a model demonstrating the interactions between the variance of the lead-times in each link of the supply chain and system inventory holding costs.

From the viewpoint of information distortion, Gavirneni et al. (1999) simulated an overall supply-chain model. Their model emphasized the value of information and extended existing inventory theory. Chen (1999) characterized optimal decision rules under the assumption that the division managers share a common goal to optimize overall performance of the supply chain.

Information distortion can lead to tremendous inefficiencies: excessive inventory investment, poor customer services, lost revenues, misguided capacity plans, and missed production schedules. They also regarded the "shortage game" as one of the major concrete causes of the bullwhip effect (Lee et al., 1997a). This can be explained as follows. When product demand greatly exceeds supply, customers might duplicate the orders with multiple retailers, buy from the first one that can deliver, and then cancel all other duplicate orders. Later, when supply exceeds real demand, backlogs suddenly will disappear. The manufacturer gets an inflated picture of the real demand for the product and places larger amounts of capital into capacity expansion

Copyright © 2008, IGI Global. Copying or distributing in print or electronic forms without written permission of IGI Global is prohibited.

based on what may be called "phantom" demands (Nehmias, 1997). Lee et al. (1997a, b) called this phenomenon the *shortage game*. Winker, Towill, and Naim (1991) emphasized the concept of total system stocks and proposed remedies for improving the performance of the entire supply chain. That work clarified and analyzed the important phenomena, the information distortion, bullwhip effect, and the shortage game for simulating supply chain dynamics in general terms.

Dynamic simulations are necessary to analyze the supply chain because it is interactive and incorporates hierarchical feedback loops or processes (Pidd, 1984). The merits of dynamic simulations are that they can combine these feedback loops with static simulations. There are two major approaches to simulating SCD. One way is to simulate movement of a supply chain by focusing on the dynamics features from the system perspective. The other approach is to demonstrate the mechanisms related to the information distortion. Many supply chain models have been built by using system dynamics.

Senge and Sterman (1992) called attention to the difference between local and global maximization and pointed out the risk of local decision-making. Paich and Sterman (1993) simulated the diffusion process of a new product and analyzed the boom and bust process. This phenomenon is caused by several influences. First, the product life cycle may be very short. Second, there is a long time lag for expanding or reducing the manufacturing facilities according to revealed demands. Finally, retailers, wholesalers, and manufacturers tend to overestimate the demand when the demand is glowing rapidly, while customers are unpredictable. Vennix (1996) has demonstrated how group model building creates a climate in which team learning can take place, fosters partnership, and helps to create acceptance of the ensuing decisions and commitments to the decision. Cheng (1996) proposed various integrated corporate models emphasizing information technology. System dynamics therefore have already proven its worth in supply chain management. For future development of supply chain, many more dynamic simulation studies integrating the various aspects are required. In the following sections, we conduct a dynamic simulation study, which combines the effects of several phenomena, information distortion, bullwhip effect, boom and bust, and multi-echelon decisions by modeling a simple but representative case. This model was designed not to reproduce the real world exactly, but rather to help decision makers in planning in situations similar to that experienced by Bandai with the *Tamagotchi*™ product.

Copyright © 2008, IGI Global. Copying or distributing in print or electronic forms without written permission of IGI Global is prohibited.

The Tamagotchi™ Case Study

The Bandai Co. introduced *Tamagotchi™* to the market at the end of November, 1996. Bandai also sells products featuring popular characters, such as POWER RANGERS™, GUNDAM™, and DIGIMON™. Table 1 is a breakdown of the sales percentages of each division. Bandai Co. classified their products into eight divisions: character goods for boys, vending machine products, video games and general toys, models, toys for girls, apparel, snacks, and others. *Tamagotchi™* is categorized in the video games and general toys. It was an egg-shaped computer game and the first simulation game of the virtual pet class. The goal of this game is to "raise" *Tamagotchi™* and the way to play is to take care of it by feeding, giving an injection, and so on. Although Bandai estimated that this toy had the potential to be a big hit, they could not accurately forecast the shift of the demand. At the beginning, they decided to place no advertisements for it in the mass media because they expected customers to buy it by word of mouth. However, the effect of word of mouth was much stronger than they had expected. Although initial target sales volume was 300,000 by the end of 1996 (for the first six weeks) in the domestic market, it became popular so rapidly that they sold about 450,000 by the end of the year and four million by the end of March, 1997. Bandai started selling it in North America, Europe, and Asia in May 1997. The total overseas sales volume exceeded 2.4 million by the end of October 1997. This demand boom outpaced Bandai's ability to meet the demand.

The shortages caused a variety of problems including actual crimes like thefts, instances of the shortage game, and counterfeit problems. Hundreds of people formed long lines at toy stores that had much smaller inventories than the demand. At the peak, Bandai received about five thousand complaints a day about the shortages by phone. Further, many robberies and aggravated assaults to acquire the toy were reported to the police. Finally, although Bandai understood that they had a high risk of overstocking and excess capacity, they had to expand their manufacturing facilities to produce two to three million units per month in July 1997. After Bandai expanded their manufacturing capability, they unfortunately soon thereafter met a sharp decline of demand. As a result, it was announced that they had 16 billion yen (US$123 million at US$1 = 130 yen) in after-tax losses in fiscal 1998 ending March 1999, mainly because huge numbers of the toy were left unsold. This case illustrates that Bandai was overly influenced by the boom and the bullwhip effects. Thus, to illustrate what happened to Bandai and to demonstrate how

Copyright © 2008, IGI Global. Copying or distributing in print or electronic forms without written permission of IGI Global is prohibited.

they might have avoided these tremendously unfortunate effects, we built a simulation model.

According to Shapiro (2001, p. 468), "systems dynamics is a well-elaborated methodology for deterministic simulation." We used the Systems Dynamics software, STELLA™ (2001), as a tool to build our supply chain model. Its merits are the following. First, it has a function that analyzes the movements of dynamic systems. It can simulate the impacts of causal relationships that have feedback loops. Second, it has a function that permits consideration of various delays and queues. These are very important elements in analyzing supply chains as noted above. Finally, STELLA™ has strong sensitivity analysis tools. Generally, sensitivity analysis is helpful in obtaining conclusions and general implications of models.

Scenarios for the Market, Retail and Factory Levels

Figure 5 summarizes our model. The model is divided into three levels: the market, retail, and factory levels, respectively. At the market level, the total demand is equal to the sum of demands for new customers, phantom demands (for example, from the shortage game), and sales for repeaters minus recycle sales in a period. We assume that the diffusion process of new products can be expressed by using the logistic curve. The logistics curve is an S-shaped curve and usually applied to the diffusion of diseases. It is given by the following differential equation, $x'_t = \alpha x_t (K - x_t)$, where x_t is the cumulative number of people who purchased by the end of time t; and x'_t is the derivative of x_t. The parameter α is a small number that controls the diffusion speed, where bigger values are associated with faster diffusion. *K* is the theoretical upper limit of the number of purchases. In our research, the logistics model is uniquely used to demonstrate the effect of shortages on the number of potential customers. If a shortage occurs, the company looses their potential customers and *K* becomes smaller. At the market level, we assumed that the diffusion speed of new product could be expressed by using the logistic curve with $\alpha = 1.5 \times 10^{-8}$ and we chose 25 million (about 15 percent of Japan's population) as the initial upper limit. To set the value of α entails two technical difficulties. First, it is very difficult to apply logistics regression to our model because the upper limit is not fixed. Second, we do not have enough data on the weekly sales. Hence, we approximated the curve visually and choose the value. We also assumed that 10 percent of the customers who could not purchase it because of shortages would withdraw from the market.

Copyright © 2008, IGI Global. Copying or distributing in print or electronic forms without written permission of IGI Global is prohibited.

Figure 5. Conceptual framework of the model

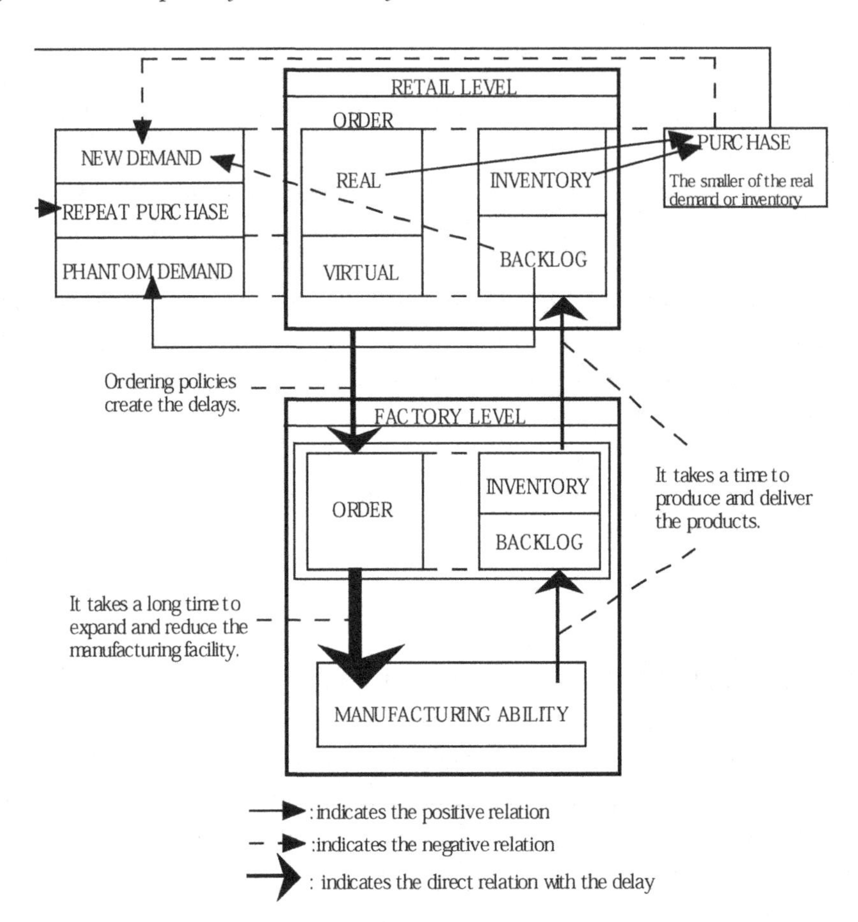

On the other hand, the shortages would create phantom demands because of the shortage game and the rate chosen for this was 20 percent. In addition, 5 percent of customers would repurchase one week later, but another 5 percent would resell four weeks after purchase due to loss of interest.

At the retail and factory levels, demands would be reviewed every week and forecasted demands are decided by using exponential smoothing. It is disputable which forecasting method fits best in this case. Forecasting methods are classified into qualitative and quantitative methods. Qualitative forecasting methods include subjective curve fitting techniques, the Delphi Method, and so on. In qualitative methods, experts play an important role to predict the future event subjectively. Usually, these methods are useful in the case of new products because there is no historical sales data (Bowerman &

Copyright © 2008, IGI Global. Copying or distributing in print or electronic forms without written permission of IGI Global is prohibited.

O'Connell, 1993). However, they are not of much use for forecasting sales with very limited information. Since *Tamagotchi*™ was the first simulation game of the virtual pet class, there was no product, which had an analogy to it. Thus, it was very difficult even for experts in the toy industry to provide reliable forecasts.

Quantitative forecasting methods also have the same difficulty in forecasting a new product like this. Gopal and Cahill (1992) note that "forecasts depend not only upon the customer, but also on the ability of the supply chain to project and respond to the product and service needs of the customer." Bandai Co. and retailers had very limited information and did not build an online information system at that time. They did not have options without simple quantitative forecasting methods based on the latest sales data and order. Among the simple quantitative methods, moving average and simple exponential smoothing are useful methods to avoid the variation inherent in the last-period technique and the variablilty in the arithmetic average. Both techniques weaken the sharp fluctuations of the demand but introduce a delay between changes in demand and their reflection in sales forecasts (Pidd, 1984). The main difference between them is that exponential smoothing is more flexible and can place greater emphasis on more recent data than does the moving average method (Barker & Kropp, 1985). Hence, we chose exponential smoothing for this illustration.

It was assumed that the plant reviews the appropriate production volume every week and the delay to increase manufacturing capacity is three weeks. The life of the manufacturing facility, essentially the machine for producing the toy, is 160 weeks, or about three years. In other words, the facility can continue to produce for 160 weeks. If the manufacturing rate is doubled, then depreciation is doubled and the period of depreciation decreases. It takes a week to ship to the customer from the factory. The manufacturer would decrease the manufacturing capacity after recognizing that the demand was declining. The initial manufacturing rate was expected to be 37,500 = 300,000/8 per week. The 300,000 comes from the initial target sales in the first six weeks; eight comes from six (first six weeks) minus one (delivery time) plus three (preparation weeks before launching). In addition, it was assumed that they could double the production volume through overtime and temporary workers without enlarging the facility. On the other hand, when the company has excess inventory, they can reduce the work rate by up to 50 percent. Their initial maximum manufacturing rate was assumed to be 75,000 = 37,500 + (450,000-300,000)/(5-1). Here, 450,000 is the number of the actual sales in the first six weeks; (5-1) represents our assumption that

Copyright © 2008, IGI Global. Copying or distributing in print or electronic forms without written permission of IGI Global is prohibited.

Table 1. Main variable in the model

Stage	Name	Feature
Market	α (SA)*	The coefficient of the logistics curve which controls the diffusion speed
	Upper Limit	Number of the potential customers in the logistics curve model - theoretical upper limit
	Diffusion Level	Number of cumulative customers at a point in time
	Total Sales	Periodic Sales
	Total Demand	Summation of periodical demands for new customers, phantom demands and sales for repeaters
	Repeat Rate (SA)*	Fraction of customers who repurchase the product
Retail	Inventory at Retail	Level of stock at the retail level at a point in time
	Backlog	Backlog of orders at the retail level (Retail shops can cancel the order after the time limit.)
	Phantom Demand	When customers experience a shortage of goods, some of them place duplicate orders repeatedly. (The amount of the phantom demand is estimated as 20 percent of backlog minus discovered phantom demands)
	Delay (SA)*	Delay to realize the phantom demands after they receive orders
Factory	Inventory in Factory	Level of stock at the retail level at a point in time
	Under Construction	Amount of the manufacturing capability which is under construction
	Manufacturing Ability	Amount of the available manufacturing capability

(SA)*: Sensitivity Analysis is conducted

Table 2. Equation used in the model

Accumalated_Order(t)	= Accumalated_Order(t - dt) + (Retail_Order - Recieved) * dt
INITIAL	= 0
INFLOW	Retail_Order = MAX(Total_Demand-Inventory_at_Retail,0)
OUTFLOW	Recieved = OF_of_IF+DELAY(Retail_Order-OF_of_IF,4)
Diffusion_Level(t)	= Diffusion_Level(t – dt) + (Periodical_Diffusion) * dt
INITIAL	= 75000
INFLOW	Periodical_Diffusion = Periodical_Sales-Sales_for_Repeater
Expected_Demand(t)	= Expected_Demand(t - dt) + (IF_of_ED - OF_of_ED) * dt
INITIAL	= 75000
INFLOW	IF_of_ED = (Retail_Order*1+Information*0)*alpha+Expected_Demand*(1-alpha)
OUTFLOW	OF_of_ED = Expected_Demand
Inventory_at_Factory(t)	= Inventory_at_Factory(t - dt) + (IF_of_IF – OF_of_IF) * dt
INITIAL	= 37500
INFLOW	IF_of_IF = Delay(Manufacturing_Ability*(2^Double)*((1/2)^Half)*(0^Stop),1)
OUTFLOW	OF_of_IF = MIN(Inventory_at_Factory,Accumalated_Order)

continued on following page

Copyright © 2008, IGI Global. Copying or distributing in print or electronic forms without written permission of IGI Global is prohibited.

Table 2. continued

Inventory_at_Retail(t)	= Inventory_at_Retail(t - dt) + (OF_of_IF - Periodical_Sales) * dt
INITIAL	= 75000
INFLOW	OF_of_IF = MIN(Inventory_at_Factory,Accumalated_Order)
OUTFLOW	Periodical_Sales = MIN(Inventory_at_Retail,Total_Demand-Phantom_Demand)
Manufacturing_Ability(t)	= Manufacturing_Ability(t - dt) + (Complete - OF_of_MA) * dt
INITIAL	= 37500
INFLOW	Complete = DELAY(IF_of_UC,3)
OUTFLOW	OF_of_MA = IF_of_IF/160
Phantom_Demand(t)	= Phantom_Demand(t – dt) + (IF_of_PD - OF_of_PD) * dt
INITIAL	= 0
INFLOW	IF_of_PD = Backlog*0.2
OUTFLOW	OF_of_PD = DELAY(IF_of_PD,Finding_Delay)
Sales_for_Repeater(t)	= Sales_for_Repeater(t - dt) + (IF_of_SR - OF_of_SR) * dt
INITIAL	= 0
INFLOW	IF_of_SR = DELAY(Periodical_Sales*Repeat_Rate,1)
OUTFLOW	OF_of_SR = DELAY(IF_of_SR,1)
Total_Sales(t)	= Total_Sales(t – dt) + (Periodical_Sales) * dt
INITIAL	= 0
INFLOW	Periodical_Sales = MIN(Inventory_at_Retail,Total_Demand-Phantom_Demand)
Under_Construction(t)	= Under_Construction(t - dt) + (Expansion – Complete) * dt
INITIAL	= 0
INFLOW	IF_of_UC = IF(Under_Construction=0) then Expansion else 0
OUTFLOW	Complete = DELAY(IF_of_UC,3)
Upper_Limit(t)	= Upper_Limit(t - dt) + (- OF_of_UL) * dt
INITIAL	= 125000000*0.15
OUTFLOW	OF_of_UL = IF(Upper_Limit>Diffusion_Level) then Backlog*0.1 else 0
A	= 0.000000015
Alpha	= 0.2
Backlog	= MAX(Total_Demand-Periodical_Sales,0)
Double	= IF(Working_Rate>1) then 1 else 0
Expansion	= IF(Upward=1) then MAX((Expected_Demand- Manufacturing_Ability-Under_Construction)/Investment_Policy,0) else 0
Finding_Delay	= 2
Half	= IF (Working_Rate<0) then 1 else 0
Information	= MAX(Total_Demand-Phantom_Demand-Inventory_at_Retail,0)
Investment_Policy	= 3
Past_Max	= MAX(DELAY(Expected_Demand,n)) n=1,2,3
Periodical_Demand	= a*Diffusion_Level*(Upper_Limit-Diffusion_Level)

continued on following page

Copyright © 2008, IGI Global. Copying or distributing in print or electronic forms without written permission of IGI Global is prohibited.

Table 2. continued

Repeat_Rate	= 0.05 Repeat_Rate
Stop	= IF(Working_Rate<-3) then 1 else 0
Total_Demand	= Periodical_Demand+Phantom_Demand+Sales_for_Repeater
Upward	= IF(Expected_Demand>Past_Max) then 1 else 0
Working_Rate	= (Expected_Demand-Inventory_at_Factory)/Manufacturing_Ability

Figure 6. Detailed schematic of the model

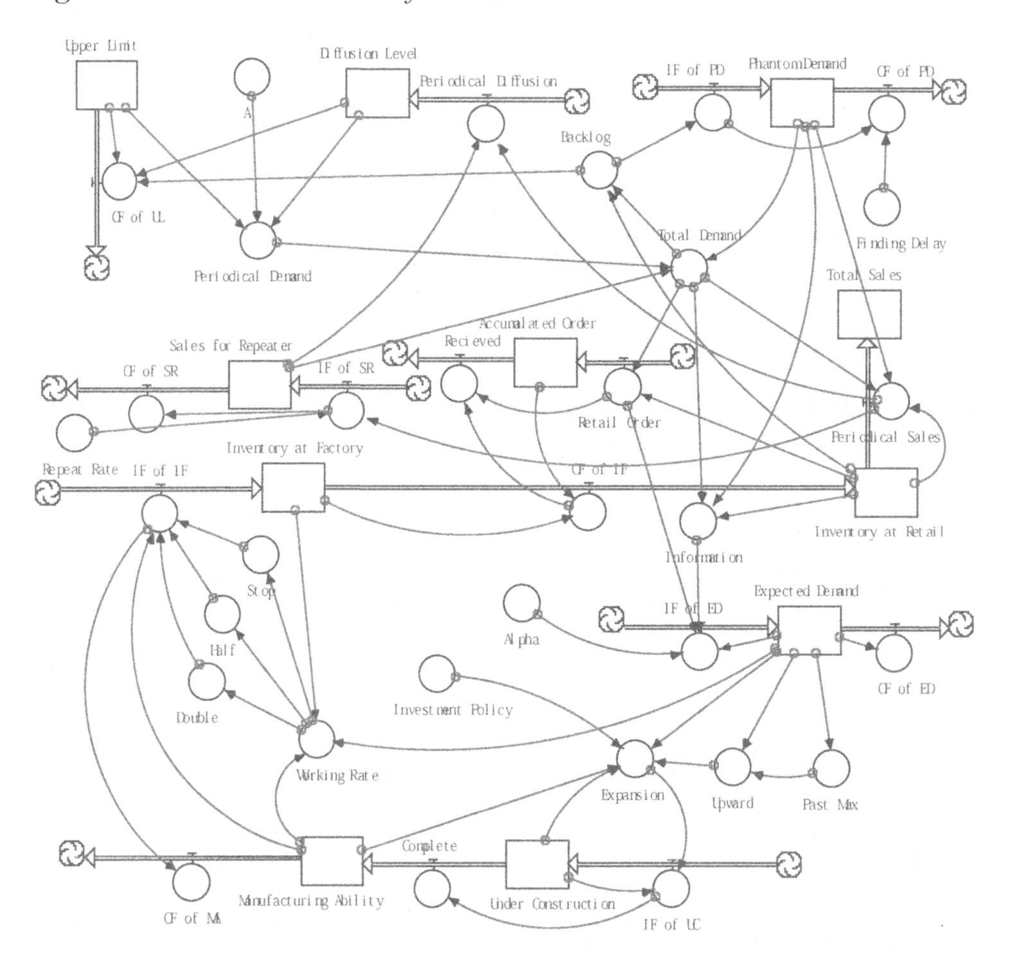

Copyright © 2008, IGI Global. Copying or distributing in print or electronic forms without written permission of IGI Global is prohibited.

they had five weeks to make most of their facility assuming the decision to do so was made by the end of the first week. Table 1 summarizes the main variables in our model. Table 2 gives the equations used in the model. Figure 6 gives a detailed schematic view of the model.

Simulation Results

The model simulates the supply chain dynamics and confirms that both the boom and bust and the bullwhip effects exert profound influences. The typical boom and bust phenomenon is shown in Figure 7, which relates the total demand and manufacturing capability. The latter had its peak enhanced by the overestimate of the demand with a delay just after the peak because of the phantom demand and construction lag (Paich & Sterman, 1993). Bandai maximized their manufacturing facilities (July 1997) just before the sharp decline of the demand because there was a lag between identifying the peak demand enhanced by the phantom demand and enlarging the facility to this level. Finally, they suffered heavy damage by the overproduction of, huge numbers of unsold toys. In our model, the bullwhip effect is combined with boom and bust. As a result, the factory level experiences a larger fluctuation in demand and much more inventory increase than does the retail level in daily operations (Lee et al., 1997a). Also, in this case, the situation becomes much worse after peak demand than before because of the accumulated information distortion and existence of the additional manufacturing facilities. This type of bullwhip effect is illustrated in Figure 8, which shows the shifts of inventory at factory and retail levels.

We see that in the case of a product with short life cycle, it is prudent to carry out more analysis and examine more potential scenarios. The demand grows faster and more capriciously. The company then faces more risk of shortages in the early stages. The shortages may create phantom demands and reduce the number of potential customers. As a result, companies may have huge inventories while still losing a certain degree of sales. The manufacturing facility might be outdated earlier than in the case of a long product life cycle. In addition, even though the repeat purchase rate is high, the company would not enjoy the usual advantages of repeat purchases. We therefore consider some additional scenarios that address different diffusion speeds and delay times in finding phantom demands, investment policies, and repeat rates. We performed sensitivity analyses on these parameters to and do what-if analysis on the information loop.

Copyright © 2008, IGI Global. Copying or distributing in print or electronic forms without written permission of IGI Global is prohibited.

Figure 7. Total demand and manufacturing ability

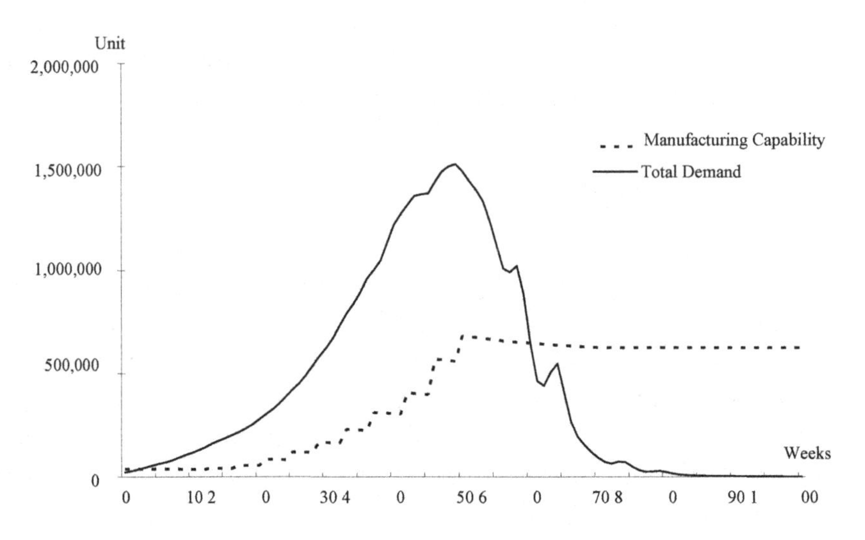

Figure 8. Factory and retail inventory

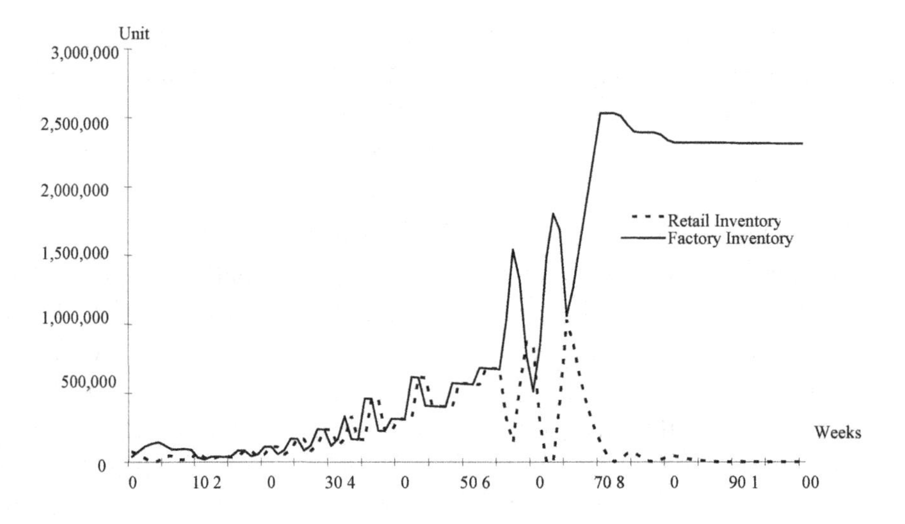

Copyright © 2008, IGI Global. Copying or distributing in print or electronic forms without written permission of IGI Global is prohibited.

The Impact of Diffusion Speed

In this model, it is assumed that the diffusion of a new product is based on the logistic curve. The derivative of the logistic curve is $\alpha \times$Diffusion Level$\times$(Upper Limit – Diffusion Level). Three sensitivity analyses were conducted on α, with values of 10^{-8}, 1.5×10^{-8}, and 2×10^{-8}, respectively. Figure 9 shows the shifts of the total and periodic demands according to the diffusion speeds. This figure suggests that, if the diffusion speed becomes faster, the total number of customers would be higher and the peak demand would be larger and sharper. Figure 10 shows the shift of the level of manufacturing capability. Although the company may face rapid shrinkage of demand after the peak with a fast diffusion speed, the maximum level of manufacturing capability would be larger. Therefore, products with short life cycles have bigger risks than those with long life cycles.

The Effect of Delay in Discovering Phantom Demand

In this model, phantom demand amplifies the variation of the demand as noted above. If it takes a long time to identify the phantom demand, then such phantom demand creates still more phantom demand. Three sensitivity

Figure 9. Total and periodic demand at different diffusion speeds

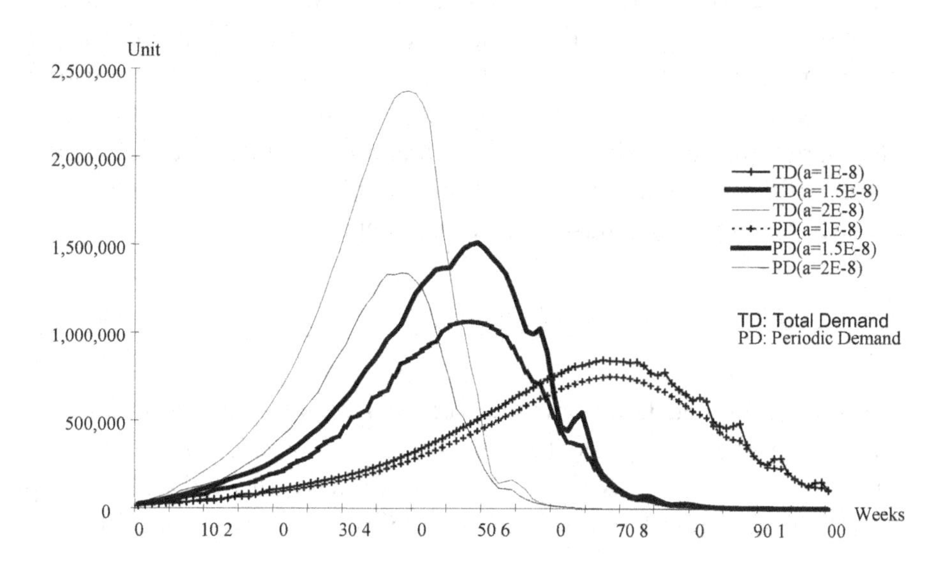

Copyright © 2008, IGI Global. Copying or distributing in print or electronic forms without written permission of IGI Global is prohibited.

Figure 10. Manufacturing ability at different diffusion speeds

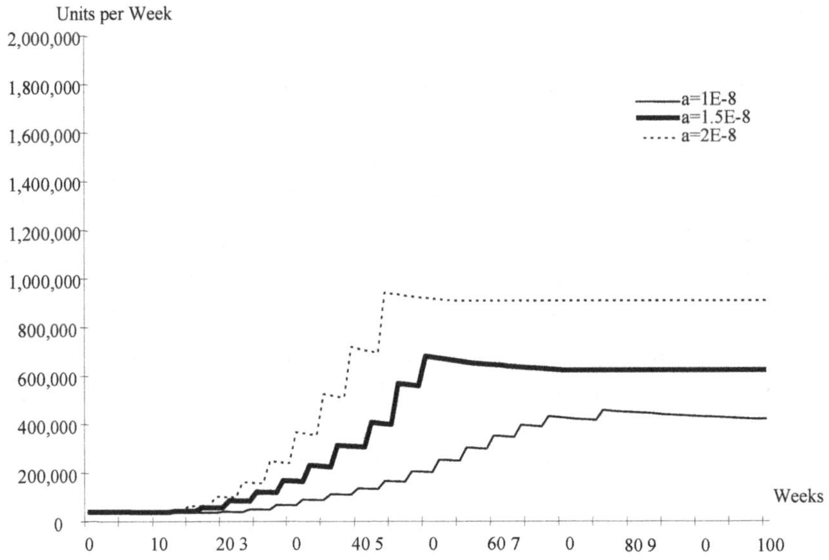

analyses have been conducted on the delay time for finding the phantom demand, with values 1, 2, and 3 weeks, respectively. Figure 11 shows the shift in manufacturing capacity of these sensitivity analyses. This figure suggests not only that the longer the delay the bigger the maximum manufacturing capacity, but also that the phantom demand has unexpected benefits. Namely, it signals the popularity of the product quickly, which promotes capacity expansion earlier. However, the most difficult part is to identify the amount of phantom demand and estimate the turning point. As a result, overestimation of this demand it might make the capacity level unnecessarily and inappropriately larger than desirable. This phenomenon was one of the main combined results of the bullwhip effect and boom and bust. Therefore, it is crucial to identify phantom demands as early and accurately as possible.

The Effect of the Investment Policy

It might be argued that companies control the level of the manufacturing capacity because they set the investment policy. We note, however, that Figure 12 is not consistent with this assertion. It shows the shifts of phantom demands caused by investment policies, which are aggressive, neutral,

Copyright © 2008, IGI Global. Copying or distributing in print or electronic forms without written permission of IGI Global is prohibited.

Figure 11. Manufacturing ability at different delays in discovering phantom demand

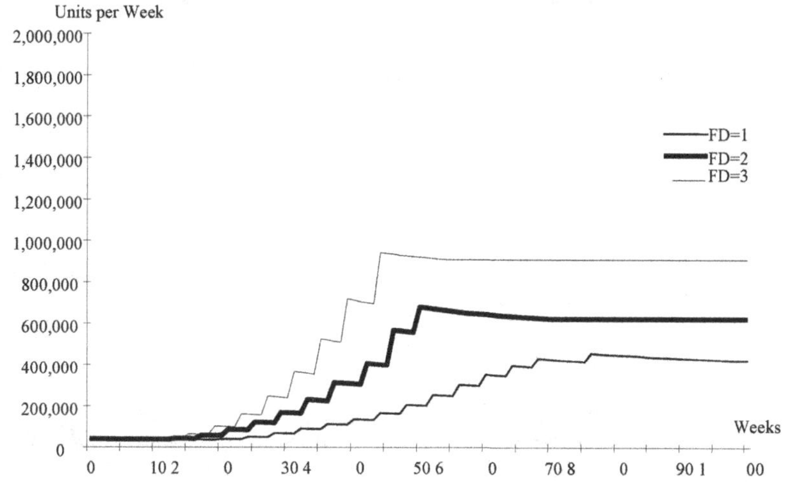

Units per Week
2,000,000
1,800,000
1,600,000
1,400,000
1,200,000
1,000,000
800,000
600,000
400,000
200,000
0
FD=1
FD=2
FD=3
Weeks
0 10 2 0 30 4 0 50 6 0 70 8 0 90 1 00

and conservative, respectively. Under the aggressive investment policy, the capacity is expanded by the difference between expected demand and the current plus planned additional capacity. Although the company could double its capacity without expansion, they in effect prepare for double the demand. Under the conservative investment policy, the size of investment is one-third of that of aggressive investment policy. Ironically, at the beginning of the boom phase, the conservative investment policy creates phantom demand most quickly among all policies, and enlarges the effect of phantom demands. They become a big driver to expand capacity to an inappropriate level. Hence, companies cannot always control the level of manufacturing capacity by investment policies alone.

The Importance of Repeat Purchases

It was assumed that 5 percent of customers in each period would repurchase the toy in the next period. It was inexpensive and simple enough to use for customers to own and handle more than two simultaneously. Kotler (1999), (see also Kotler & Keller, 2006), proposed three total sales patterns from the viewpoint of repeat purchasers, one-time purchase, infrequently purchased, and

Copyright © 2008, IGI Global. Copying or distributing in print or electronic forms without written permission of IGI Global is prohibited.

Figure 12. Phantom demands and manufacturing ability

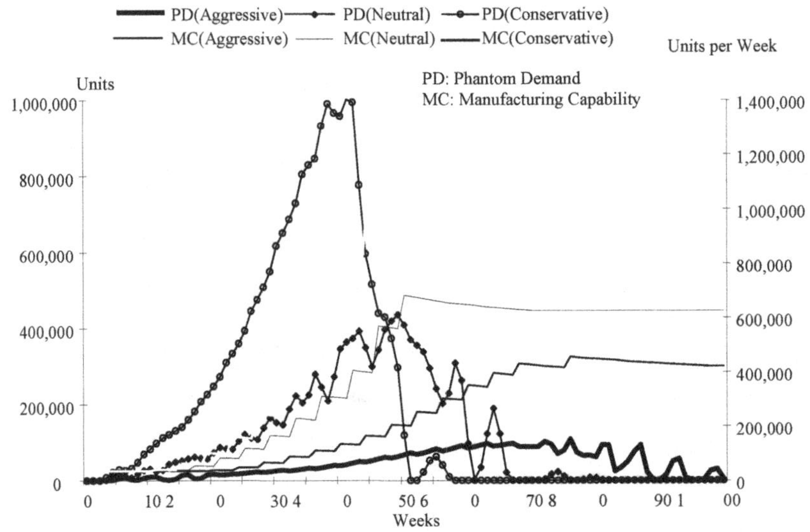

frequently purchased. Our result is similar to that of the frequently purchased case. Generally, repeaters serve as a buffer because whether the company can minimize the damage from shrinkage after the peak demand depends to a large degree on the repeat rate. However, from Figure 13, which shows the shift of the total demand for the different repeat rates (0, 5, 10 percent), it is very difficult to conclude that repeat purchases always play a role as a buffer and help avoid busting. In our model, 5 percent of buyers would buy another in the next week. The toy was evidently so attractive that repeaters want to have another as people would want actual pets like dogs, but it is too simple for repeaters to keep interest in it for a long time. This type of higher repeat rate might not enhance the effect of the buffer rather increase the level of the peak and sharpen it. As a result, even though the repeat rate is high, companies may be unable to avoid sharp demand shrinkage after the peak unless stable repeat purchases continue long term.

What-If Analysis: Value of Information

As noted earlier, information distortion is known to be one of the biggest problems in supply chains. What-If analysis was conducted to contrast two different cases of whether the phantom demands are identified or not.

Copyright © 2008, IGI Global. Copying or distributing in print or electronic forms without written permission of IGI Global is prohibited.

Figure 13. Total demand for different repeat purchase rates

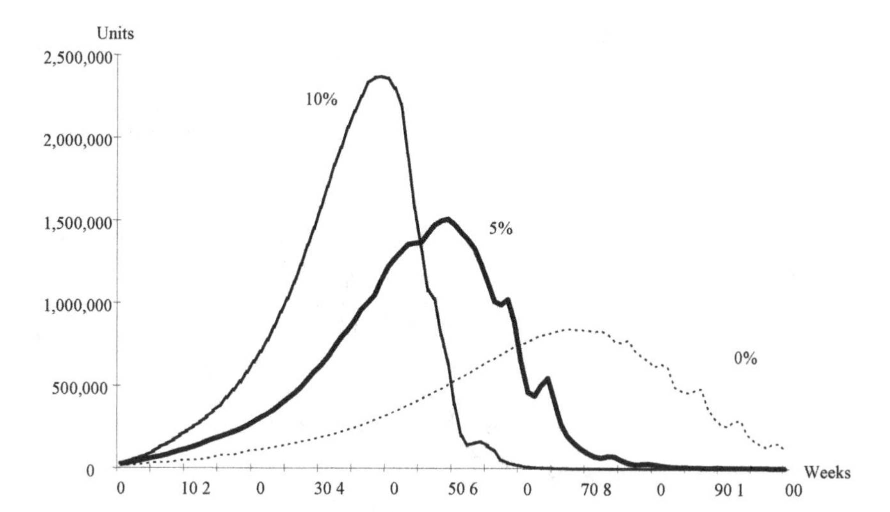

Originally, in this model, investments in manufacturing capacity were done based on orders from retail shops including the phantom demands. We also considered the case that expected demand at the factory is calculated by exponential smoothing of the market demand minus the sum of inventory at the retail level and the phantom demand. Figure 14 shows the results of the change. This figure demonstrates that, in the case that the phantom demands can be identified, the variance and final unsold of inventory at the factory level would be much smaller.

Simulation Conclusions

In this chapter, we reviewed some of the principal dynamics features of supply chains and focused on supply chain dynamics in the short product life-cycle case. In that case, setting the product and supply chain specifications is more important than improving them later because the supply chain might not have enough time to improve them and enjoy the benefits under the short life-cycle assumption. Often, supply chains require a large number of members to provide a wide variety of products at desired cost and service levels. The competitiveness of the supply chain has a close relation with its intrinsic dynamics, with both positive and negative aspects. These dynam-

Copyright © 2008, IGI Global. Copying or distributing in print or electronic forms without written permission of IGI Global is prohibited.

Figure 14. The effect of information

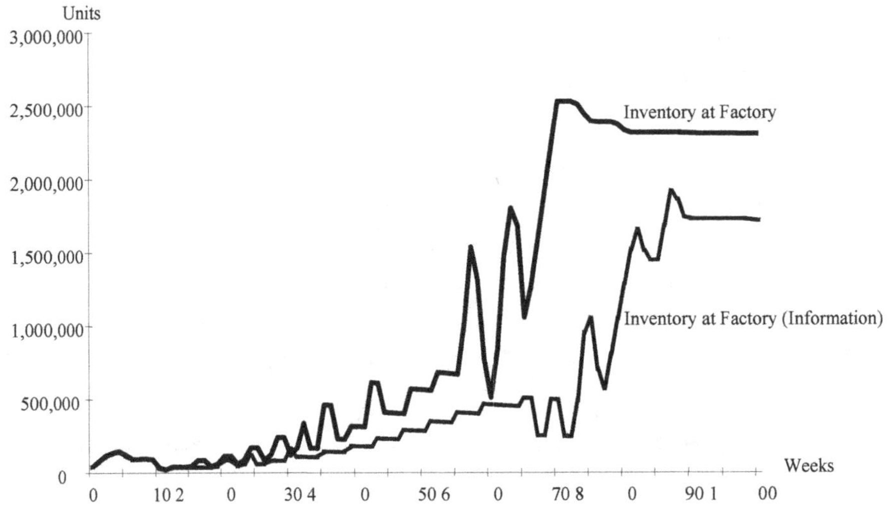

ics make the supply chain not only flexible and agile, but also complicated and unpredictable. Management of supply chains should therefore generally attempt to capitalize on the positive aspects, while minimizing the effects of the negative aspects. The dynamic modeling approach contributes to decision-making such as on the levels of manufacturing capacity and advertising, as well as the timing to foreign market. Based on the simulation experiments, the following recommendations emerged.

Control of Diffusion Speed

From the results, it may be concluded that faster diffusion is not always beneficial. It is true that in the product introduction period fast diffusion is important. However, fast diffusion might sharpen and increase the peak demand. To stabilize the demand variation and minimize shortages and phantom demands, it is desirable to control the diffusion speed. Otherwise, due to acquisition lag, the supply chain might experience a large loss, as Paich and Sterman (1993) demonstrated. To slow the diffusion speed it is imperative not to advertise in the mass media and to entreat customers not to discuss the product on the Internet as Bandai did. In reality, manufacturing and distribution capacity must continue to be the ultimate constraints in the supply

Copyright © 2008, IGI Global. Copying or distributing in print or electronic forms without written permission of IGI Global is prohibited.

chain. Therefore, not only the demand side but also the supply side should be involved with the decision processes related to diffusion strategy.

The Importance of Repeat Purchases as a Buffer

The importance of repeat purchases is recognized in every business. Generally, after the peak, most products face shrinkage of demand. If the repeat rate is higher, the peak becomes flatter and the shrinkage after the peak becomes smaller. A high repeat rate, however, does not always play a significant role because certain types of repeat purchases enhance and sharpen the peak demand and offset the advantages of repeat purchases. For example, *Tamagotchi*™ was a very novel but simple game. Most repeaters bought another immediately after they had bought the first one. However, very few repeaters continued to buy it in the long term because they soon lost interest in it. Unlike high repeat rates under long product life cycles, those of short product life cycles may entail big risks and high peaks of demand.

Identifying Phantom Demand

Phantom demands amplify the peak of the demand. If it takes a long time to detect phantom demands, they tend to re-amplify themselves. Therefore, even if a company controls its diffusion speed and capacity effectively, it would suffer amplified peaks without early discovery of phantom demands. Furthermore, the information distortion in supply chains is one of the main causes of the boom and bust and the bullwhip effects. If the information that was processed without recognizing the phantom demands at each stage is used for forecasting the demand, the expected demand has greater variance. Lowson, King, and Hunter (1999) assert that timely and accurate information flows enable quick and accurate response, and contribute to minimize the inventories and make optimal use of contractors. Corbett, Blackburn, and Van Wassenhove (1999) concluded that, through the more open, frequent, and accurate exchange of information, companies could eliminate the bullwhip effect and ensure ongoing improvement. Many researchers demonstrated the effect of information sharing by using simulation models (Forrester, 1961; Lee et al., 1997a, b; Paich & Sterman, 1993).

Some more general conclusions emerge as well. Profitability is considered to be the primary goal. From the logistical perspective, total system stocks (Winker et al., 1991) and throughput (Goldratt & Cox, 1992) are the keys for

Copyright © 2008, IGI Global. Copying or distributing in print or electronic forms without written permission of IGI Global is prohibited.

improving the performance of the entire supply chain. Both in practice and theory, various phenomena, such as information distortion, the bullwhip effect, and the shortage game are observed. To increase supply chain profiility, synchronization of the players and minimization of information distortion within the supply chain is essential. To achieve these, commitment, coordination and information sharing play important roles.

Commitment to the Supply Chain

Commitment to the supply chain encourages the players to recognize the overall strategy and to see the importance of their assigned roles in the context of the whole system. Without recognition of these, each player may behave inconsistently with each other and the results may be far from the plan. It is natural for each player to be tempted to sub-strategies. For example, even though a supply chain (integrator) has a strategy to sell a product at high price using a brand, some retailers may decide to sell it as a loss leader. Such behavior increases the chance of negative effects, such as business tricks, the bullwhip effect and phantom demands. As a result, a supply chain fails to execute its strategy. Hence, enhancing the commitment to the supply chain is an important first step to increasing profitability.

Coordination within the Supply Chain

Coordination is fundamental to the effective operation of a successful supply chain. Its function is to embody commitment and to smooth the collaboration within the supply chain. To make sure that each player takes a consistent direction, sharing of the strategy, profits and risks is very useful. In addition to the direction, adjusting the speed (agility) is also essential to collaboration within the supply chain. Even though some players are very agile, the throughput is decided by the slowest part (stage) in the chain. In this case, a supply chain might suffer from excessive inventories or backlogs. So coordination of both direction and speed is the next step. We note next that coordination also plays an important role in information sharing.

Information Sharing Among Supply Chain Players

Information distortion in the supply chain can start from and be amplified by independent decision-making. Multi-echelon systems tend to create time lags in knowing about the latest market demand and unnecessary phantom demand order swings in the supply chain. For minimizing information

Copyright © 2008, IGI Global. Copying or distributing in print or electronic forms without written permission of IGI Global is prohibited.

distortion, the coordination process becomes very important in the sharing of information effectively and efficiently and by use of an appropriate information system. Bowersox and Closs have noted that "Coordination is the backbone of overall information system architecture among value chain participants" (1996, p. 37).

A potential limitation of the simulation illustration is that exact parameter values were not available from the *Tamagotchi*™ case. Thus, it was not possible to assess model validity with respect to the original situation. However, results followed the general patterns of that case very closely in the qualitative sense. The proposed scenarios covered the most likely parameter ranges and the results are likely to have bracketed the actual dynamics of the case. In reality, the value of a model of this kind will be to permit what-if analyses by planners who will be themselves uncertain of the most valid choices of parameter values to use. What-if analyses should be critically important in observing possible results from various choices of parameter settings. For example, it may be possible that conservative (large) estimates of phantom demand are generally advisable. However, such choices may not be appropriate if repeat purchases are high or capacity decisions are conservative at the same time.

Future Trends

By observing a large number of scenarios, more informed decisions can be made based on the likely impacts of the interactions among the parameter choices. In addition, for simplicity, we have focused the model on logistics. The model might be expanded to include a financial component reflecting costs and profitability. One avenue for future research is the development of modular components for simulation models of this kind. Such a modular system might usefully include flexible options for demand model and forecasting model choices.

Conclusion

In this chapter, we first discussed the central role of the intrinsic dynamics of the supply chain, setting the stage and rationale for systems dynamics-based

Copyright © 2008, IGI Global. Copying or distributing in print or electronic forms without written permission of IGI Global is prohibited.

simulation. We then described an application of the systems dynamics simulation tool STELLA™ (2001). The application was developed from a case study of a short life-cycle example, the well-known case of the *Tamagotchi*™ toy. Several features of that case were well illustrated by the simulation results and some general features and recommendations emerged. Dynamic simulation was found to be a useful and appropriate tool for supply-chain analysis. We believe it promises to be increasingly useful for what-if policy analysis for issues such as pricing, capacity change, and advertising. Enabling stochastic variables and optimal control features would further expand its value.

References

Abernathy W. (1978). *The productivity dilemma*. Baltimore: The Johns Hopkins University Press.

Abernathy, W., Clerk, K., & Kantrow A. (1983). *Industrial renaissance*. New York: Basic Books.

Ballou R. (1992). *Business logistics management*. Englewood Cliffs, NJ: Prentice Hall.

Barker, K. R., & Kropp, D. H. (1985). *Management science*. New York: John Wiley & Sons.

Becheck B., & Brea, C. (2001). Deciphering collaborative commerce. *Journal of Business Strategy, 22*(2), 36-38.

Bowerman, L. B., & O'Connell, R. T. (1993). *Forecasting and time series* (3rd ed.). Belmont, CA: Duxbury Press.

Bowersox, D., & Closs, D. (1996). *Logistical management: The integrated supply chain process*. New York: McGraw-Hill.

Bowersox, D., Closs D., & Helferich, O. (1986). *Logistical management: A systems integration of physical distribution, manufacturing support, and materials procurement* (3rd ed.). New York: Macmillan Publishing Company.

Bradley P., Thomas, T., & Cooke, J. (1999). Future competition: Supply chain vs. supply chain. *Logistics Management and Distribution Report, 39*(3), 20-21.

Chen, F. (1999). Decentralized supply chains subject to information delay. *Management Science, 45*(8), 1076-1090.

Copyright © 2008, IGI Global. Copying or distributing in print or electronic forms without written permission of IGI Global is prohibited.

Cheng, H. (1996). *Enterprise integration and modeling: The meta base approach.* Massachusetts: Kluwer Academic Publishers.

Chopra, S., & Meindl, P. (2001). *Supply chain management: Strategy, planning and operation.* Upper Saddle River, NJ: Pearson Prentice Hall.

Corbett, C. J., Blackburn, J. D., & Van Wassenhove, L. N. (1999). Partnerships to improve supply chains. *Sloan Management Review, 40*(4), 71-82.

Cox, J., & Spencer, M. (1998). *The constraints management handbook.* Boca Raton, Florida: St. Lucie Press.

Dettmer, H. (1997). *Goldratt's theory of constraints: A systems approach to continuous improvement.* Milwaukee: ASQC Quality Press.

Forrester, J. (1961). *Industrial dynamics.* Cambridge,Massachusetts: MIT Press.

Gavirneni, S., Kapuscinski, R., & Tayur, S. (1999). Value of information in capacitated supply chains. *Management Science, 45*(1), 16-24.

Goldratt E. (1990a). *The theory of constraints.* Croton-on-Hudson, NY: North River Press.

Goldratt E. (1990b). *The haystack syndrome: Sifting information out of the data ocean.* Great Barrington, MA: North River Press Publishing Corporation.

Goldratt E. (1990c). *What is this thing called the theory of constraints and how is it implemented?* Croton-on-Hudson, NY: North River Press.

Goldratt, E., & Cox J. (1992). *The goal: A process of ongoing improvement* (2nd Ed.). Croton-on-Hudson, NY: North River Press.

Goldratt E., & Fox, R. (1986). *The race.* Croton-on-Hudson, NY: North River Press.

Gopal, C., & Cahill, G. (1992). *Logistics in manufacturing.* Homewood, IL: Richard D. Irwin, Inc.

Gossain, S. (2002). Cracking the collaboration code. *Journal of Business Strategy, 23*(6), 20-25.

Gunasekaran, A. (2005). The build-to-order supply chain (BOSC): A competitive strategy for the 21st century. *Journal of Operations Management, 23*, 419-422.

Higuchi, T., & Troutt, M. D. (2004). Dynamic simulation of the supply chain for a short life cycle product - Lessons from the *Tamagotchi* case. *Computers & Operations Research, 31*(7), 1097-1114.

Copyright © 2008, IGI Global. Copying or distributing in print or electronic forms without written permission of IGI Global is prohibited.

Higuchi, T., & Troutt, M. D. (2005).Understanding and managing the intrinsic dynamics of supply chains. In *Successful strategies in Supply chain management* (pp. 174-193). Chan, C. K., & Lee, H.W.J. (Eds). Hershey, PA: Idea Group Publishing Co.

Kendall, G. (1998). *Securing the future: Strategies for exponential growth using the theory of constraints.* Boca Raton, FL: St. Lucie Press.

Kotler, P. (1999). *Marketing management, the millennium edition.* Englewood Cliffs, NJ: Prentice Hall.

Kotler, P., & Keller, K. L. (2006). *Marketing management* (12th ed.). Upper Saddle River, NJ: Pearson Prentice Hall.

Lee, H., Padmanabhan, V., & Whang S. (1997a). The bullwhip effect in supply chains. *Sloan Management Review, 38*(3), 93-102.

Lee, H., Padmanabhan, V., & Whang S. (1997b). Information distortion in a supply chain: The bullwhip effect. *Management Science, 43*(4), 546-558.

Lowson, B., King, R., & Hunter, A. (1999). *Quick response*. New York: John Wiley & Sons

Luebbe, R., & Finch, B. (1992). Theory of constraints and linear programming: A comparison. *International Journal of Production Research, 30*, 1471-1478.

Mabin, V., & Balderstone, S. (2000). *The world of the theory of constraints: A review of the international literature*. Boca Raton, FL: St. Lucie Press.

Mabin, V., & Gibson, J. (1998). Synergies from spreadsheet LP used with the theory of constraints: A case study. *Journal of Operational Research Society, 49*(9), 918-927.

Magee, J., Copacino, W., & Rosenfield, D. (1985). *Modern logistics management: Integrating marketing, manufacturing, and physical distribution*, New York: John Wiley & Sons.

McMullen, T., Jr. (1998). *Introduction to the theory of constraints (TOC) management system.* Boca Raton, FL: St. Lucie Press.

Nehmias, S. (1997). *Production and operations analysis* (3rd ed.). New York: McGraw-Hill.

Nersesian, R. L., & Swartz, G. B. (1996). *Computer simulation in logistics.* London: Quorum Books.

Copyright © 2008, IGI Global. Copying or distributing in print or electronic forms without written permission of IGI Global is prohibited.

Paich, M., & Sterman, J. (1993). Boom, bust and failures to learn in experimental markets. *Management Science, 39*(12), 1439-1458.

Pidd, M. (1984). *Computer simulation in management science* (2nd ed.). Chichester: John Wiley & Sons.

Schragenheim, E. (1999). *Management dilemmas: The theory of constraints approach to problem identification and solutions*. Boca Raton, FL: St. Lucie Press.

Schwarz, L., & Weng, K. (1999). The design of JIT supply chains: The effect of leadtime uncertainty on safety stock. *Journal of Business Logistics, 20*(1), 141-163.

Senge, P. M., & Sterman, J. D. (1992). System thinking and organizational learning: Acting locally and thinking globally in the organization of the future. *European Journal of Operational Research, 59*, 137-150.

Shapiro, J. F. (2001). *Modeling the supply chain*. Belmont, CA: Duxbury Press.

STELLA. (2001). STELLA and STELLA Research software Copyright 1985, 1987, 1988, 1990-1197, 2000, 2001. High Performance Systems, Inc.

Takeda, K., & Kuroda, M. (1999). Optimal inventory configuration of finished products in multi-stage production/inventory system with an acceptable response time. *Computers & Industrial Engineering, 37*, 251-255.

Tompkins, J. (2000). *No boundaries*. Raleigh, NC: Tompkins Press.

Troutt, M., White, G., & Tadisina, S. (2001). Maximal flow network modeling of production bottleneck problems. *Journal of the Operational Research Society, 52*, 182-187.

Vendemia, W., Patuwo, E., & Hung, M. S. (1995). Evaluation of lead time in production/inventory systems with non-stationary stochastic demand. *Journal of the Operational Research Society, 46*, 221-233.

Vennix, J. M. (1996).*Group model building*. Chichester: John Wiley & Sons.

Winker, J., Towill, D., & Naim, M. (1991). Smoothing supply chain dynamics. *International Journal of Production Economics, 22*, 231-248.

Zeleny, M. (1981). On the squandering of resources and profits via linear programming. *Interfaces, 11*(5), 101-117.

Zeleny, M. (1986). Optimal system design with multiple criteria: De novo programming approach. *Engineering Costs and Production Economics, 10*, 89-94.

Copyright © 2008, IGI Global. Copying or distributing in print or electronic forms without written permission of IGI Global is prohibited.

Endnote

[1] In 1989, an electronic information system replaced the "kanban."

Copyright © 2008, IGI Global. Copying or distributing in print or electronic forms without written permission of IGI Global is prohibited.

Section II

Innovation Aspects

This part discuses the development of the product and the industry from the view of innovation, the product innovation, and the process innovation. Generally the process innovation starts after the product innovation advances very much. The emergence of the product design is a good indicator to measure the advancement of the product because the first product which satisfies the prerequisite conditions for the commercializing and which widely diffuses is called the dominant design. While the process innovation increases the quality of products and reduces the production cost, the room for further improvements is diminishing. It is called the productivity dilemma. Under the productivity dilemma, some companies try to de-mature by launching a new series of products and others aim to create a new market by alternative products.

Chapter V

Prerequisite Conditions for Commercializing

This chapter focuses on the period that precedes commercialization of the product and is based on the VCR case study. Only one or a few companies can succeed, usually after many attempts, in developing a sufficiently excellent product to achieve the dominant design. In this chapter, first, the time until the emergence of the dominant design is separated into three parts: the embryo period, the fetus period, and the birth period. Next, the basic requirements for achieving commercial success are discussed. It is very important for any new category of products to satisfy all the minimum consumer requirements rather than to improve a critical factor because those products which can not satisfy some minimum requirements are defective and will not diffuse widely. To complement the discussion in this chapter, the movement of Sony and JVC in this period and the emergence of the video software industry are explained.

Copyright © 2008, IGI Global. Copying or distributing in print or electronic forms without written permission of IGI Global is prohibited.

Periodization until the Emergence of the Dominant Design

Embryo Period (In the 1950s and 1960s)

It took almost two decades to succeed in commercializing the home-use VCR since the emergence of the VCRs for broadcasting. The demand for VCRs for broadcasting was born in the US because broadcasting stations needed to broadcast the same program at different times, taking into account the time differences within the country. Ampex and RCA developed most essential technologies for the VCR in the 1950s. Ampex monopolized the VCR for broadcasting in the 1950s and RCA invaded the area in the 1960s. At that time, Japanese companies demonstrated excellence in mass production in the preceding industries of cassette recorders and TVs.

During the 1950s, a few companies succeeded in developing experimental VCRs. In 1950, RCA, which had the latest technologies in TV, developed the prototype of the monochrome VCR. In 1951, Bing Crosby Enterprises demonstrated the monochrome TV recorder (12 heads). In 1953, RCA succeeded in prototyping a three-head color VCR. BBC also developed a monochrome VCR, VERA, in 1956. These models all had the fatal defect of utilizing a stationary head that recorded the signals in a longitudinal direction with AM sound. They wasted videotapes and required the mechanism to run the tape at high speed. This waste of videotape made them unstable in operation. The models were huge and heavy.

Ampex developed the essential technologies of FM (Frequency Modulation) recording and rotary head recording[1] for the VCR. These technologies reduced the volume of signals and increased the utilization of videotape. As a result, the recording capacity of videotapes expanded dramatically. Subsequently in the mid-1950s, Ampex presented the first practical VCRs, in the form of the VR-1000, and the MARK III and IV (VRX-1000). They each cost $75,000 and recorded the signals on the bias of the surface of two-inch tape with AM sound. CBS started using the VCR on the air on November 30, 1956. *Douglas Edwards and the News* was a delayed broadcast for the West Coast. NBC and ABC also started using VCRs in 1957. Ampex monopolized the VCR for broadcasting and sold about 600 units until RCA entered the market at the beginning of the 1960s.

Copyright © 2008, IGI Global. Copying or distributing in print or electronic forms without written permission of IGI Global is prohibited.

Ampex made a cross license agreement with Sony in 1960 to reinforce their weak points. Ampex received the latest semiconductor technologies from Sony and in return gave their latest VCR technologies to Sony. While Sony made remarkable contributions to the development of the home-use VCR, Ampex could not. Ironically, the change of the president from George Long to William Roberts because of the deficit in 1961 due to the excess inventory of VCRs, delayed the utilization of semiconductor technologies.

In the early 1960s, Japanese companies whose background was quite different from that of US companies, aggressively started developing the home-use VCR. In 1959, Toshiba completed a helical scan model with two-inch tape. In October of 1959, JVC[2] took out a patent on the rotary two-head helical scan, for which Matsushita and Sony applied within ten days after JVC, and which became the mainstream VCR design. After making a cross license with Ampex in 1960, Sony led the home-use VCR in technology. In 1961, JVC made a prototype of a helical scan color VCR.

In the 1960s, many companies launched and attempted to commercialize home-use VCR products. In 1962, Sony developed and exported the PV-100, the smallest VCR that was an open-reel half-inch helical scan VCR for business use. American Airlines adopted them. In 1964, Matsushita and Sony sold helical scan monochrome VCRs (at 250,000 Yen = US $2,130 in 2007) and the CV-2000 (at 198,000 Yen = US $1,687 in 2007), for home use. Some companies sold stationary head VCRs. Japanese companies had an established expertise in the miniaturization, cost-cutting, and mass marketing of tape recorders and television sets.

Ampex also dove into the VCR business and education markets. From Chicago and California, two departments that specialized in VCRs for broadcasting

Table 1. Alternative products circa 1970

Product	EVR	SV (PREVS)	U Format
Year	1968	1969	1971
Inventor	CBS	RCA	Sony
Material	Film (Silver Chloride)	Plastic Tape (Hologram)	Magnetic Tape
Weight	25-26kg	-	13-25kg
Recording Function	Unavailable	Unavailable	Available
Recording Time (Playing Time)	(50 min. monochrome) (25 min. color)	60 min. (monochrome)	60 min. (color)
Tape Speed	152.4mm/s	-	190.5mm/s
Horizontal Resolution	430-450		220-240 (300 monochrome)

Copyright © 2008, IGI Global. Copying or distributing in print or electronic forms without written permission of IGI Global is prohibited.

Table 2. Details of alternative products

	Company	Model	Recording Function	Price (Yen)	Size (cm)			Weight (kg)
					W	D	H	
Open-reel	Matsushita	NV-3120	Available	298,000	40.2	43.5	43.5	17
		NV-3110	None	190,000	21.7	34.9	21.7	14
	Toshiba	GV-211C	Available	305,000	43.3	42.7	23.3	17
		GV-216C	None	240,000	43.3	34.9	23.3	15
		GV-616C	None	190,000	35.7	37.5	22.0	13
	JVC	FV-3500	Available	285,000	40.5	38.0	24.7	15
		FV-1500	None	190,000	40.5	38.0	24.7	15
	Sony	AV-5100A	Available	330,000	46.2	39.7	27.6	25
		AV-1700C	None	198,000	39.0	38.0	23.7	15
	Sanyo	VTR-2000C	Available	278,000	35.7	35.7	22.0	14
		VTR-2000CP	None	190,000	35.7	35.7	22.0	13
	Shiba	SV-520	Available	298,000	42.0	35.6	24.8	16
	Ikegami	VTR-321C	Available	360,000	43.4	39.0	22.6	25
U Format	Matsushita	NV-2125	Available	368,000	61.0	45.0	19.8	27
		NV-2120	Available	338,000	61.0	45.0	19.8	27
		NV-2110	None	248,000	51.0	44.8	19.8	24
	JVC	CR-7000	Available	355,000	81.5	45.8	32.0	28
		CR-6000	Available	328,000	52.4	44.9	19.2	26
		CP-5000	None	248,000	52.4	44.9	19.2	26
	Sony	VO-1700	Available	358,000	61.6	46.5	20.5	27
		VP-1100	None	238,000	49.1	46.5	20.5	22
EVR and others	Hitachi	EV-1000	None	268,000	56.5	49.0	19.1	25
	Mitsubishi	VP-200	None	276,000	51.6	47.6	22.3	26
	Toshiba	EVR-101	None		54.6	49.8	22.0	25
	JEOL	EBR	None	250,000	62.2	50.5	23.4	25
	Fuji Film	CVR	None	270,000	46.3	30.3	22.6	18
		TM-40	None	70,000	40.0	27.0	23.0	6.5

and consumer electronics, separately developed the VR-303 and the VR-7000, home-use VCRs. The VR-7000 was superior to the VR-303 because of the rotary recording and the helical scan. Ampex chose the VR-7000 for delivery to the market. However, the model was more expensive and lower in quality than those of competitors. Although US companies owned critical VCR technologies, they lacked the techniques for mass-production. In the

Copyright © 2008, IGI Global. Copying or distributing in print or electronic forms without written permission of IGI Global is prohibited.

Table 3. Road to U format VCR (3/4" Cassette)

Type of Format	Year	Fast and Last Model
Open-reel (monochrome)	1964	CV-2000 (Sony)
	1968	NV-1010 (Matsushita)
Unification Type I (monochrome)	1968	KV-810 (JVC)
	1973	NV-3040 (Matsushita)
Unification Type I (color)	1969	GV-201C (Toshiba)
	1972	AV-5700 (Sony)
New Color Recommendation Standard (color)	1970	FV-3500 (JVC)
	1974	GV-215C (Toshiba)
U Format (color)	1971	VP-1100 (Sony)

1960s, the size, price, quality, and complex operability of VCR products were still far removed from the requirements of home-use products.

In the late 1960s, having too many heterogeneous devices was disadvantageous to end users and the VCR industry. In 1967, Ampex introduced the HS-100, a color magnetic disk recorder with rapid playback. CBS developed EVR (Electric Video Recording), which used a film in a cassette[3]. EVR was superior to the U Format VCR introduced later in the vividness of the image, the controllability of the still frame and the mass-productive capacity of the software. In 1969, RCA demonstrated SV (SelectaVision) that used a holographic tape but could not record. In 1971, Sony introduced the 3/4" U Format[4] one-hour cassette tape, whose use was somewhat widespread because Sony allowed others to sell machines that used their cassette. Table 1 is the catalogue of the alternative products circa 1970 and Table 2 explains them in detail. Table 3 summarizes the road to U Format VCR which was the predecessor of Betamax and VHS.

Fetus Period (In the Early 1970s)

In the early 1970s, the minimum and essential functional requirements of the home-use VCR became clear. These requirements consisted of a long recording time (more than 120 minutes), the price (approximately less than US $1000), the appropriate level of size and weight to be moved easily by one person (approximately less than 50 pounds). However, the VCR industry would go through a process of trial and error in order to achieve these requirements.

Copyright © 2008, IGI Global. Copying or distributing in print or electronic forms without written permission of IGI Global is prohibited.

With the exception of Sony and Matsushita, companies such as Ampex, Philips, and Matsushita also tried to commercialize the home-use VCR market, but most of them failed. Although companies could make a prototype, most of them could not establish a mass production system. Even though some established it, their price was too high for consumers to purchase. In addition, the high rate of service problems resulted in a huge number of returns and repairs. Matsushita was the first company to develop a mass-production system for the home-use VCR in the early 1970s, but the demand was much lower than expected. The lack of demand made Matsushita reluctant to pursue the home-use VCR market.

In addition to VCRs that used magnetic tape, EVR and SV entered the market. Hitachi, Mitsubishi and Toshiba chose the EVR over the VCR. Matsushita took part in both systems. Hitachi started selling their EVR products on October 10, 1971, the same day Sony introduced the U Format VCR into the market. The price was 268,000 Yen and their monthly output was 2000 units. Mitsubishi followed Hitachi in late 1971. However, at that time CBS announced that they had withdrawn from the EVR because of shortcomings of the EVR approach. Nevertheless, Hitachi, Mitsubishi, Toshiba and Matsushita continued to produce EVRs frugally.

In 1972, CTI (Cartridge Television Inc.), one of the subsidiaries of AVCO, introduced Cartrivison, a videocassette recorder that used the skip field recording technology, in the market. It offered the following novel points:

1. It was a TV/VCR combination unit.
2. It used a smaller 1/2 inch cassette, rather than the 3/4 inch U Format cassette tape.
3. CTI collaborated with Columbia Pictures Industries to provide sufficient software.
4. They exploited the distribution channels of Sears and Wards for sales promotion.

However, CTI sold about 2500 Cartrivision units, only 5 percent of the expected sales. At that time, the demand for combination products was very small because people who purchased the VCR already owned a TV. In addition, a great number of units were recalled because of defective tapes. The tapes deteriorated easily and were the major source of head problems. As a result, CTI went bankrupt in 1973.

Copyright © 2008, IGI Global. Copying or distributing in print or electronic forms without written permission of IGI Global is prohibited.

In the early 1970s, many companies tried to commercialize the home-use VCR market through their unique products. Most of them were cancelled before reaching the market and others had major problems with recalls and excessive inventory. However, there was a great step forward toward the commercialization of the home-use VCR. In 1970, a cross-license agreement among the Japanese companies Sony, Matsushita, and JVC heralded great contributions in the progress of the home-use VCR market. Sony, Matsushita, and JVC agreed to adopt the U Format, a format developed by Sony in 1969. Later, while US and European companies decentralized their R&D and competed based on the different formats, disks, cartridges, and tapes, Japanese companies concentrated on R&D according to the U Format and therefore competed within the same field. This allowed acceleration of the R&D activities and was eventually the cause of the market victory of the Japanese companies.

Table 4. Activities just after the emergence of the U format VCR (1972-1973)

Year	Month	Event
1971	10	Sony introduced VP-1100 (U Format 3/4" cassette VCR).
	10	Hitachi launched EVR in Japan.
	10	Telefunken, Teldec and Decca exhibited TeD (Television for Disk). (TeD players were introduced on 17 March 1975)
	11	Mitsubishi launched EVR in Japan.
	12	CBS announced withdrawal from EVR.
1972	1	Matsushita launched a U Format VCR.
	3	RCA developed SelectaVision (SV).
	4	JVC launched a U Format VCR.
	6	Cartridge Television (America) exhibited Cartrivision.
	7	Shiba Electric (Shibaden) agreed with Philips to start selling the Philips-type VCR in 1973.
	8	Fuji Photo Film (Fujifilm) launched CVR.
	11	Teijin, Hitachi, Mitsubishi, etc. formed Nippon EVR.
	11	Philips developed Video LP System (Laserdisc).
1973	1	Matsushita launched 1/2" cartridge VCR.
	2	MCA developed DiscoVision (the third Video Disc).
	7	Nippon EVR started manufacturing at Mihara Factory.
	10	Sanyo and Kingrecords joined TeD group.

Copyright © 2008, IGI Global. Copying or distributing in print or electronic forms without written permission of IGI Global is prohibited.

During this period, manufacturers need a huge amount of investment in R&D for the creation of a new category of products. Although it may be very profitable in the future, it involves great risks. It is necessary for companies to develop certain key technologies by themselves and to integrate them with others essential for the development of new products. Some companies hesitate to develop the new category of products and wait for the time when a standard technology or design becomes stable in order to avoid grave consequences in the event key technologies change.

Birth (In the mid-1970s)

There were too many formats using the 1/2 inch magnetic videotape cassette in the mid- 1970s. Between 1974 and 1977, Japanese companies launched five different machines, including Betamax and VHS, which used the magnetic videotape. All of them used the 1/2 inch cassette instead of the cartridge. The difference between the cassette and the cartridge is the number of reels. While the cartridge incorporates one reel, the cassette has two. Although the cartridge had a big advantage in that it was compatible with some open-reel machines, it could not be ejected on the way. It became clear that the mainstream home-use VCR would use the 1/2 inch cassette format.

In 1974, Toshiba and Sanyo introduced the KV-3000 and the VTC-7230 except the TV tuner. They were compatible because both used the same *V-Code* format, which was developed by Toshiba and Sanyo. After Sony introduced Betamax in 1975, Matsushita introduced the VX100 in 1975 and the VX2000 in 1976. In 1976 Toshiba and Sanyo launched the KV-4000 and the VTC-8200, both of which incorporated the newly updated V-Code II format. JVC also introduced the HR-3300 with the VHS format. In 1977, Sony also upgraded Betamax to BetaXII.

Betamax and VHS were two prominent home-use VCR formats in the mid-1970s. In 1975, Sony introduced the Betamax at 229,800 Yen. It was the first product that satisfied the essential functional requirements of the home-use VCR market. Betamax could record signals 3.5 times greater than the U format in the same area. Sony was the only household appliance manufacturer that knew the importance of the magnetic tape in the VCR and had strength in the magnetic tape. As a result, Sony could downsize Betamax dramatically due to the reduction of the mechanical parts and the utilization of smaller cassettes. In 1976, Sony announced the birth of the post color TV era, the video age.

Copyright © 2008, IGI Global. Copying or distributing in print or electronic forms without written permission of IGI Global is prohibited.

Table 5. New release of 1/2 inch VCR formats during 1973-1976

Year	Month	Model (Company)	Recording Time	Price (Yen)
1973	1	1/2" cartridge		
		NV-5120 (Matsushita)	30 minutes	310,000
1974	9	V-Code		
		KV-3000 (Toshiba)	30 minutes	325,000
		VTC-7230 (Sanyo)	30 minutes	338,000
1975	4	**Betamax**		
		SE-6300 (Sony)	60 minutes	229,800
		SE-7300 (Sony)	60 minutes	298,000
	6	**VX**		
	10	VX-100 (Matsushita)	-	198,000
1976	6	VX		
		VX-2000 (Matsushita)	100 minutes	210,000
	10	**V-Code II**		
		KV-4000 (Toshiba)	60 minutes	279,000
		VTC-8200 (Sanyo)	(120 minutes)	329,000
	10	**VHS**		
		HR-3300 (JVC)	120 minutes	256,000
1977	4	BetaXII		
		SE-7100 (Sony)	120 minutes	268,000

Undoubtedly, Betamax was regarded as the dominant design in home-use VCRs at that time. Sony disclosed its critical home-use VCR technologies to JVC and Matsushita in order to make Betamax the de facto standard in the home-use VCR market. However, in 1976 JVC introduced the HR-3300 model with the VHS format at 256,000 Yen (about US $892 at US $1 = 287 Yen) and thereby competed with Sony for the de facto standard. The severe competition between Betamax and VHS was to be expected given that they developed the VCR based on the same U Format.

Betamax and VHS are similar from a technological viewpoint and are quite comparable in performance. However, they are incompatible because of the different sizes of cassettes, the Betamax cassette being smaller than the VHS cassette. On the other hand, VHS surpassed Betamax with its lower weight and a longer recording time capacity. The Model HR-3300 VHS was lighter than Betamax by 11 pounds and could record for two hours, while the Betamax could record for just one hour at a time. This longer recording time was considered to be the critical factor that appealed to customers in the United States, Europe, and Japan.

The main risk in this period is whether a technology or a product design can become a standard in a new category of products or not. In the case that the products are almost comparable in performance and price, faster is better.

Copyright © 2008, IGI Global. Copying or distributing in print or electronic forms without written permission of IGI Global is prohibited.

The level of risk is determined collectively by the number and strength of the competitors, the degree of difficulty of the technology, the range of essential technologies, and the alliance with others.

Basic Requirements

Playback and Recording Functions

The SL-6300 (Betamax) and the HR-3300 (VHS) satisfied some functions that were prerequisite requirements for the commercial success of the home-use VCR. The recording function was a most critical factor for the home-use VCR to be widely adopted. In the 1970s, not nearly enough software appeared in the market, and the little that was available was very expensive. Users needed to make their own software by recording TV programs, such as a movie, a drama, a baseball game, and the like. The early home-use VCRs had to improve operation stability, image quality, and the length of videotaping. The application of transistor technology and high-density recording on the videotape by Sony contributed to increased stability of operation. The agreement on U Format and the cross license between Sony, JVC, and Matsushita, accelerated the sophistication of the image. They focused on the format and accumulated their performance.

Some functions associated with playback such as still shots and slow motion playback, were features available on the SL-6300 (Betamax) and the HR-3300 (VHS). Although these functions were necessary for commercial success, their order of priority was much lower than that of the actual recording function. Technically, these functions were already established in the 1950s. In addition, users would have to videotape TV programs because of an underdeveloped software market including video rentals.

Size and Weight

The critical differences between the VCR for broadcasting and the home-use VCR were the size and the weight. For commercial success in the home-use VCR, the products would have to be reduced in size and weight. The broadcasting stations had ample space and budget for the VCR. In the 1950s and

Copyright © 2008, IGI Global. Copying or distributing in print or electronic forms without written permission of IGI Global is prohibited.

60s, they purchased the products placing importance on the basic functions, image quality, and machine stability at a reasonable price. On the other hand, consumers are strictly subject to space. There was a great divide between VCRs for broadcasting and home-use VCRs. VCRs for business and education bridged this gap.

In 1959, JVC launched the KV-1[5], a VCR for broadcasting with a size of 720,000 cubic centimeters and a weight of 300 kg. Afterward, JVC introduced a few VCRs with a built-in TV tuner for business and education use. These included the KV-200 (in 1963) and the CR-6100 (in 1972), with respective sizes of 74,100 and 52,933 cubic centimeters and respective weights of 67 and 29 kg. In 1976, JVC launched the first VHS VCR for home-use, the HR-3300, whose size and weight were 20,910 cubic centimeters and 13.5 kg respectively. For initial commercial success in the home-use VCR industry, products would have to be small enough and light enough to be carried by the average person.

Some products launched before 1975 were as light as the HR-3300 and the SL-6300. For example, in 1965 Sony launched the CV-2000, a monochrome product using an open-reel design and weighing in at 15kg. In 1970, JVC launched the 16 kg open-reel KV-340. However, these products were lacking some critical functions such as colorability and recording. As a result, they could not become the dominant design or de-facto standard in the home-use VCR.

Sony introduced the semiconductor and other technologies for use in downsizing the home-use VCR. The huge mechanical parts had various negative effects on the commercialization of the VCR. For instance, they accounted for most of the size and weight of the VCR. Sony downsized parts and reduced their number by using semiconductors. Although it required additional cost in the short run, it contributed to cost cutting in the long run because of the reduced costs of materials, transportation and installation. On the other hand, they improved the efficiency of the videotape. Before the 1970s, VCR manufacturers wasted videotape because of the requisite high tape speed. As a result, the maximum recordable time was very short and the tape's reliability was very low. Without the efficient use of the videotape through new technologies in the signal and magnetic head, it would have taken a much longer time to commercialize the VCR market. Efficient use of the videotape increased the stability of operation because it removed the heavy burden of the mechanical parts. Supply chain partners responded to the changes either partially or totally because all mechanical parts were linked together. They would have collaborated as a whole to downsize the VCR products in accordance with the

Copyright © 2008, IGI Global. Copying or distributing in print or electronic forms without written permission of IGI Global is prohibited.

design change. Sony accomplished the mass-production system and supply chain of their promising product Betamax, for the first time.

Cassette

Prior to the emergence of the cassette tape, the open-reel tape was the standard. The open-reel tape was inferior to the cassette tape in handling, storage, and durability. The open-reel tape machines were divided into three types. The first and oldest one was the monochrome model without unification. It started with the CV2000, which Sony introduced in 1964 and ended with the NV1010, which Matsushita introduced in 1968. The next one was the monochrome one with unification, Unification Type I. It started with the KV810 JVC introduced in 1968 and ended with the NV3040 Matsushita introduced in 1973. The third type was the color one based on Unification Type I. However, this one did not have compatibility with the others. It started with the GV201C and ended with the AV5700 Sony sold in 1972. The last one was the New Color Recommendation Standard, which started with the FV3500 JVC introduced in 1970, and ended with the GV215C Toshiba introduced in 1974.

Even after Sony introduced the VP-1100 (the first U Format VCR) in 1971, the other non-cassette type products took the offensive because the U Format VCR could not become the de-facto standard. In addition to the EVR and SV, Philips introduced the Video LP System (laserdisc) in 1972. However, at that time, the U Format VCR was closest to achieving success in the market because the recording function was critical while the video software industry was undeveloped.

Table 6. Comparison of cassette tapes

Cassette Type	Size (Width, Height, Depth)
U-Matic	22.1×3.2×14.0
½" Cartridge*	13.0×2.9×12.8
V-Code	15.6×2.5×10.8
Betamax	15.6×2.5×9.6
VX-2000	21.3×4.4×14.6
VHS	18.8×2.5×10.4
8mm	9.5×1.5×6.25

* While the cassette incorporates two reels, the cartridge had one.

Copyright © 2008, IGI Global. Copying or distributing in print or electronic forms without written permission of IGI Global is prohibited.

Table 7. Price down of Betamax and VHS cassette tapes

		1976.10	1977.6	2006.9
30 min.	Betamax	2600	-	-
	VHS	2800	-	-
60 min.	Betamax	3800	2850	-
	VHS	4000	3500	300-400
	VX	5000	-	-
120 min.	Betamax	-	4000	-
	VHS	6000	4800	200-300

Many Japanese companies believed that the cassette type VCR would become mainstream and they improved the U Format VCR. They launched 1/2" cassette VCRs. The cassette was so practical in those days that consumers became satisfied with the size, the price, and the storage of TV programs and itself. It took more than two decades before the variety and volume of DVD software caught up with those of VHS, and the performance/cost of DVD and HDD recorders became comparable with that of VHS.

Price Down

Price is one of the most critical factors of commercialization. In 1963, Sony and JVC introduced the PV-100 and the KV-200, priced respectively at 2,480,000 and 2,000,000 Yen. Both machines used an open-reel tape, were monochrome and at more than 60kg each, were fairly heavy. Sony and JVC reduced the cost of this type of machine and sold the CV-2000 at 198,000 Yen in 1965, and the KV-800 at 200,000 Yen in 1966. Although these prices were cheaper than those of the SL-6300 (Betamax, in 1975) and the HR-3300 (VHS, in 1976), they could not succeed in the market because consumers experienced much lower performance than expected. In 1972, Sony and JVC introduced the VO-1700 and the CR-6100, the U Format color VCRs with recording function at 358,000 Yen and 395,000 Yen respectively. U Format VCRs evolved into Betamax and VHS.

The markdown of the first series of products, the monochrome and open-reel machines, was much bigger than that of other series. It was a result of the cost reduction and the strategy that a company developed a new series first and afterward, sought to discern the appropriate price. If they could not settle on a price, in other words, the product was not promised. They added

Copyright © 2008, IGI Global. Copying or distributing in print or electronic forms without written permission of IGI Global is prohibited.

new functions and launched a new series. Finally, the price and performance of the product, the Betamax CL-6300, satisfied the innovators' needs. This was critical for the start up.

Sony's Contribution in the Development of the Home-Use VCR

Functions pertaining to the Betamax SL-6300 and the VHS HR-3300 were gradually put to practical use by the 1950s. Sony was a leading company in the development of the home-use VCR. In 1958, Sony produced a four-head video recorder prototype based on the Ampex standard and which was the first of its kind in Japan. In 1961, Sony developed the first transistor-based VCR in the world; the SV-201 with a total weight of about 200kg, offered still shots and slow motion playback. In 1963, Sony launched a 1.5 head PV-100 for business and education use. Although it still used two-inch tape, it had a reduced weight of approximately 60kg, 30 percent that of the SV-201. In 1965, Sony launched the first all-transistor VTR for home use, the one-inch tape CV-2000 at a total weight was about 15 kg, which was able to record and playback 90 minutes continuously. In 1971, the VP-1100 was launched as the first U Format color videocassette 3/4-inch tape player. In 1972, the VO-1700 was launched as the first U Format color videocassette 3/4-inch tape recorder with a TV tuner. In 1975, Sony succeeded in introducing the Beta system models SL-6300 and SL-7300, which used half-inch cassette tape. The SL-6300 was advertised as a *time-shift machine* and known as *Betamax*. The SL-7300 incorporated a TV tuner. It took almost two decades for Sony to meet all prerequisite conditions for commercializing the home-use VCR.

The critical differences between the VCR for broadcasting and the home-use VCR were the size, the weight, and the price. The VCRs for broadcasting in the 1950s were so huge and expensive that only broadcasting stations would buy them because of their strong need and available space. For commercializing the home-use VCR, downsizing and price down were required. The VCR was a facility or fixture and it needed to become an appliance.

For downsizing, Sony had a big advantage in the technologies of the transistor and the videotape. The transistor changed VCR products dramatically. The transistor needs a minute amount of electricity and is much smaller and stronger than the vacuum tube. In addition, it was suitable for mass produc-

Copyright © 2008, IGI Global. Copying or distributing in print or electronic forms without written permission of IGI Global is prohibited.

tion and possessed great potential for advancing VCR products. Transistor technology contributed to downsizing, price down, the sophistication of products and mass production.

The videotape also played a very important role in the development of the VCR. The tape must be durable and capable of recording vast amounts of signals. Durability is an essential factor for VCRs because repeated use of the tape over a long period of time is expected. The density is closely related to the area (width and length) of the tape, which in turn has a direct impact on the size of the cassette and consequently an indirect impact on the size of the VCR and the durability of both the VCR and the cassette. While the PV-100 launched in 1963 used two-inch tape, the SL-6300 in 1975 used half-inch tape. As a result, the cassette tape for Betamax was smaller than that of VHS. The high-density tape contributed to a reduced running speed of the tape, which alleviated the burden on the mechanical parts of VCR and saved energy consumption. The development of the videotape resulted in the downsizing of the VCR.

JVC's Challenge

JVC, one of the subsidiaries of Matsushita, was founded as the Victor Talking Machine Company of Japan, Limited in 1927 and later revised its name to JVC (Victor Company Japan, Limited) in 1945. JVC garnered a big advantage in TV technology after they succeeded in projecting an image, the first character of the traditional Japanese alphabet, on a cathode ray tube for the first time worldwide. Kenjiro Takayanagi led the project and was called the father of TV. Under his control, JVC focused on the development of the two-head helical scan VCR to project a still frame. In 1959, they took out the patent on it and succeeded in launching the first two-head VCR, the KV-1, which became the mainstream VCR. In 1963, they introduced the KV-200, the smallest two-head VCR in the world at that time. In 1967, they developed a DFC (Direct FM Combine) color VCR, which corresponds with TV. In 1969, they succeeded in developing the cartridge VCR by themselves. In 1970, JVC introduced the KV-340, a two-head helical scan VCR that used half-inch tape and which was small and cheap (models W427, H228, D390, at 16kg and 185000 Yen each). They tried to make it the de-facto standard in the home-use VCR but were unsuccessful.

Copyright © 2008, IGI Global. Copying or distributing in print or electronic forms without written permission of IGI Global is prohibited.

In 1971, JVC sold the U Format VCR, a format developed by Sony. In the early 1970s, they promoted to home-users the U Format VCR that could record color images. However, the U Format still had a lot of problems to be overcome before being readily adopted by home-users. These problems included being very expensive, heavy, and huge, and still having low image and sound quality. In addition, the recording time was very short. During 1971-1976, the VCR department accumulated a deficit of 900 million Yen. In addition, they were in debt for 1.5 billion Yen for facility investments. JVC suffered a major setback and reduced the number of people in the VCR department in the early 1970s. They owned huge inventories and were pushed for the repair service. Frugally, they continued research and development in the home-use VCR. When Sony launched the Betamax in 1975, JVC fell so far behind Sony that few people believed JVC could ever make a comeback.

A project team for developing VHS was created in the VCR department in February of 1976. Their first mission was to complete the basic design within half a year. JVC had several reasons to be able to catch up quickly. First, JVC, Sony, and Matsushita made a cross license in VCR technology. All of them could use patents others owned at no cost. Second, JVC and Sony carried out product development based on the U Format. Their products were already similar. Finally, JVC had unique technologies and excellent concepts for the home-use VCR. They especially enjoyed excellence in the application of TV technologies. They listed the following requirements for commercial success:

a. Requirements related to Specifications
    1. The VCR should be able to be connected with other companies' TVs.
    2. Image and sound quality at playback should be equal to original broadcast signal.
    3. The recording time should be two hours.
    4. Technology should be compatible with other companies' VCRs (Common size cassette tapes).
    5. VCR should offer extensive features.
b. Requirements for Commercial Success
    1. The price should be reasonable.
    2. The machine should be very easy to operate.

Copyright © 2008, IGI Global. Copying or distributing in print or electronic forms without written permission of IGI Global is prohibited.

    3. The running cost should be low.

c. Requirements related to Manufacturing
    1. Productivity should be high.
    2. Ability to easily manufacture various models.
    3. JVC should offer high quality service.

d. Requirement to be widely accepted
    1. VCR should be a medium for cultural exchange.

JVC conducted R&D from the viewpoint of the home-use requirements. Although they fell behind Sony in the technology, JVC owned 451 patents related to the VCR by 1976. They had the essential technologies for commercialization, such as a perfectly round magnetic head at micron level. JVC developed the technologies when they launched a theatrical projector in the middle of the 1930s. Good concepts for the home-use VCR and other factors made it possible for JVC to catch up with Sony quickly, and to introduce the HR-3300, the first VHS in 1976.

Emergence of the Video Software Industry

Midway in 1960, Sony, JVC and others launched some low cost products without a recording function. At that time, manufacturers should have been preparing consumer software simultaneously. Instead they focused on education and business use only. Sony explored the education market. The school ownership rate of VCRs more than tripled during the late 1960s. 20 percent of high schools owned at least one VCR by 1969. VCRs had various uses in business such as orientation classes, public relations to promote products, and so forth. Movies about deep-sea fishers triggered their use on distant voyages. Hotels and motels have also been good customers since the late 1960s. They offered movies and adult films on VCRs with a timer.

Many Japanese video software companies were founded around 1970 due to the expectation of market expansion in the near future. Video software companies included many companies that did not actually produce video software (CATV Journal, 1972). About a hundred video software companies could be divided into four groups. The leader of the first group was Pony

Copyright © 2008, IGI Global. Copying or distributing in print or electronic forms without written permission of IGI Global is prohibited.

Canyon which is a member of Fujisankei Communication Group, a media group. The leader of the second one was Pack-In-Video Co., Ltd., which was founded by JVC, Tokyo Broadcasting System, Matsushita, and others. The third one was Video Pack Nippon Co., Ltd., which was supported by NET (Nippon Educational Television Co., Ltd), Oubunsha Co., Ltd., Asahi Shimbun Company, and others. This company specialized in agricultural and medical programs. The last one was NESCO whose base was print and movie companies. Video software companies formed a group to establish a low cost, low risk distribution channel.

In 1970, Pony Canyon started Pony Video, a monochrome video series (Unification Type I) for home users that included documentary movies and educational programs. The price of a 30-minute program was 30,000 Yen. The price was too expensive for the average consumer because at that time the average starting pay of new college graduates was about 40,000 Yen. This price structure continued to be the standard for a long time in Japan. Visual Communication Journal (August 15, 1970) estimated the cost of prerecorded videotape at 10,500 Yen (the videotape 3,500 Yen; the case 650 Yen; the dubbing fee 2,000 Yen; the production cost including royalty and copyright 4,000 Yen; the distribution cost, etc. 250 Yen)[6].

The software market in the early 1970s was very lopsided. Visual Communication Journal (September 9, 1971) said that 98 percent of software for sale was for business use and just 2 percent was for home use. The adult movies made available at hotels and motels accounted for 60 percent of the software for business use, while 25 percent was used for orientation classes in companies. 1.5 percent of the home use software (of the 2 percent for home use) was used for demonstration purposes and a mere 0.5 percent was actually used in the home.

In late 1971, Sony introduced the U Format VCR, which used the 3/4-inch cassette. Matsushita and JVC also introduced U Format VCRs in the first half of 1972. In the early 1970s, the U Format cassette dominated the software industry. However, in the mid-1970s, EVR software temporarily gained market strength. After that, Betamax and VHS became very powerful, with VHS ultimately dominating the software industry by the middle of the 1980s. Video software companies had difficulty in their choice of machine. They already possessed a wide assortment of various prerecorded videos on open-reel tape. The emergence of the U Format VCR compelled them to decide to maintain their existent products or to treat them as obsolete. They decided to service both products because customers hesitated to change. This was temporary until Betamax and VHS dominated the home-use VCR market.

Copyright © 2008, IGI Global. Copying or distributing in print or electronic forms without written permission of IGI Global is prohibited.

In addition, Japanese video software companies had other options in EVR and SV, which offered them some desirable aspects. EVR and SV were much more suitable for the mass-production of software than the VCR, and they could not work without the prerecorded software. After CVS declared their withdrawal from EVR, a joint venture named Japan EVR was founded to produce the software in Japan. The monthly output of prerecorded cassettes was increased from 5,000 to 25,000 units during the period 1973 to 1975. EVR stopped producing prerecorded cassettes by 1977. It was a significant burden for the video software companies to prepare for multiple formats.

Role of SCM for Prerequisite Conditions

Manufacturers, such as Sony and JVC, formed their supply chain with an eye for commercialization. Even in the R&D phase, they needed a lot of suppliers and close communication among them. They need to overcome many technically difficult problems through trial and error in order to achieve the prerequisite conditions for commercializing the home-use VCR. As an example, the precise processing of the magnetic head was essential to launch the VHS. JVC collaborated with domestic machine tool and die makers for the development of high-tolerance equipment. In addition to the difficulty of these problems, speed was also a very important factor for the supply chains. Otherwise, another supply chain would achieve the prerequisite conditions first and leave all the others far behind.

Sony and JVC's ideas for the essential functional and minimum requirements for the home-use VCR became clear by the early 1970s. It was no surprise that the playback image and sound of the home-use VCR were almost the same as those of TV. The recording time was more than 120 minutes, the price was approximately 250,000 Yen (about US $900), and the appropriate level of size and weight were approximately less than 50 pounds (a person could move it alone).

VCRs for broadcasting could record and playback high quality image and sound stably in the 1950s. For commercializing the home-use VCR market, downsizing and price down were required. In addition, a long recording time was desired by using the cassette. Even the leading companies, such as Sony and JVC, could not solve these problems alone quickly and efficiently. They had formed a supply chain through the_VCRs for business use. Effective

Copyright © 2008, IGI Global. Copying or distributing in print or electronic forms without written permission of IGI Global is prohibited.

ways for downsizing without downgrading performance were the reduction and downsizing of parts through improvement of the design or the development of innovative parts. These methods had an enormous impact on part or all of the supply chain. Suppliers should deal with the switching cost when they are concerned about the future and the volume of transaction per unit is decreasing. Quick responses are very much required because there will be many attempts before achieving success in the market.

In addition to downsizing, which in the short run raised the cost because of the switching and precision processing costs, suppliers were required to simultaneously cut cost in other ways. They achieved both by establishing a mass-production system and accumulating their experience in the long run. The sharing of information and risk was so important within the supply chain that they invested in a mass-production system, the mold, at the right time. Before the 1990s, supply chains in Japan were formed exclusively based on the long transaction and the capital relation. That type of supply chain was suitable under an environment that changed very slowly because they could steadily adapt to the situation while sharing information and risk within the supply chain.

Conclusion

This chapter focused on the commercialization process. At the beginning of a durable goods industry, many companies have the opportunity to attempt commercialization because their products were developed by their own unique approaches. It is natural that many or most of such initial product attempts will have had fatal competitive deficits. Thus it generally takes a long time for the emergence of the dominant design or de facto standard. This period can be subdivided into three parts, the embryo period, the fetus period, and birth. During these sub-stages some companies can hope to achieve a viable candidate for the de facto standard by use of their unique technologies.

The basic requirements for achieving commercial success are discussed in this chapter. It is very important for any new category of products to satisfy all the minimum requirements, such as the product performance, the cost, the size, the weight, and the functions, simultaneously. Those products which do not satisfy some minimal requirements are defective and would not diffuse widely.

Copyright © 2008, IGI Global. Copying or distributing in print or electronic forms without written permission of IGI Global is prohibited.

In the VCR case study, US companies such as AMPEX and RCA tapped the VCR market for the broadcasting industry and set the technological foundation for the home-use VCR in the 1950s. Many companies including Japanese and European companies developed a variety of prototypes and some of them introduced their own products into the market. Most of these ended in failure. Finally, Sony and JVC succeeded to develop an excellent product which satisfied all minimal requirements after many attempts.

References

Abernathy, W. J. (1978). *The productivity dilemma.* Baltimore: The John Hopkins University Press.

Abernathy, W. J, Clark, K.B., & Kantrow, A.M. (1983). *Industrial renaissance*. New York: Basic Books.

Anonymous. (1967). Ampex shows home video. *Business Week*, June 17, 164-172.

Anonymous. (1974). Japanese high band VCR. *CATV Journal*, *October*, 66-69.

Camras, M. (1985). *Magnetic tape recording*. New York: Van Nostrand Reinhold.

Christensen, C. M. (1997). *The innovator's dilemma*. Boston: Harvard Business School Press.

Economic and Social Research Institute, Cabinet Office, Government of Japan. Retrieved March 15, 2006 from http://www.cao.go.jp/index-e.html

EITA (Japan Electronics and Information Technology Industries Association). (2005). Minnseiyou Dennsi kiki Deta Shyu. Tokyo: JEITA.

Electronics Industries Association of Japan. Retrieved March 15, 2006 from http://www.jeita.or.jp/eiaj/english/index.htm

Ernest, B., & Macdonald, S. (1978). *Revolution in miniature: The history and impact of semiconductor electronics*. New York: Cambridge University Press.

Funai. Retrieved March 15, 2006 from http://www.funaiworld.com/index.html

Copyright © 2008, IGI Global. Copying or distributing in print or electronic forms without written permission of IGI Global is prohibited.

Hayakawa, S. (1975). Betamax shutugen niyoru VCR kai no hamon, *CATV Journal, June*, 48-50.

Higuchi, T, Troutt, M. D., & Polin, B. A. (2004). Life cycle considerations for supply chain strategy. In C.K. Chan & H.W.J. Lee (Eds.), *Successful strategies in supply chain management* (pp. 67-89). Hershey, PA: Idea Group.

Hitachi. Retrieved March 15, 2006 from http://www.hitachi.com/

Itami, H. (1989). Nihon no VTR sanngyou naze sekai wo seiha dekitanoka. Tokyo: NTT Publishing.

Japan Electronics and Information Technology Industries Association. Retrieved March 15, 2006 from http://www.jeita.or.jp/english/index.htm

Japan Video Software Association. Retrieved March 15, 2006 from http://www.jva-net.or.jp/en/index.html

JNHK (Japan Broadcasting Corporation.). (2000). Project X: Challengers 1. Tokyo: Japan Broadcast Publishing.

Marlow, E., & Secunda, E. (1991). *Shifting time and space: The story of videotape.* New York: Praeger.

Matsushita. Retrieved March 15, 2006 from http://panasonic.net/

Mitsubishi Electric. Retrieved March 15, 2006 from http://global.mitsubishi-electric.com/

National Science Museum, Tokyo. Retrieved March 15, 2006 from http://sts.kahaku.go.jp/sts/

Nmungwun, A. F. (1989). *Video recording technology: Its impact on media and home entertainment*. Hillsdale, NJ: L. Erlbaum Associates.

Osterholm, J. R. (1994). *Bing Crosby: A bio-bibliography*. Westport, CT.: Greenwood.

Rosenbloom, R. S. & Abernathy, W. J. (1982). The climate for innovation in industry. *Research Policy*, *11*(4), 209-225.

Sato, M. (1999). *The story of a media industry*. Tokyo: Nikkei Business Publications.

Schoenherr, S. Retrieved March 15, 2006 from http://home.sandiego.edu/~ses/

Sony Corporation. Retrieved March 15, 2006 from http://www.Sony.net/

Toshiba. Retrieved March 15, 2006 from http://www.toshiba.co.jp/index.htm

Copyright © 2008, IGI Global. Copying or distributing in print or electronic forms without written permission of IGI Global is prohibited.

Utterback, J. M. (1994). *Mastering the dynamics of innovation: How companies can seize opportunities in the face of technological change*. Boston: Harvard Business School Press.

Victor Company of Japan. Retrieved March 15, 2006 from http://www.jvc-victor.co.jp/english/global-e.html

Endnotes

[1] The patent on rotary head recording was filed by Luigi Marzocci, an Italian inventor, in 1938. Ampex utilized this technology for the VCR and changed the direction for recording electrical signals. As a result, the consumption of videotape was dramatically reduced.

[2] JVC (Victor Talking Machine Company Japan, Limited) was established in 1927 as a subsidiary of Victor Talking Machine Company. The parent companies changed many times. Matsushita is the parent company of JVC but gave it the freedom to work hard by competing with itself.

[3] EVR was invented by Peter C. Goldmark who first succeeded in the practical use of TV and who also invented the LP record.

[4] U Format uses 3/4 inch tapes that are wider than the mainstream 1/2 inch tapes used today.

[5] JVC could not sell the KV-1 at all because of its incompatibility with Ampex products.

[6] While US companies' pricing policy was based on customer needs, Japanese policy was based on the cost plus.

Copyright © 2008, IGI Global. Copying or distributing in print or electronic forms without written permission of IGI Global is prohibited.

Chapter VI

Struggle for De Facto Standard

In this chapter, we discuss the struggle for the de facto standard based on the Betamax versus VHS case in the late 1970s. In the middle of the 1970s, Sony and JVC introduced Betamax and VHS, both of which satisfied minimum requirements for commercialization. Although Sony had a first mover advantage in the market, JVC and VHS group turned the face about by 1980. The capabilities for the first Betamax and VHS, SL-6300 and HR-3300, were almost equal. The first VHS excelled the first Betamax in the lighter weight, the longer recording time, the compatibility. The consumers preferred HR-3300 because of the longer maximal recording time to videotape the movies, the baseball games, and the football games. Most videotaped programs by all VHS machines can playback except for a few models. On the other hand, Sony made a disconnection to the first Betamax when they launched Beta II in 1977. In addition, JVC formed the VHS group to catch up with Sony. This group worked effectively to enhance the product capability and the sales promotion.

Copyright © 2008, IGI Global. Copying or distributing in print or electronic forms without written permission of IGI Global is prohibited.

Emergence of Betamax and VHS

It was two almost decades after Ampex and RCA developed the VCR for broadcasting in the 1950s that two prominent home-use VCRs appeared on the market. In 1975, Sony introduced the first Betamax, SL-6300. In the following year, JVC introduced the HR-3300 model with the VHS format. Both products satisfied the essential functional requirements of the home-use VCR market, such as the functions. These requirements were the quality, weight, size, price, and operability. They were quite equal in performance and capabilities except for the maximal recording time limitation. At that time, however, it was expected that Sony (Betamax) would win against JVC (VHS) because Sony had achieved fame as a pioneer of the Home-use VCR and Sony had a great lead over JVC when JVC introduced the VHS into the market.

There were two major reasons why JVC was able to catch up with Sony within a relatively short time. First, their target product concepts were almost the same. Both focused on a version which used a magnetic cassette tape, the U Format. While companies other than those in Japan tried to commercialize according to a variety of formats, most Japanese companies developed a 1/2-inch magnetic tape VCR in a plastic cassette body. Second, a cross-license agreement among the Japanese companies, Sony, Matsushita, and JVC, in 1970, accelerated the technological progress of the home-use VCR. They shared the latest techniques. Even though Sony had a substantial lead in the home-use VCR, JVC and Matsushita could catch up quickly because they also had enough experience in a magnetic tape VCR and they could use the latest technologies Sony owned without payment.

As a result, the recording time should have been more than 120 minutes. While the SL-6300 could record only 60 minutes, the HR-3300 could record 120 minutes. Many more people preferred the ability to record 120 minutes because most movies and sports, such as baseball and football games, require more than 60 minutes. At least, we can say without doubt, that the HR-3300 was superior to SL-6300 in the maximal recording time feature.

A comparison between the features of the SL-6300, the first Betamax, and the HR-3300, the first VHS, is shown in Table 1. Betamax was introduced one year earlier than VHS. In addition, it was cheaper than VHS. Furthermore, its tape speed was slower than that of VHS. The slower the tape speed the better because mechanical parts then have a lighter burden, thus leading to more reliability and a longer life. As a result, Betamax could utilize the tape efficiently and operate stably. However, the Betamax machine was much

Copyright © 2008, IGI Global. Copying or distributing in print or electronic forms without written permission of IGI Global is prohibited.

Table 1. Comparison of Betamax and VHS

	Betamax	VHS
Release	1975	1976
Model	SL-6300	HR-3300
Size	W 45.0cm, H 20.5cm, D 40.0cm	W 45.3cm, H 14.7cm, D 31.4cm
Weight	18.5kg	13.5kg
Price	229,800Yen	256,000Yen
Cassette Size	W 15.6cm, H 9.6cm, D 2.5cm	W 18.8cm, H 2.5cm, D 10.4cm
Tape Speed	1.9cm/s	3.3cm/s (2.4cm/s for PAL)
Cassette Price	3800 Yen	6000 Yen (T-120)
Recording Time	60 min.	120 min.
Horizontal Resolution	Approx 250 lines	Approx 240 lines

bigger and heavier than the VHS. In addition, although the price of Betamax cassette tape was cheaper than that of VHS cassette tape, the VHS cassette tape was more economical than Betamax cassette tape because of the recording time. It was very difficult to tell which was superior technically because both had good points and bad points.

Formulation of the Betamax and VHS Groups

Betamax and VHS are similar from a technological viewpoint and are quite similar in performance. However, they are incompatible because of the difference in the recording format and the cassette size. The Betamax cassette is smaller than the VHS cassette. On the other hand, VHS surpassed Betamax with its lower weight and a longer recording time capability. The Model HR-3300 VHS was lighter than Betamax by 11 pounds (approx 5 kg) and could record for two hours, while the Betamax could record for just one hour at a time. Thus, these two versions were locked in a deadly struggle for dominance for a decade.

In the late 1970s, Japanese consumer electronics companies were divided into two groups, the Betamax group and the VHS group (NHK, 2000). Sony started forming the Betamax group earlier than the VHS because they launched the Betamax earlier than did JVC. Toshiba, Sanyo, NEC, Aiwa, and Pioneer

Copyright © 2008, IGI Global. Copying or distributing in print or electronic forms without written permission of IGI Global is prohibited.

adopted the Betamax format directly after Sony launched it. Sony had expressed a desire for Matsushita and JVC to adopt the Betamax format and lent them a prototype of SL-6300 for the persuasion. Unfortunately, the loan of the Betamax prototype to JVC boosted the development of the VHS.

The VHS (JVC) was considerably far behind Sony (Betamax group) toward the de facto standard due to their precedence in the market. When JVC introduced the model HR-3300 (first VHS VCR) in 1976, no one but JVC believed that they could catch up with Sony (Betamax group). To do so, JVC started making strategic alliances with other consumer electrics companies. In 1976, JVC called on Matsushita, Hitachi, Mitsubishi, Sharp, and Akai to adopt the VHS format and all of them agreed to accept it. This VHS group was quite effective in recovering from their lagging position far behind Sony in both technical and commercial aspects of VCR development and production. For example, Sharp contributed the front-loading function, Mitsubishi the fast forward function, and Hitachi the integrated circuit (IC) technology. Matsushita has had the largest distribution channel in Japan.

All VHS members except for Matsushita made Original Equipment Manufacturing (OEM) contracts with JVC in the middle1970s. Hitachi, Mitsubishi, Sharp, and others did not have enough technology to develop their own VCR products and wanted to reduce their risks until the battle between Betamax and VHS was resolved. At the beginning, they asked JVC to supply the products and studied the product for manufacturing by themselves. However, JVC did not have a sufficient manufacturing capacity. JVC had about 50 workers in 1976 and the products were almost handmade. Although their target monthly production was 10 thousand units, they could produce only 144 units in September 1976. It took a year to be able to produce 10,000 units in a month. By the late 1970s, Japanese companies such as Hitachi, Mitsubishi, and Sharp, started manufacturing VCRs by themselves, Hitachi in 1977, Mitsubishi in 1978, and Sharp in 1979. However, no U.S. or European companies have ever manufactured VCRs.

The struggle for the de facto standard expanded around the world. The United States was considered to be the crucial battleground for the de facto standard. In 1977, OEM contracts were made between Sony and Zenith Electronics Corp., and between Matsushita, a member of the VHS Group, and RCA. Matsushita succeeded in doubling the recording time and in cutting the production cost to about US $1000 by acceding to the requests of US companies. As a result, other major US companies such as GE and Magnavox, adopted the VHS format. In the same year, Matsushita also supplied OEM to US major shop-

Copyright © 2008, IGI Global. Copying or distributing in print or electronic forms without written permission of IGI Global is prohibited.

pers and the VCR produced by Matsushita achieved 60 percent of US sales. On the other hand, JVC focused on the European Market and made OEM contracts and technology transfer contracts with major European electronics companies in the late 1970s. Finally, VHS overtook Betamax domestically and globally and established itself as the de-facto standard—the dominant design— in the home-use VCR market.

The VHS group surpassed that of the Betamax in production of VCRs by 1980. JVC had much less power than did Sony to control its group. Companies that had some strength in the VCR technologies wanted the influence and freedom. JVC functioned as a coordinator and developed the VHS with strong members including Matsushita, Hitachi, Sharp, and Mitsubishi. In addition, the design of the VHS VCR was much simpler than that of the Betamax because JVC had a chance to improve their model by investigating Betamax machines. The simpler design was very favorable to manufacturability. Many electronics firms therefore expected that they could reproduce the VHS faster than the Betamax. Thus, within a few years, most Japanese companies started producing VHS products.

The year 1983 was an important one for the VHS group because new two groups joined them. First, Philips and Grundig, which had pursued the development of their own original VCR versions, decided to adopt the VHS format. Second, Toshiba, Sanyo, NEC, and General, who were members of the Betamax group, decided to adopt VHS for exports. Thus, even rivals recognized VHS as a winner, the de-facto standard. As a result, more than 0.1 billion VHS VCRs had been sold all over the world by 1986.

Table 2 summarizes the formation details for the Betamax and VHS groups. From the viewpoint of a greater number of members, the VHS caught up with the Betamax very quickly. It appears that Sony fought alone against the VHS group. Sony did excel over the other VHS group in terms of the technology and marketing. Aiwa is a subsidiary of Sony. On the other hand, the VHS group had two core members and many strong supporting members. JVC developed VHS and Matsushita steered the VHS group to victory. Matsushita had the greatest sales force in Japan. In addition, other members, such as Hitachi, Sharp, and Mitsubishi, had excellent technologies. By the 1980s, VHS surpassed the Betamax group. JVC did not control other group members except for the observance of the VHS format. JVC emphasized the guarantee of the compatibility among VHS machines. JVC had a less power to control the other companies and the market than did Sony. Other members who had unique technologies had an incentive to be a VHS group

Copyright © 2008, IGI Global. Copying or distributing in print or electronic forms without written permission of IGI Global is prohibited.

member. Furthermore, JVC encouraged other members toward greater quality, especially the importance of precision instruments and measurements, for the mass production of the VCR. The gap between VHS group and Betamax group expanded so rapidly that most Betamax group members had switched to the VHS format by the middle of 1980s.

Table 2. Formation of the Betamax and VHS groups

		Betamax (Sony)	**VHS (JVC)**
1975	May	Sony launched SL-6300. NEC, Aiwa, Pioneer, and General adopted.	
1976	Sep.		JVC launched HR-3300.
	Dec.		Hitachi and Sharp sold VT-3000 at 258,000 and VC-5000 at 256,000 Yen.
1977	Jan.		Matsushita sold HV-1100 at 268,000 Yen.
	Feb.	Zenith adopted and Toshiba sold V-5200 at 255000 Yen.	
	Mar.	Sanyo sold VTC-9000 at 268000 Yen.	RCA, GE and Magnavox adopted.
	Jun.		Matsushita produced and sold NV-8800 at 266,000 Yen
	Dec.		Saba (West Germany) adopted.
1978	Feb		Hitachi started producing.
	Mar.		Thomson (France) adopted.
	Jun.		Thorn-EMI (UK) adopted. Mitsubishi started producing.
	Sep.		Rank (Austria) adopted.
	Dec.		Telefunken adopted.
1979	Jul.		Sharp started producing.
1980	Sep.		Standard Electric Lorenz adopted.
1981	Sep.		Tokyo Sanyo adopted and started producing.
			Olympus made alliance with Matsushita.
1982			Cannon adopted.
1983	Dec.		Philips and Grundig adopted. Toshiba, Sanyo, NEC and General adopted for exports.
1984	Jan.		Zenith made an OEM contract with JVC.

Copyright © 2008, IGI Global. Copying or distributing in print or electronic forms without written permission of IGI Global is prohibited.

Product Advancement Under the Struggle

Table 3 shows the product advancement of Betamax and VHS during the phase of the struggle for dominance. The other notable movement in the late 1970s was the advancement of the Betamax and VHS format. Both groups upgraded their products by adding new functions and extending recording time. SL-6300 (Betamax) and HR-3300 (VHS) were essentially equal in performance and capabilities except the allowable recording time. While Sony dominated the Betamax group and extended the recording time, JVC coordinated the VHS group flexibly and pursued the goals which users would need.

To extend the recording time, Sony developed Betamax II in 1977 (two hours recording capacity) and Betamax III (three hours) in 1978. When Sony extended the recording time, they were in a dilemma whether to tolerate deterioration of the image quality or to give up the compatibility (Sato, 1999). They chose the latter. Sony introduced a three-hour cassette because they had strength in the R&D on cassettes. On the other hand, Matsushita introduced four-hour VHS VCRs in the United State in 1977. In 1977, JVC developed the triple-recording-time mode. But from the basic or original design perspective, VHS was better designed to extend the recording time than Betamax.

In addition, there were substantial differences with respect to compatibility among upgraded versions between Betamax and VHS group. The Betamax group did not emphasize compatibility. Betamax II and Betamax III were not compatible with the basic Betamax. On the other hand, the JVC group stressed compatibility. All VHS models are compatible with each other except one, a four-hour VHS model that Matsushita introduced in 1977. The coherent compatibility with other versions obviously pleased VHS users and made the VHS version stronger.

The other functions incorporated in this period were the double-speed playback mode and the addition of HiFi (High Fidelity) sound. In 1977, JVC introduced the HR-3600, which had a new function, the double-speed playback mode with monaural sound. In 1983, Sony introduced the HiFi VCR first. Sony lost the coherent compatibility again because they developed the European HiFi model. JVC followed Sony and introduced their HiFi model about a half year later. While Sony had excellent technology, JVC had an idea ahead of the times, which led VHS to the rapid and long product diffusion.

Copyright © 2008, IGI Global. Copying or distributing in print or electronic forms without written permission of IGI Global is prohibited.

Table 3. Product advancement under the struggle

		Betamax	VHS
1975	May	Sony started selling Betamax VCRs.	
1976			JVC introduced VHS VCR.
1977	Mar.	Sony launched Beta II (2 hours).	Matsushita introduced 4 hour recordable VCR.
	Dec.	Sony introduced a 3 hour cassette.	
			JVC introduced the double-speed playback model.
1978	Dec.	Sony launched Beta III (3 hour).	JVC introduced 6 hour model.
1982	Mar.		JVC developed 8 hour model.
	Sep.	Sony introduced 5 hour model.	
1983	Apr.	Sony introduced HiFi model.	
	Nov.		JVC introduced HiFi model.
1985	Feb.	Sony introduced a high band model.	

OEM Under the Struggle

In many industries, including especially electronics (Chopra & Meindl, 2004), there are very few major component manufacturers, whose customers, or receivers of these components are called original equipment manufacturers (OEMs). The term is slightly confusing and perhaps a term more like final product producers or assemblers might have been better. Chopra and Meindl observe that the 80:20 rule holds. This rule is often also called the Pareto Principle (Troutt & Acar, 2005) but apparently is based on Lorenz curve considerations. Fewer than 20 percent of the OEMs were using over 80 percent of the components according to Chopra and Meindl (2004). OEMs therefore tend to sell directly to their large customers and thus are upstream suppliers to set makers. However, Chopra and Meindl (2004) also note that intermediate distributors can still improve supply chain performance in serving smaller OEMs.

OEMs played an important role for both VCR groups in the late 1970s. Sony and JVC (Matsushita) provided OEM for other manufacturers and shoppers. When they requested other manufacturers to adopt their format, they served their original products until others started producing. Otherwise, they did not abandon their original model and developed their original format. During the OEM period, the receivers could sell VCRs and earn profitable returns. In addition, the other manufacturers except Sony, JVC, and Matsushita, had

Copyright © 2008, IGI Global. Copying or distributing in print or electronic forms without written permission of IGI Global is prohibited.

the patent problem. Sony, JVC, and Matsushita made a cross license among them. But others did not. However, neither group exercised the patent right because the group members were the hostage of the other group and the exercising the patent right would have discouraged the diffusion of the VCR. The OEM periods were different depending on the manufacturers. Matsushita was an exception, which did not receive OEM products from the beginning and which started OEM to the US companies from 1977. They had enough experience in the mass-production of VCRs and wanted to get a higher margin from their own products rather than OEM products.

Others took much more time than Matsushita because of their inexperience in mass-production and the aversion to risk. A recent literature review on risk management in supply chains has been given by Chen, Li, and Wang (2006). These concerns were especially critical for the VHS group members who desired to produce VCRs which were compatible with each other to a large degree. Ideally, all VHS VCRs should be able to playback all prerecorded video cassettes by any VHS manufacturer with adequate image and sound quality. Some manufacturers waited for the time until the winner was settled. It was natural for most companies to avoid the risk of the double investment in producing both versions. The receivers asked OEM suppliers for the products based on their design. When they started producing by themselves, they added some originality. For example, Hitachi and Sharp added the remote controller by making the most of their competence in micro-computers and Mitsubishi added the timer function.

All US companies and customers, such as RCA, Zenith, and GE, continued to receive OEM products from Japanese companies. The reasons why they could not produce VCRs by themselves were the following. First, the VCR was a very complex machine combining both mechanics and the electronics. To produce VCRs, companies needed to own the latest technologies in both areas. While Japanese household appliance manufacturers were generalized, US manufacturers were specialized. Second, the VCR was developed exclusively in Japan. Foreign manufacturers could not access the latest information quickly. In addition, the development speed was too fast for any foreign manufacturers to catch up and overcome the time lags with the Japanese companies.

OEM continued to play a very important role into the late 1970s. OEM contributed to the formation of the groups by guaranteeing that receiver companies could sell products and begin earning profits in advance of being ready for producing components on their own. It also boosted the diffusion

Copyright © 2008, IGI Global. Copying or distributing in print or electronic forms without written permission of IGI Global is prohibited.

of the VCR to the world. The VHS was designed to be applicable to all TV formats, NTSC, PAL, and SECAM. After JVC announced the VHS at an international exhibition in West Germany in August 1952, JVC started OEM for the European makers. By that time, JVC had established their mass production system. If JVC and Matsushita had not made most of the OEM, the diffusion speed would be very slow. Betamax would have had a bigger chance to be a winner or even another machine might have a chance to dominate the home-use VCR markets. The OEM functioned as the catalysis to decide the winner by speeding up product diffusion.

VHS's Great Victory

JVC introduced HR-3300, the first VHS VCR, after Sony (Betamax) achieved prominence in technological excellence and a significant lead in the home-use VCR market. Contrary to expectations, the VHS did come from behind

Figure 1. Shift of domestic market share of Betamax and VHS (others) (Source: Yano Research Institute Ltd., Market Share in Japan 1978-1987)

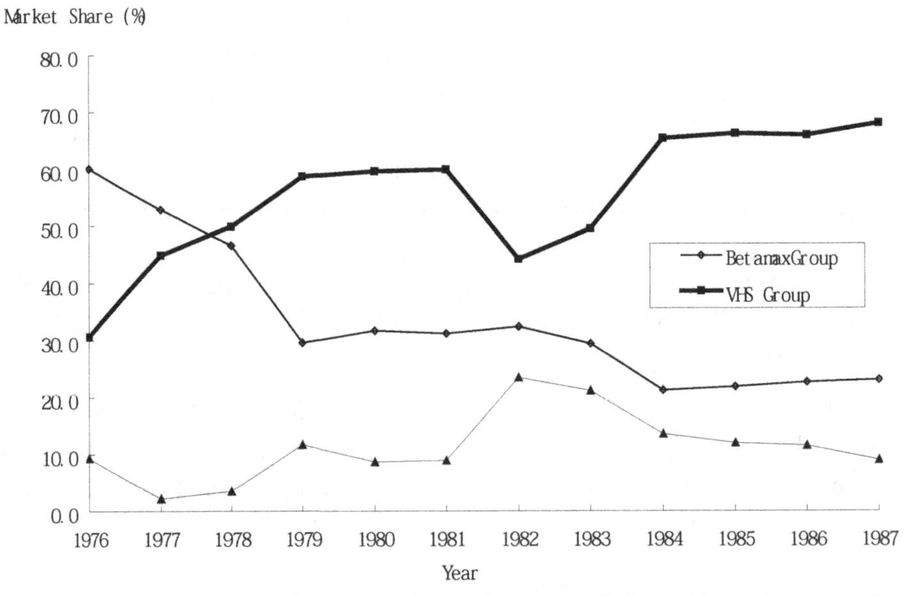

Copyright © 2008, IGI Global. Copying or distributing in print or electronic forms without written permission of IGI Global is prohibited.

to achieve victory as in Figure 1. VHS caught up with Betamax very quickly and, on the other hand, Betamax was able to retain a level of the market share, in fact, more than 20 percent. This was because the software industry was not well enough established at that time. When customers chose the products, they did not care about externalities. They mainly recorded TV programs. After the software industry developed sufficiently, the VHS finally completely dominated the Betamax. VHS was able to overcome adversity due to the following reasons in the 1970s.

First, JVC had a visionary leader, Shizuo Takano, who headed the VCR division. JVC downsized the division again and again because of the large declines before JVC introduced the HR-3300. Although the sales of VCRs remained at a very low level, JVC spent huge amounts of money and labor on customer aftercare and the returns of goods because of the poor quality of the products. It was natural for JVC to cut the R&D budget on the home-use VCR, under the assumption of the inferiority to Sony at the time. However, Takano decided to continue it because he had a clear vision that the home-use VCR would be virtually a necessity to the consumer in the near future. He designed the VHS to be very applicable to consumer needs. For example, it was made possible for the latest VHS machines to playback the video recorded by earlier HR-3300. Takano had an excellent vision and his distinguished consistency was in large part responsible for the great success of VHS. Later, Takano was called Mr. VHS.

Second, the cross license among Sony, JVC and Matsushita, made it possible for JVC to catch up with Sony quickly. In addition, Sony did not interfere with other group members, such as Hitachi, Mitsubishi, and Sharp to protect Betamax group members. As a result, VHS and Betamax had a lot of similarities and their performances were almost the same.

Third, JVC accumulated their unique technologies along with the excellent vision. JVC had strength in TV technology and grasped the future of consumer needs for the home-use VCR. As an example, they emphasized the projection of a still shot, that is, a freeze or pause, and took out a patent for the rotary two-head helical scan in 1959, which was necessary for a still shot. Another example is that JVC developed a new function, the double-speed playback mode with the monaural sound in 1977. The last and most critical one is that the VHS had a longer recording time than the Betamax. This addressed an important consumer need as they wanted to record the TV programs such as movies and sports events, which were typically longer

Copyright © 2008, IGI Global. Copying or distributing in print or electronic forms without written permission of IGI Global is prohibited.

than one hour. Thus JVC developed the VCR products with future consumer needs clearly in mind.

Fourth, the VHS group was much more effective than the Betamax group. Although JVC developed VHS format, JVC did not have an enough power to control the VHS group. All JVC did was to protect and develop the VHS format and to coordinate the strong members of their group. Members brought their technologies to VHS products and the performance of VHS products improved rapidly. Furthermore, the VHS group made the most of OEM. JVC did OEM for other group manufacturers while they were in preparation and Matsushita did it for shoppers in the United States.

Finally, they carefully protected the original VHS format. Compatibility was one of the most critical factors in the home-use VCR. The VHS format was

Table 4. Domestic market share (1976-1985)(Source: Yano Research Institute Ltd., Market Share in Japan 1978-1987)

	1st (Market Share)	**2nd**	**3rd**	**4th**	**5th**	**6th**	**Production (Thousands of Units)**
1976	**Sony* 60.1%**	JVC 18.5%	Matsushita 12.2%	-	-	-	286
1977	**Sony 44.1%**	Matsushita 21.8%	JVC 21.0%	Shibaura 6.2%	**Sanyo 2.6%**	Hitachi 2.1%	131,041
1978	**Sony 35.3%**	Matsushita 23.0%	JVC 22.1%	Shibaura 8.1%	Hitachi 4.8%	**Sanyo 3.2%**	211,162
1979	Matsushita 27.0%	JVC 21.2%	**Sony 17.3%**	**Sanyo 8.0%**	Hitachi 6.7%	Shibaura 4.3%	296,168
1980	Matsushita 30.0%	**Sony 19.4%**	JVC 19.2%	**Sanyo 8.4%**	Hitachi 6.1%	Sharp 4.3%	562,825
1981	Matsushita 30.1%	**Sony 19.9%**	JVC 19.3%	**Sanyo 7.7%**	Hitachi 6.0%	Sharp 4.5%	1,086,799
1982	Matsushita 23.9%	**Sony 22.6%**	JVC 12.1%	**Sanyo 7.0%**	Hitachi 4.8%	Sharp 3.4%	1,284,987
1983	Matsushita 26.2%	**Sony 20.8%**	JVC 14.3%	**Sanyo 6.4%**	Hitachi 5.2%	Sharp 3.8%	1,513,991
1984	Matsushita 24.6%	JVC 17.0%	Hitachi 15.0%	**Sony 11.4%**	Sharp 8.7%	**Toshiba**** 5.2%	2,019,919
1985	Matsushita 25.0%	JVC 16.0%	Hitachi 15.5%	**Sony 13.0%**	Sharp 9.7%	**Toshiba** 5.0%	1,790,383

*Beta group members are written in the boldface.

**Toshiba started manufacturing VHS in 1983.

Copyright © 2008, IGI Global. Copying or distributing in print or electronic forms without written permission of IGI Global is prohibited.

developed protecting the compatibility except for a four hour recordable VCR which Matsushita introduced into the US market to accommodate requests from US companies in 1977. Compatibility meant that previously recorded material, that is, older videotapes, did not become obsolete. Perhaps more importantly, although prices had declined substantially, typical consumers found VCR prices sufficiently high that purchase of new machines for new formats was not a practical reality.

Table 4 shows the domestic market share of the VCR during 1976-1985. Sony was a leading company during 1976-1978. However, their share declined from about 60 percent to about 20 percent during 1976-1979 and remained about 20 percent in the early 1980s. After that they lost market share again as the total amount of domestic production increased. Sanyo, a member of the Betamax group, kept their share at about 8 percent in the early 1980s and then lost further. Contrary to the Betamax group, the VHS group members grew very quickly. Matsuhsita excelled over JVC and Sony very quickly and Hitachi caught up with JVC by the middle of 1980s. The contributions of the group members may be regarded as amounting to the main cause of the great victory of VHS.

Betamax's Fate

Sony, originator of Betamax, was a leading company in the development of the home-use VCR. A cross license agreement with Ampex in 1960 made it possible for Sony to fuse the latest technology on the VCR which Ampex owned and the semiconductor technology which Sony owned. Sony also had a big technical advantage in video tape. The semiconductor and video tape technologies are the keys to downsizing and stabilizing the VCR. Sony drew Japanese companies toward the development of VCR which used the magnetic tape, U Format. In 1976, Sony introduced the SL-6300, the first Betamax VCR, which was the first practical machine for home-use and opened up the home-use VCR industry.

When JVC introduced the VHS in 1976, Sony already had a big lead in the market. Sony had formed the Betamax Group and was increasing sales. However, the home-use VCR market in the middle of 1970s was so small that there was a big opportunity for the tables to be turned. At that that point in

Copyright © 2008, IGI Global. Copying or distributing in print or electronic forms without written permission of IGI Global is prohibited.

the diffusion curve, the VCR was mainly for business-uses or was purchased by the high-tech maniacs. These types of purchasers tested the products and would pursue further tries as new products became available to test. Sony was therefore caught up to by JVC during the period when VHS had a chance to reverse the market.

In the early struggle for the de facto standard, Sony could not form an effective Betamax group. It seemed that Sony alone fought against the VHS group. Sony had a big technical advantage and the Betamax group was under its exclusive control. In 1977, they made a fatal mistake when they disconnected the compatibility of the formats by developing a new format, the Betamax II. Users feared that the disconnection might happen again. In addition, the manufacturing capacity of the Betamax group also was surpassed by the VHS group by 1978. While Sony and Sanyo could produce 45,000 and 15,000 units in a month, respectively, Matsushita, JVC, Toshiba, and Hitachi could produce 100,000, 20,000, 10,000, and 6,000.

Sony apparently had believed that they had time until the others caught up in the middle of the 1970s. However, JVC launched the HR-3300 which was a very high quality product and Matsushita launched the VX-2000 at a very reasonable price, 210,000 Yen, in 1976. Most Japanese companies had focused on the magnetic tape recording method and the cross license among them made others made it possible for others to catch up with Sony very quickly. The SL-6300 did not have some essential functions that the HR-3300 had, such as 120-minute recording, slow motion playback and the still frame projection. In addition, Matsushita promoted a price war so that Sony could not gain enough profit as a first mover (Fujioka 1978). As a result, Sony acknowledged that they were losing their field by 1980.

Even though clearly having the less desirable product version, Sony continued to upgrade Betamax toward the longer recording time and HiFi sound. Sony did not give up hope because they had many Betamax fans and users. Unfortunately, the struggle was settled so slowly and gently that many people decided to purchase Betamax machines and owned the tapes recorded by Betamax. Some of them continued to use Betamax machines for a few decades because of the high switching cost and the strong attachment. Sony continued to produce Betamax for them until 2002, when most of the VHS group stopped producing VHS machines altogether.

Copyright © 2008, IGI Global. Copying or distributing in print or electronic forms without written permission of IGI Global is prohibited.

Marketing Interface and Strategy Issues

The VCR case history also illustrates important points and general principles on the relationship of marketing and production, and manufacturing strategy, respectively. These principles were developed by operations theorists but apply with equal or greater force to the supply chain environment. With regard to the marketing and production interface Hill, Berry, and Klompmaker (1994) observed that:

The basic tasks of a business are straight-forward – getting and keeping customers and making profits, both set within the context of the short and long term. Explicit in these is not only a need for marketing and manufacturing to work well in themselves but to work well together. Companies, however, typically fail in this provision. To move to a more competitive response in today's difficult markets requires more than co-existence, it requires a shared partnership between the core parts of the firm. Shared understanding of marketing and manufacturing's approach is the starting point. The other is building on a shared understanding of the market itself. In such a way, a move to being customer-driven starts to take shape.

In the case of JVC and the VHS, this integrated view of marketing and production appears to have materialized mainly through the vision of Shizuo Takano. It makes sense that a single top leader might most easily embody a shared understanding the two needs. He was clearly able to stay focused on customer product requirements and needs while bringing to bear the technologies, alliances, and strategies for the most effective production decisions.

Manufacturing strategy is a term used for the work of Hill (2000). See also and Vollman, Berry, Whybark, and Jacobs (2005). This work stressed the marketing and production relationship with a fresh perspective. Products might have a just-sufficient set of features to be competitive, but should go beyond these to become "order winners." Additional or added new features beyond these were called "exciter/delighters." In the early stages of the VCR, it can be seen that the Betamax and VHS designs were essentially equivalent. To overcome Betamax the VHS group needed to develop order winner and exciter/delighter features. Many of the VHS features clearly qualify as one or the other of these kinds of customer requirements or wants. Clearly the longer recording time capability was an important order winner kind of feature. At

Copyright © 2008, IGI Global. Copying or distributing in print or electronic forms without written permission of IGI Global is prohibited.

the same time, the production process must support the order winner features, which might be regarded as cost and quality in the VHS case.

In the maturity and de-maturity stages, further cost reduction, globalization, and location of production facilities become of dominant interest (Higuchi, Troutt, & Polin, 2005; Polin, Troutt, & Acar, 2005). However, in the case of the Betamax versus the VHS supply chains, it is clear that the VHS had won by the time the maturity phase had arrived.

Supply Chain vs. Supply Chain

During the dominance struggle, Sony and JVC built their own supply chains. They put greater emphasis on product development than on the efficiency of operations. They competed in product performance and on downsizing within a more or less certain price zone. Consumers in this period, the innovators of the product life cycle, did not pay much attention to the slight price differences. Anyway, both supply chains needed to establish mass-production systems. Otherwise, they could not serve enough volume of quality products to the market. Unfortunately, Sony and JVC (including Matsushita) completed their mass-production systems for the Betamax and the VHS with which consumers were satisfied, within a relatively short time frame. The role of supply chains in this period was to serve the products which customers wanted as promptly and stably as possible. These supply chains for the home-use VCR simply repeated a process, adapted to the technical changes, and to the declining prices, while improving product performance until they could serve the products that consumers wanted in sufficient quantities and quality.

Also during this struggle, Betamax and VHS competed in product performance features such as the basic functions (the recording function and the built-in TV tuner), the quality of the image (the number of horizontal resolutions or the pixels and the level of the noise), the recordable time, the size and weight, and the price. Sony was first to satisfy the minimum requirements from the consumers at that time. Betamax had achieved those of the basic functions and the quality of image. Sony evaluated the consumer needs for the size, weight, and cost, and their supply chain partners contributed to reduce these to improved levels. They started mass-production, based on the machine design choice after they felt assurance of future success. However,

Copyright © 2008, IGI Global. Copying or distributing in print or electronic forms without written permission of IGI Global is prohibited.

Betamax was followed by VHS whose recording time was longer, and size was smaller, than that of the Betamax.

In addition to the product specifications, manufacturing capabilities played very important role for the growth of sales. Manufacturing capability includes the quality and cost of production and the maximum yield. In the case of the VCR, it was a critical factor for manufacturers to guarantee compatibility among other VHS or Betamax machines. The price decline also had a great impact on sales. Furthermore, the demand for the VCR increased dramatically after the SL-3600 and HR-3300 models emerged. The manufacturing capacity was a constraint to rapid growth. To achieve these goals, collaboration among supply chain partners was necessary. Without it, the various bottlenecks appeared quickly and the growth speed was impeded. The expansion of the manufacturing capability of each group member decided the winner of the struggle for the de facto standard.

All the upstream supply chain partners could do was to deliver the required number of parts with the desired quality. It was a hard task because the demand for the products increased very rapidly. However, upstream partners of both the Betamax and VHS groups operated so efficiently that the winner was decided by the recording time (the product design) and the activities in the downstream. A consumer-oriented view led JVC to a superior product design. Along with Matsushita's distribution power and the global strategic alliances with other strong distributors, the global diffusion of the VHS accelerated. The battlefield had changed from the upstream to the downstream portions of the supply chains.

Conclusion

The importance of the de facto standard has been recognized since the late 1970s. The VCR case demonstrated how to set the de facto standard and how advantageous the de facto standard actually can be. Japanese companies exclusively manufactured VCRs in the 1970s and 1980s because they could launch new generation products with good timing for them. As a result, no companies except the Japanese ones could follow. Many countries learned this lesson and set de jure standards. Nowadays, many international de jure standards exist in the global market such as the digital TV and the cell phone.

Copyright © 2008, IGI Global. Copying or distributing in print or electronic forms without written permission of IGI Global is prohibited.

A struggle for the de facto standard between Betamax and VHS in the late 1970s is very famous. Sony was a leading company in the VCR in 1960s and 1970s and launched the first Betamax, the SL-6300, which satisfied minimal requirements for commercialization. When JVC launched the first VHS, HR-3300, whose capability was almost equal to the first Betamax, Sony had a first mover advantage in the market. On the other hand, VHS surpassed the Betamax by having lighter weight, a longer recording time, and compatibility. Many consumers preferred the VHS to Betamax because of the longer maximal recording time and the compatibility. In addition, JVC formed the VHS group, which worked so effectively that the VHS topped the Betamax in product capability and sales promotion by 1980.

References

Abernathy, W. J. (1978). *The productivity dilemma*. Baltimore: The John Hopkins University Press.

Abernathy, W. J, Clark, K.B., & Kantrow, A.M. (1983). *Industrial renaissance*. New York: Basic Books.

Anonymous. (1967). Ampex makes a play for the home market. *Business Week, 1972*, 164-174.

Chen, J., Li, J., & Wang, S. (2006). Risk management in supply chains: Literature review. In *Proceedings of the international workshop on successful strategies in supply chain management*, *January 5-6*, (pp. 73-84). Hong Kong: The Department of Applied Mathematics, The Hong Kong Polytechnic University.

Chopra, S., & Meindl, P. (2004). *Supply chain management: Strategy, planning, and operation* (3rd ed.). Upper Saddle River, NJ: Pearson Prentice Hall.

Christensen, C. M. (1997). *The innovator's dilemma*. Boston: Harvard Business School Press.

Economic and Social Research Institute, Cabinet Office, Government of Japan. Retrieved March 15, 2006 from http://www.cao.go.jp/index-e.html

Electronics Industries Association of Japan. Retrieved March 15, 2006 from http://www.jeita.or.jp/eiaj/english/index.htm

Copyright © 2008, IGI Global. Copying or distributing in print or electronic forms without written permission of IGI Global is prohibited.

Fujioka, T. (1978). Sony shinwa no saisei ha kanou ka? *The Economist, 56*(43), 112-117.

Funai. Retrieved March 15, 2006 from http://www.funaiworld.com/index.html

Higuchi, T, Troutt, M. D., & Polin, B. A. (2004). Life cycle considerations for supply chain strategy. In Chan, C. K., & Lee, H. W. J. (Eds.), *Successful strategies in supply chain management* (pp. 67-89). Hershey, PA: Idea Group.

Hill, T. J. (2000). *Manufacturing strategy: Text and cases*, 3rd ed. Homewood, IL: Richard D. Irwin.

Hill, T. J., Berry, W. L., & Klompmaker, J. E. (1994). Customer-driven manufacturing, *International Journal of Operations & Production Management, 15*(3), 4-12.

Hitachi. Retrieved March 15, 2006 from http://www.hitachi.com/

Itami, H. (1989). *Nihon no VTR sanngyou naze sekai wo seiha dekitanoka.* Tokyo: NTT Publishing.

Japan Electronics and Information Technology Industries Association. Retrieved March 15, 2006 from http://www.jeita.or.jp/english/index.htm

Japan Video Software Association. Retrieved March 15, 2006 from http://www.jva-net.or.jp/en/index.html

JEITA (Japan Electronics and Information Technology Industries Association). (2005). *Minnseiyou Dennsi kiki Deta Shyu.* Tokyo: JEITA.

Matsushita. Retrieved March 15, 2006 from http://panasonic.net/

Mitsubishi Electric. Retrieved March 15, 2006 from http://global.mitsubishi-electric.com/

National Science Museum, Tokyo. Retrieved March 15, 2006 from http://sts.kahaku.go.jp/sts/

NHK (Japan Broadcasting Corporation.). (2000). Project X: Challengers 1. Tokyo: Japan Broadcast Publishing.

Polin, B. A., Troutt, M. D., & Acar, W. (2005). Supply Chain Globalization and the Complexities of Cost Miminization. In C.K. Chan & H.W.J. Lee (Eds.), *Successful strategies in supply chain management* (pp. 109-143). Hershey, PA: Idea Group Publishing Co.

Rosenbloom, R. S., & Abernathy, W. J. (1982). The climate for innovation in industry. *Research Policy, 11*(4), 209-225.

Copyright © 2008, IGI Global. Copying or distributing in print or electronic forms without written permission of IGI Global is prohibited.

Sato, M. (1999). *The story of a media industry*. Tokyo: Nikkei Business Publications.

Schoenherr, S. Retrieved March 15, 2006 from http://home.sandiego.edu/~ses/

Sony Corporation. Retrieved March 15, 2006 from http://www.Sony.net/

Toshiba. Retrieved March 15, 2006 from http://www.toshiba.co.jp/index.htm

Troutt, M. D., & Acar, W. (2005). "A Lorenz-Pareto measure of pure diversification", *European Journal of Operational Research, 167*(2), 543-549.

Utterback, J. M. (1994). Mastering the dynamics of innovation: How companies can seize opportunities in the face of technological change. Boston: Harvard Business School Press.

Victor Company of Japan. Retrieved March 15, 2006 from http://www.jvc-victor.co.jp/english/global-e.html

Vollman, T. E., Berry, W. L., Whybark, D. C., & Jacobs, F. R. (2005). *Manufacturing planning & control for supply chain management* (5th ed.). New York: McGraw-Hill.

Copyright © 2008, IGI Global. Copying or distributing in print or electronic forms without written permission of IGI Global is prohibited.

Chapter VII

Development of Products

This chapter explains the advancement and the price decline of products based on the VCR case. After the dominant design emerges, the product advances incrementally or cumulatively because the dominant design sets a standard design of the product and a framework for the competition. Many new generation products appeared in the market with innovative functions to spur sales. Some of them became popular and others did not. In the VCR case, most consumers bought a monaural VHS machine and, then later, a HiFi VHS machine. On the other hand, most consumers did not purchase S-VHS, D-VHS, and other advanced machines because those were too expensive in comparison with their performance. As a result, the alternation of generations of the VCR occurred only once, from the monaural to the HiFi machine.

Copyright © 2008, IGI Global. Copying or distributing in print or electronic forms without written permission of IGI Global is prohibited.

Innovations and Productivity Dilemma

The industry-wide product development process can be explained from the viewpoint of innovation (Abernathy, 1978; Abernathy, Clark, & Kantrow, 1983). Innovations are mainly classified into two types, based on their characteristics, that is, product and process innovations. Usually, product innovations take place frequently in the early stages of the life cycle after the dominant design has emerged. They contribute to considerable enhancement of performance of a product. They are sometimes also destructive because they make obsolete old technologies and existing products by changing the basic design, components, production process, and usage of a product, such as cellular phones and portable games. In the middle stage, while product innovations gradually abate, process innovations frequently occur. Process innovations make the production system more efficient by sophisticating the product designs, components, and production processes. In the later half of the life cycle, a product innovation sometimes creates a brand-new category of products that makes an existing category of products obsolete. This phenomenon is called *de-maturity.*

Figure 1 shows the declining prices of three Product Series (PSs), 1 to 3. PS1 is the first series of the product which includes *the dominant design.* The dominant design is the first product to be widely adopted to satisfy latent customer needs. The Model T (developed in 1908) in the auto industry and the HR3300 (VHS format: developed in 1976) in the VCR industry, are good examples of a dominant design. The PS2 and PS3 series, which appear later, are more advanced versions of PS1. In time, the prices of the PS1 decline because of process innovations, the experience curve, and competition. At some point in time, a subsequent series is more expensive to produce than a previous one. However, the variation in price of subsequent series is less than that of preceding series. The impact of product innovations continues to further decrease; specifically, the process innovations, standardization, and modularization advance so much that it becomes possible to localize the impact of the change in the product design.

Figure 2 shows the progress of the product development. The vertical axis is an index, performance/cost ratio. At the beginning of each series, the cost (price) is very high. However, as the new functions are gradually incorporated, the production system becomes more efficient. As a result, performance/cost is enhanced as time elapses. However, after a certain point in time, the enhancement becomes regressive. Abernathy (1978) called the situation the

Copyright © 2008, IGI Global. Copying or distributing in print or electronic forms without written permission of IGI Global is prohibited.

Figure 1. Product series (PSs) and price (cost)

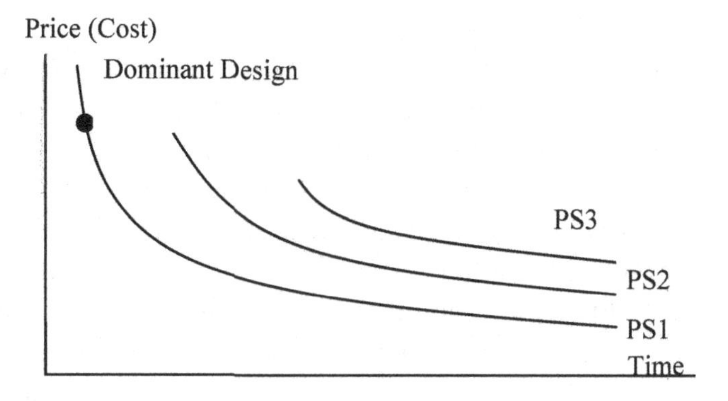

Figure 2. Product development

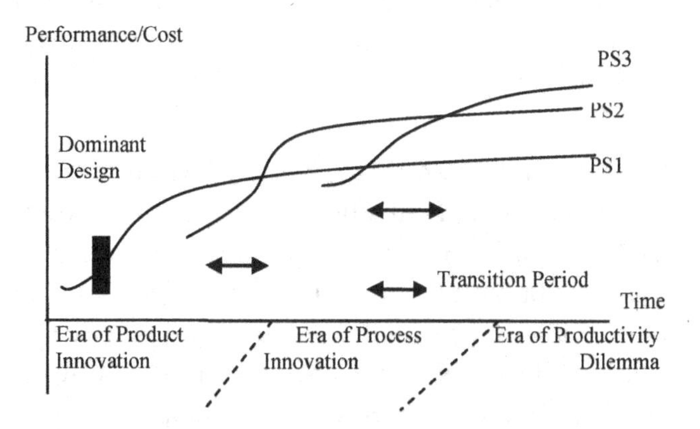

productivity dilemma. Although one expects the performance of a subsequent series to be higher than that of a preceding series, performance/cost does not always realize this expectation, depending on the point in time in the product development. Abernathy et al. (1983) called the period during which a previous series is superior to a subsequent one *the transition period.* A subsequent series is not always characterized by substantial enhancement, as PS3 shows in Figure 2. Sometimes, new series are launched for a commercial reason without any remarkable technical advance.

Copyright © 2008, IGI Global. Copying or distributing in print or electronic forms without written permission of IGI Global is prohibited.

Evolution of VHS Formats

JVC upgraded VHS formats several times, maintaining the compatibility among them. They were the platform and new functions were added to them. In other words, upgrading the formats was a major change and adding new functions a minor change. Whenever JVC upgraded the formats, the image or sound was improved very much and the usability of the VCR was expanded. Earlier formats had a limitation, which the formats incorporated originally. To overtake the limit, JVC should develop another format technically. On the other hand, to stimulate consumer interest, they launched new formats.

The first series of VHS was called the standard VHS or the monaural VHS. HR-3300 was a fair machine for the innovative customers such as high-tech aficionados and the business people. Although those segments spend much more money than ordinary people on the high-tech goods, they are small in numbers. Generally, the primary products had considerable room for improvement and needed to be further refined to satisfy ordinary people. But this was the only VHS format during 1976-1983, except the formats for the video camera.

In 1983, JVC introduced HR-D725, the first VHS machine that included the HiFi (High Fidelity Sound) function. HiFi functions were put on both the VHS and Betamax in 1983. Sony put them on Betamax slightly sooner than JVC and Matsushita put them on VHS. JVC added another magnetic head for the exclusive use of the sound. Thus, a high-quality FM sound was added to the standard VHS. This major change was very effective for customer satisfaction. This feature has been so widely adopted by other manufacturers and consumers that most VCRs came to incorporate this function.

In 1987, JVC launched the HR-S7000, the first S-VHS (High Resolution Super VHS). This new format featured high quality images as good as the one inch tape VCR in the broadcasting industry. This first S-VHS product's price was 220,000 yen. Even in the EP (long play) mode, the image was much finer than that of broadcasting. JVC believed the new format would stand for the next decade. JVC also developed the high quality cassette tape, S-VHS Cassette (Standard Recording Time was 120 minutes), for the S-VHS and the recording method which reduced noise. S-VHS adopted a terminal, S-Video, which is used widely in image equipment. However, S-VHS had a much smaller impact on the market than HiFi because of the performance/cost ratio. In addition to the fact that most customers were satisfied with HiFi,

Copyright © 2008, IGI Global. Copying or distributing in print or electronic forms without written permission of IGI Global is prohibited.

the price of the machine (220,000 Yen) and cassette (3,000 Yen, Standard Recording Time; 120 minutes) remained expensive.

In 1994, JVC introduced the HR-W1, the first W-VHS. It corresponded to the HDTV. Although it could use other VHS cassettes, JVC also developed a cassette which could make the most of the W-VHS's ability. The price of the machine was 620,000 Yen. The tape, WT-180, cost 5,900 Yen and could record about three hours in HD mode (nine hours in SD mode). The recording method was the same as S-VHS. Users could select the sound, HiFi sound or S-VHS Digital Audio. The sales of W-VHS machines were very low because of the expensiveness and the low diffusion rate of HDTV.

Figure 3 illustrates the price decline pattern of JVC products by each format. It was natural that the lowest price of the monaural (standard) VHS was cheapest among others at any time. However, the difference among them became smaller and smaller. In addition, and of especial interest, the lowest price of the each VHS format became lower cheaper than that of others. This demonstrates the advancement of the productivity and the productivity dilemma.

Figure 4 shows the price decline of the monaural products. There are two groups of companies. The first group, JVC and Matsushita, entered into the

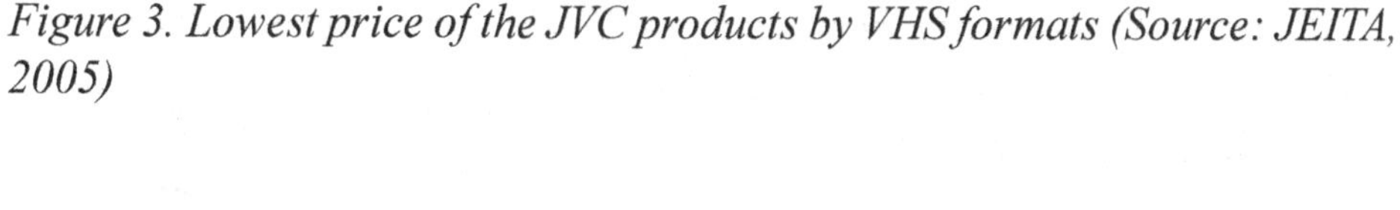

Figure 3. Lowest price of the JVC products by VHS formats (Source: JEITA, 2005)

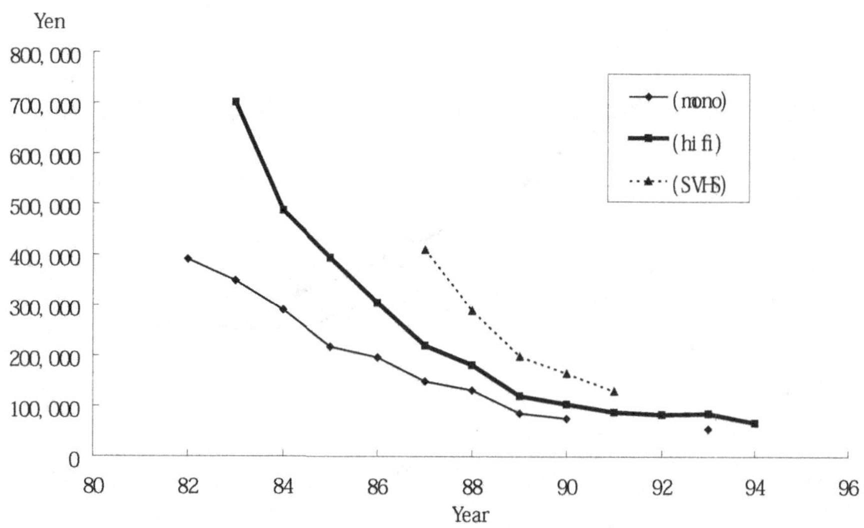

Copyright © 2008, IGI Global. Copying or distributing in print or electronic forms without written permission of IGI Global is prohibited.

market early and decreased their price gradually. On the other hand, the second group of companies, Funai and Aiwa, entered into the market late but sold the products at a very cheap price from the beginning. The price was so low that they did not have very much room for further cost reduction and that the first group did not want to compete with them because of their high cost structure. The first group tried to keep the price higher and the gap between the actual price and the theoretical lowest price gave the cost cutters the chance to enter into the market. Sanyo, a follower, entered into the market just behind the first group and the price of their products were close to that of the first group. However, they did not keep a wide selection. Sanyo launched the VZ-T100 experimentally at 120,000 Yen in 1993. This model could do random access playback. Figure 5 demonstrates the price decline of HiFi VHS. Figures 4 and 5 are very similar. The cost cutters could enter into the HiFi VHS market just after they entered into monaural VHS market and the price difference among them became smaller.

In 1997, JVC introduced the HM-DSR100DU, the first D-VHS (Digital VHS). It corresponded to digital TV broadcasting, such as MPEG-2, and could record TV programs as a bit stream. D-VHS kept compatibility with

Figure 4. Price decline of monaural VHS (Source: JEITA, 2005)

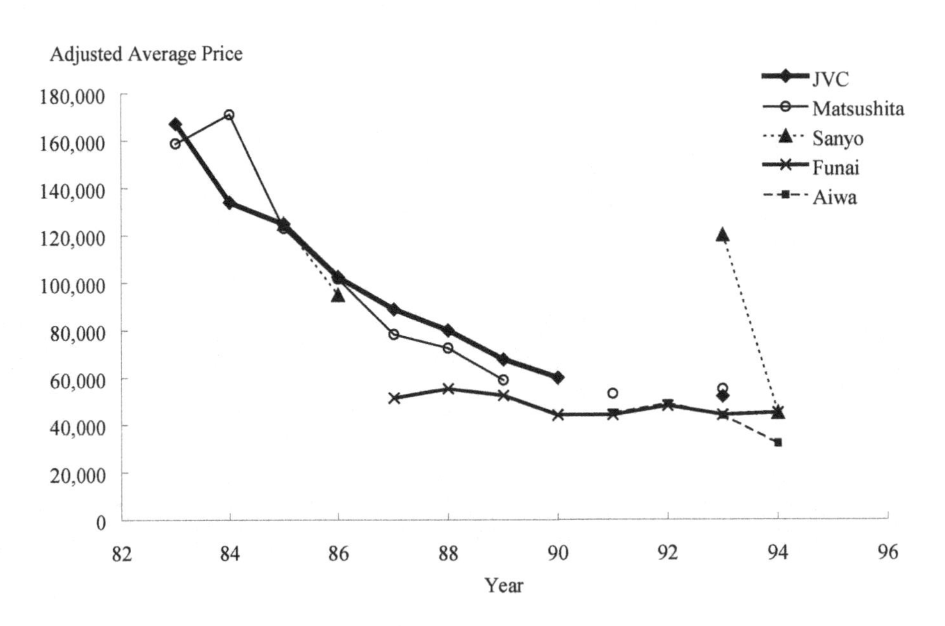

Copyright © 2008, IGI Global. Copying or distributing in print or electronic forms without written permission of IGI Global is prohibited.

Figure 5. Price decline of HiFi VHS (Source: JEITA, 2005)

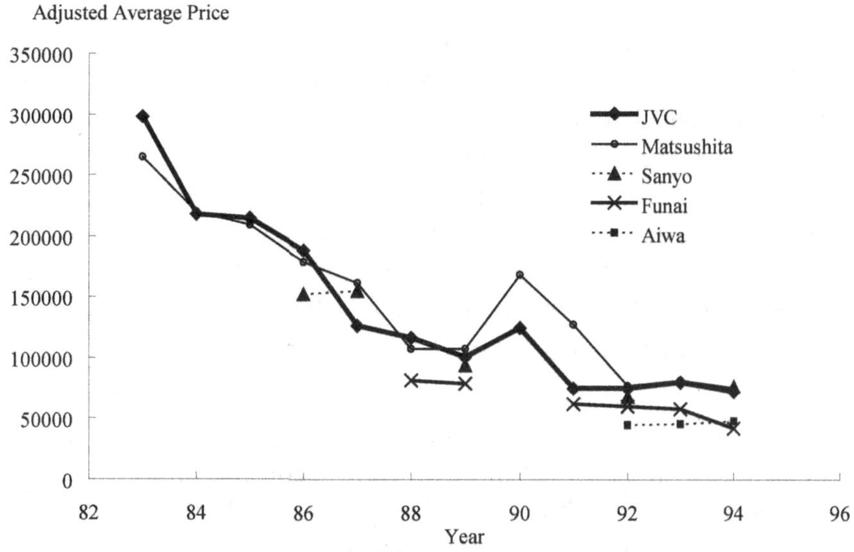

Table 1. Advancement of VHS format

Format	Year	Model	Image	Sound
Monaural VHS	1976	HR-3300	Standard	Monaural
HiFi VHS	1983	HR-D725	Standard	HiFi
S-VHS	1987	HR-S7000	S-VHS	HiFi
W-VHS	1994	HR-W1	S-VHS	HiFi / S-VHS Audio
D-VHS	1997	HM-DSR100DU	Digital / S-VHS	Digital Audio

other analogue VHS formats and extended the recording time dramatically. Table 1 summarizes the development of VHS format to D-VHS. JVC also developed a D-VHS Cassette, DF-480 (6,200 Yen; 480 minutes in the standard mode) and DF-300 (2,000 Yen; 300 minutes in the standard mode), which was the same size as other VHS formats. Unfortunately, other digital media, such as the DVD and the HDD recorder, became predominant gradually in the next generation recorders. This digital technology is therefore finishing off the VHS, which was analogue medium.

Copyright © 2008, IGI Global. Copying or distributing in print or electronic forms without written permission of IGI Global is prohibited.

New Functions

The HR-3300, the first VHS machine, satisfied the minimum requirements for commercialization. These may be regarded as the quality of image and sound, stableness, the durableness, the recording function, adequate recording time, the right size, weight, and price. Although HR-3300 was an excellent product at that time, the performance was simple. VHS had a big margin for the improvement. VCR manufacturers added new functions to the VHS format for development. Some of these functions belonged to the product innovations, such as the new versions of VHS and the extension of the recording time, and others were complementary.

In the 1970s, companies added new functions, which were adopted widely. To the VHS format, JVC added the double-speed playback mode with monaural sound in 1977 for the first time. The triple-recording-time (6-hour) mode was developed in 1978, while maintaining compatibility among other versions. The triple-recording-time mode with the high quality image saved the use of the video cassettes. These functions were essential to the development of VHS and different from other complementary functions the other JVC group members contributed.

In the 1980s and 1990s, to differentiate their products from other company's products and to avoid price wars, the leading companies and followers continued to add new functions. These included 24-hour quick programming, the jog shuttle control, the full-function remote control, the auto index, child lock, and others. The trend started with HR-3500 and HR-3750 which launched in 1979. The HR-3500 could playback in slow-motion with a clear image and the HR-3750 coped with the sound multiplex broadcasting. VHS products became multi-functional and opened to utilization in various ways. However, as time passed, the impact and necessity of these new functions became smaller and smaller. Some models were equipped with a certain function and others were not. Furthermore, cost cutters eliminated the extra functions and simplified the product for cost reduction and the consumer's convenience.

In the 1990s, JVC added new functions to keep pace with the advancement of the digital world. In 1992, JVC introduced the HR-Z1, which could record and playback digital sound from digital audios, which was a precursor of the D-VHS. Then, JVC added some functions, such as HDMI (High Definition Multi-Media Interface) and MPEG-2 CODEC encoding. They reinforced the coordination with the PC and the home theater performance connecting to

Copyright © 2008, IGI Global. Copying or distributing in print or electronic forms without written permission of IGI Global is prohibited.

HDTV and high resolution projector systems. JVC also improved the S-VHS Cassette and introduced XG120 for HR-Z1. However, consumers chose the DVD and HDD recorders for digital broadcasts.

Advancement of Key Parts

The minimum requirements for commercialization were considered to be: the quality of image and sound, the stability of the image, the durability, the recording function, the recording time, the size, the weight, and the price. To satisfy these, companies advanced and improved individual parts and united them repeatedly. In case of the VCR, companies upgraded the functions and the quality of image and sound first, then reduced the size, weight and the price, and finally increased the stability and durability. In the VCR, the price declined to the target level earlier than the weight as Figure 6 (a) and (b) show. The weight downsizing took a longer time than the price reduction because the down sizing process included the design changes, the reduction of and downsizing of parts, and required feedback and cooperation from the various supply chain partners. The VCR products approached to the de facto standard, VHS, step by step.

Before the modular manufacturing system became established, the advancement of the key parts brought about major changes of the product design and increased the risk of obsolescence of the existing facilities. After considering the advantages companies promoted modular approach. With this advance they could keep a wide variety of products efficiently and reduce the impact of the changes of the parts simultaneously because the parts were exchangeable. On the other hand, it set the basic design and regulated or constrained the possibilities. It is very important for any product to start the modular manufacturing system at the right time. If it was too early, companies should change it later. If it is too late, companies would suffer from the inventory to keep the wide variety of products.

After the modular manufacturing system became established, companies could upgrade parts individually. In particular, without requiring the nonmodular or full coordination among all parts, companies could advance the key parts individually. As a result, manufacturers could shift from monaural VHS to HiFi VHS smoothly. They replaced some parts related to the monaural sound to those of HiFi sound. However, many companies did not shift from HiFi

Copyright © 2008, IGI Global. Copying or distributing in print or electronic forms without written permission of IGI Global is prohibited.

Figure 6 (a). Shift of the price of JVC products before HR-3300

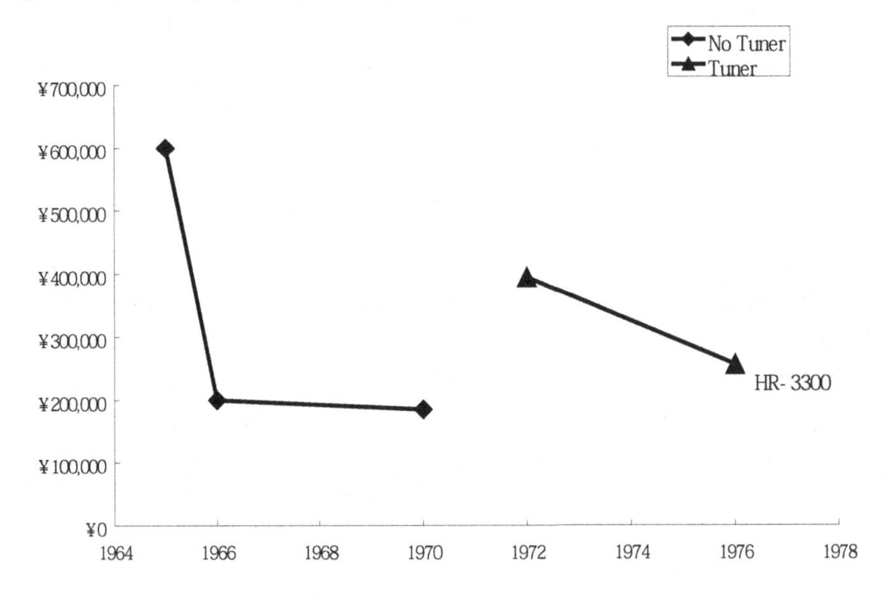

Figure 6 (b). Shift of the weight of JVC products before HR-3300

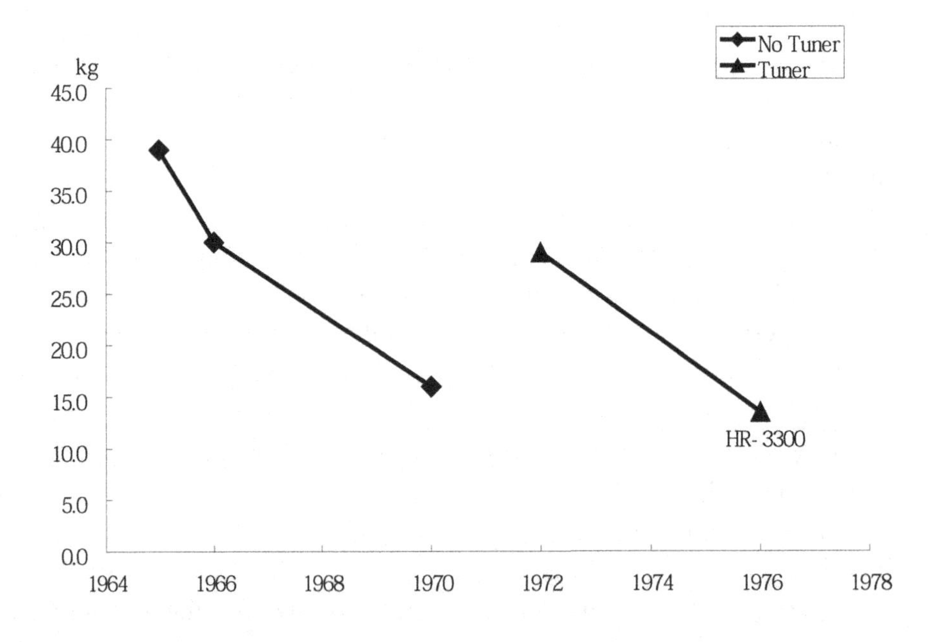

Copyright © 2008, IGI Global. Copying or distributing in print or electronic forms without written permission of IGI Global is prohibited.

VHS to S-VHS because the high resolution images required a lot of specific parts and major changes. Therefore, the price of the S-VHS was so expensive that the majority of customers did not want to adopt it as well as the D-VHS. In fact, the modular manufacturing system guaranteed the advancement of individual parts to some extent. For example, Sanyo, who introduced S-VHS latest among the major companies, introduced and sold IC chips. Sanyo reduced the price of parts LA7317 and LA7322/7322M to 500 Yen in 1987 and 1988. The numbers of pins in these were reduced very much to simplify the assembly process. They also introduced the LA7380 at 900 Yen which consolidated many chips into one. Aiwa made efforts to reduce the number of print boards by simplifying the design. Funai decided to focus on becoming an excellent assembler. Their R&D force concentrated on the simplification of the design for easy assembly and high quality production. The reduction of the number of parts linked with the simplification of the assembly process and the cost reduction were attained. Modular manufacturing thus made it possible for many companies to compete in the market.

In addition, the modular manufacturing system supported a wide variety of products. Companies created new parts with new functions or better performance to differentiate their products from others. They could propose and support a wide range or scope of products at reasonable costs due to the modular approach. They could save parts and products inventory costs by utilizing sharable parts and components, as well as using postponement strategies. The specification of the products could be determined just before shipment, thereby permitting an improved degree of customization. Companies could make various combinations by picking up the required parts and assembling them on the same line. Otherwise, the lead time would have been longer because it took a time to complete a product from the beginning or they would need a huge inventory to shorten such lead times.

Progress of Process Innovation

After a specific product design was determined, innovations occur frequently. The change of the basic design sometimes has a great detrimental effect on the production system. Once the process innovation starts, companies could not change the basic design easily. Generally, it leads to the mass-production system first and, then, the flexible manufacturing system. It promotes

Copyright © 2008, IGI Global. Copying or distributing in print or electronic forms without written permission of IGI Global is prohibited.

manufacturing cost reduction and then severe competition by drawing new competitors.

The increase of the yield is a good indicator to measure to what degree the process innovation progresses or how far the mass-production system perfects. Companies do not make a big investment into the manufacturing facility after they establish their standard process and establish the supply chain. 1980 was the first year that production volume of the home-use VCR exceeded the deci-peak (10 percent of its peak in 1986). The establishment of the mass-production system is a primary key word in the beginning of the process innovation.

The other keyword is the localization of the impact of change. After the completion of the manufacturing facility, companies add new functions for differentiation. If they can not localize the impact, they should decide whether they continue to manufacture the same product or whether they should change the manufacturing facility. With some new functions, such as the double speed playback and the triple-recording-time mode, in the late 1970s, the manufacturing facilities could absorb the impact of change. When the HiFi sound function appeared, the manufacturing facilities could work so smoothly that many manufacturers started producing HiFi machines quickly because it had little impact on the production system in spite of the addition of the extra head.

The performance of the process innovation appears also in the reduction of production cost. The production costs of home-use VCRs decreased rapidly due to the multiple effects of the process innovation, the economies of scale, and the learning curve effects. The production cost per unit (deflated by the GDP deflator) continued to show a period of tremendous decline and had been decreased by half by the early 1980s.

After the process innovation progresses to some extent, new competitors can enter into the market easily. The production process has been standardized so that new competitors can start manufacturing quickly and get parts though the component market. Most of them focused on the low-end products or the niche market. They boost the growth of the production volume rapidly and the price war by driving other advanced companies out of the low-end market. By the early 1980s, 20 companies manufactured VCRs. Cost-cutters such as Funai and Aiwa entered the market after the production system had been specified and focused on the low-end products. As a result, the top three companies, JVC, Matsushita, and Sony, who collectively maintained 90 percent of the market share in 1976, lost about 30 percent of their market share.

Copyright © 2008, IGI Global. Copying or distributing in print or electronic forms without written permission of IGI Global is prohibited.

Advanced companies should have a wide variety of goods in the middle range and high-end products. Otherwise, they would lose against newcomers and other advanced competitors. The average numbers of models JVC introduced in a year were 2.2 during 1976-1980, 3.2 during 1981-1985, 8.4 during 1986-1990, 6.2 during 1991-1995, 13 during 1996-2000, and 14.8 during 2001-2005. These figures show the necessity and manufacturability of the wide variety of products depending on the market condition. In the later half of the process innovation, the key activity is the establishment of the flexile and efficient manufacturing system for the variety of products and the variance of the demand.

Effect of Entry Timing on Price and Width of Selection

Entry timing has a great effect on company behavior. It is true that a company who owns the latest technologies and great market power tends to exploit the new market, but the company strategy determines the entry timing. The strategy realized depends on the price of the products and the width of selection for the consumers.

Figure 7 shows the relation between entry timing and the weighted average price. The entry timing was calculated by the following way. First, we iden-

Figure 7. Entry timing and weighted average price

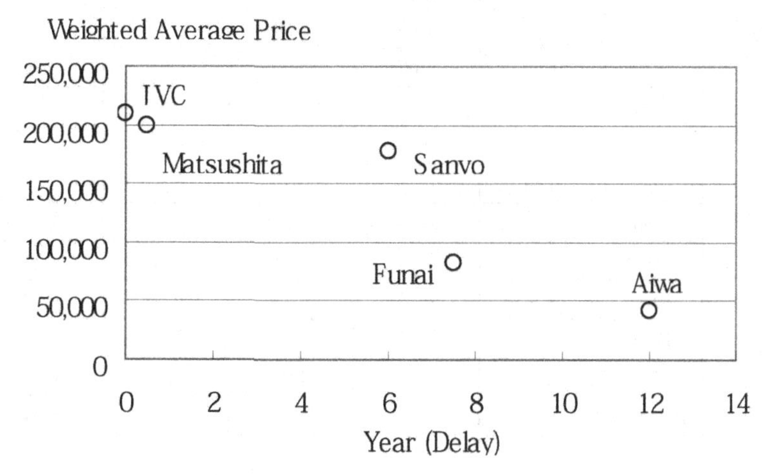

Copyright © 2008, IGI Global. Copying or distributing in print or electronic forms without written permission of IGI Global is prohibited.

tify the first year in which companies introduced the monaural (standard) VHS and HiFi VHS. Second, the delay, the difference to the first one, was calculated in each company and category. Finally, the delays of two categories are averaged by each company. The weighted average price of the products during 1983-1994 is calculated for each company, which is adjusted by the GDP deflator. A company that entered into the market earlier tends to have a higher value of the index because the initial products are generally expensive. However, this figure shows the unique strategy of cost cutters. The weighted

Figure 8. Entry timing and width of selection

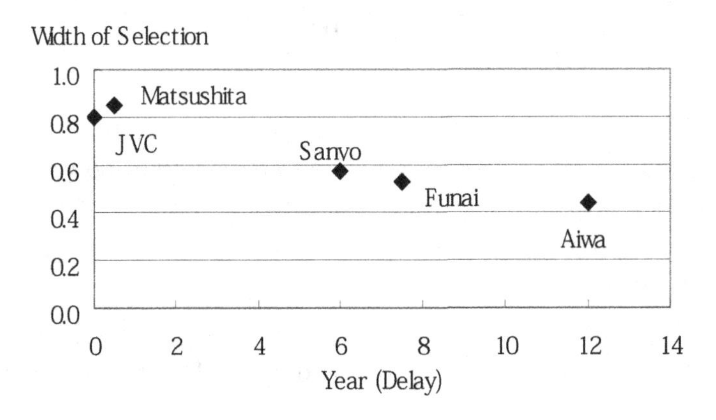

Table 2. Cover rate of four segments

	JVC	Matsushita	Sanyo	Funai	Aiwa
1985	1.00	1.00	0.50	-	-
1986	1.00	1.00	1.00	-	-
1987	0.75	0.75	0.75	0.50	-
1988	1.00	0.75	0.25	0.50	-
1989	0.75	0.75	0.50	0.50	-
1990	1.00	0.75	0.25	0.50	-
1991	0.75	1.00	0.25	0.50	0.25
1992	0.50	0.75	0.50	0.50	0.50
1993	0.75	1.00	0.75	0.50	0.50
1994	0.50	0.75	1.00	0.75	0.50
Average	0.80	0.85	0.58	0.53	0.44

Copyright © 2008, IGI Global. Copying or distributing in print or electronic forms without written permission of IGI Global is prohibited.

average prices for this group were much lower than those of the others. They waited until they could get most parts though the market and entered the market with the low-end products instead of the high-end ones because they focused on price competition rather than the brand image.

Figure 8 demonstrates the relation between market entry timing and the width of product selection. The width of selection is calculated in the following the way. First, all products were sorted into four categories: monaural VHS, HiFi VHS, S-VHS, and BS-HiFi VHS. Second, we identified the cover rate of all segments in each year and company as Table 2. Finally, we calculated the average of each company. From Figure 8, we confirmed the following things. First movers covered a much larger segment than others, but they did not cover all segments. It was not reasonable for them to keep the low-end products when the price competition became severe. Second, Matsushita tried to keep a wide selection of products because they wanted to emphasize sales promotions. Finally, a company who entered into the market later tended to either focus on the low-end or squeeze the width of selection.

Productivity Dilemma

After the middle of the 1980s, some signs of the productivity dilemma in the home-use VCR industry could be seen, wherein the opportunities for product and process innovations of the product diminish. Although it is very difficult to make a distinct line when the productivity dilemma occurred, the impact of new products and functions became very weak.

First, the beginning of the productivity dilemma is the suspension of product development. The innovative new functions, which had novelty and were popularized, seldom were added to the product after the middle of 1980s. That is, most innovative functions had been added by the middle of 1980s, the timer (1976), the double-speed playback with the sound (1977), the triple-recording-time mode (1979) and HiFi sound (1983). Almost all home-use VCRs incorporate these functions. Functions added after the middle of 1980s can be divided into two categories: the extension of existing functions and the minor functions. For examples of the former functions, JVC introduced some products, which could make eight reservations for two weeks. In 2001 and 2002, JVC introduced two products which could make 32 reservations for a year. However, the average models introduced in the early 2000s could

Copyright © 2008, IGI Global. Copying or distributing in print or electronic forms without written permission of IGI Global is prohibited.

Figure 9. Shift of domestic shipment during 1985 to 2003 (Source: JEITA, 2005)

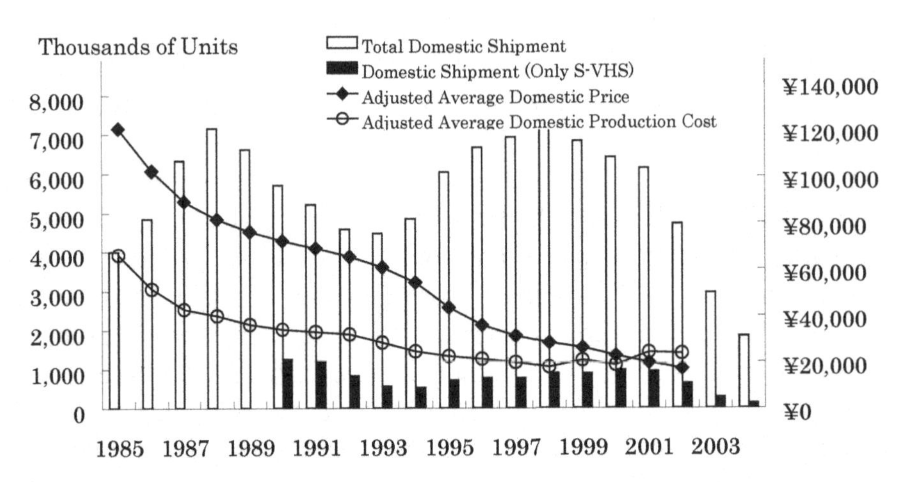

make eight programs for a month or a year. The double speed playback with the sound function upgraded to the seven-times speed playback. Most DVDs adopt up to the double speed playback because more than double speed playback is too fast for us to understand. Although S-VHS, W-VHS, and D-VHS improved the quality of the image and the sound and extended the recording time, these could not be popularized as well as HiFi VHS (Figure 9). HiFi VHS has occupied more than 80 percent of the total domestic shipments since 1991. Examples of the latter functions were G-code and Auto CM Cut, which emerged in the early 1990s. These were fresh but not necessary because some users used them and others did not. Average products which were introduced in the middle of the 1980s incorporated the all essential functions for the home-use VCR.

Second, the middle of the productivity dilemma is the slowdown of the cost reduction rate. The reduction rate of the manufacturing cost of the home-use VCR slowed down in 1987. While the domestic production cost per unit (in Japan) decreased more than 20 percent before 1987, the rate fluctuated 6 percent to 15 percent after 1987. In addition, the savings by further cost cutting became very small, because the total cost per unit had become small. This suggested that the limitation of the efficiency of the production system had been essentially achieved.

Third, the last of the productivity dilemma is when foreign production in the low cost operation areas becomes the mainstream. In 1994, foreign produc-

Copyright © 2008, IGI Global. Copying or distributing in print or electronic forms without written permission of IGI Global is prohibited.

Table 3. Development of magnetic videotape recorders during 1950-1987

	Company	Event
1950	Crosby Enterprises	Crosby Enterprises developed a magnetic TV recorder at first.
1951	Crosby Enterprises	Crosby Enterprises demonstrated an experimental 12-head VTR.
	Ampex	Ampex started R&D on VTR.
1953	RCA	RCA demonstrated a 3-head VTR.
1956	Ampex	Ampex succeeded in developing a first practical VTR.
	CBS	CBS utilized an Ampex VTR for broadcasting at first.
1957	Ampex & RCA	Ampex and RCA pooled VTR patents.
1958	NHK	NHK succeeded in prototyping a 4-head VCR.
	Sony	Sony developed a 4-head VCR.
1959	Ampex	Ampex launched the first mobile VTR
	Toshiba	Toshiba demonstrated the helical scan 1-head VTR.
	Shibaura	Shibaura started selling a broadcasting station VTR.
1960	Ampex & Sony	They made a cross license about VTR technologies and the transistorized circuit technologies.
	JVC	JVC introduced helical scan 2-head VTR.
1961	JVC	JVC introduced helical scan color 2-head VTR.
	Ampex	Ampex launched 1-head VTR and undersized VTR.
1962	Sony	Sony developed a 1/2 inch miniature VTR.
	Sibaura	Sibaura made a transistor VTR for broadcasting experimentally.
1963	Sony	Sony launched PV100 at 3,350,000 Yen
	JVC	JVC launched KV-200 at 2,000,000 Yen
1964	Matsushita	Matsushita developed transistor VTR whose size was almost same as a trunk.
	Shibaura	Shibaura developed a VTR which could playback slow-motion pictures.
1965	Sony	Sony launched CV2000, a home-use 1/2 inch VTR, at 198,000 Yen.
	Philips	Philips introduced a compact cassette audio recording and playback.
	Ampex & Toshiba	Ampex and Toshiba developed VR-2000, a high-band color VTR.
	Shibaura	Shibaura developed SV1000, a stationary head VTR, whose price was 90,000 Yen.
1966	Matsushita & Sony	Matsushita and Sony developed a home-use color VTR separately.

continued on following page

Copyright © 2008, IGI Global. Copying or distributing in print or electronic forms without written permission of IGI Global is prohibited.

Table 3. continued

1967	EAIJ	EAIJ (Electric Industries Association of Japan) started working on the standardization of the VTR.
1968	Hitachi & Sanyo	Hitachi and Sanyo started R&D on the VTR separately.
1969	EAIJ	EAIJ standardized the format of the monochromatic VTR.□
	Sony	Sony developed the first 3/4 inch color VCR, Video Cassette Recorder, whose format was called U-Format.
	Matsushita	Matsushita developed a 1/2 inch magazine VTR.
1970	Ampex & Toshiba	Ampex and Toshiba introduced the Instavision, a color cartridge VTR.
	Philips	Philips introduced a VCR in Europe.
	AVCO	AVCO introduced a CartriVision, a VCR.
1971	JVC & Matsushita	They reached an agreement on the adoption of U Format VCR Sony had developed.
	EAIJ	EAIJ proposed a recommended format for the color VTR.
1972	Matsushita & Sanyo	Matsushita and Sanyo separately developed a 1/2 inch cartridge VTR based on the format EAIJ recommended.
	Sears and Wards	Sears and Wards started selling CartriVision.
1973	EAIJ	EAIJ acknowledged three formats produced by Sony, Philips and Sanyo.
	Matsushita	Matsushita sold NV5120, 1/2 inch cartridge VTR at 310,000 Yen.
1974	Sony	Sony launched VC3900, portable U Format VCR.
1975	Sony	Sony launched SL-6300, the first Betamax.
	Matsushita	Matsushita sold their original VCR model, VX100, in the Shikoku district.
	Sanyo	Sanyo launched their original VCR model, VTC7225.
1976	JVC	JVC launched HR-3300, the first VHS.
1977	Sony	Sony upgraded Betamax to extend the recording time to 120 minutes.
	Matsushita	Matsushita delivered 4-hour VHS VCR to RCA.
1978	Pioneer	Pioneer developed a laser disk.
	Sony	Sony introduced a portable consumer camcorder, SL3100 and HVC1100.
1979	Sony	Sony introduced the Betascan which could project the fast-forwarding mages.
	Philips & Grundig	Philips and Grundig jointly developed 8-hour recording time VCR.
1980	Funai	Funai developed 1/4 inch portable VCR.
	Matsushita	Matsushita three U Format VCR for broadcasting.
	Sharp	Sharp sold a combine product of the VCR and TV at 268,000 Yen.

continued on following page

Copyright © 2008, IGI Global. Copying or distributing in print or electronic forms without written permission of IGI Global is prohibited.

Table 3. continued

1981	Matsushita	Matsushita developed a 240-hour recording time VCR.
1982	Toshiba	Toshiba developed an 8mm VTR.
1983	Sony	Sony introduced, SLHF77, Beta HiFi VCR
	JVC	JVC introduced, HRD725, HiFi VHS VCR.
1984	Kodak	Kodak introduced an 8mm VTR.
1985	JVC	JVC introduced HRD565, a high quality model.
	Sony	Sony introduced SLHF900, a high band Beta model.
	Toshiba	Toshiba introduced a high quality model with a digital memory.
1986	Matsushita	Matsushita introduced NUG21 which could reserve the programs by the bar code.
1987	JVC	JVC introduced HR-S7000, the first S-VHS VCR.

tion exceeded the domestic production in total units. To further reduce costs, while Japanese companies started reducing the number of factories in the world, they increased or reinforced the factories in Asia, where almost all of them manufactured the majority of parts and assembled the VCR units. This shows that it is very difficult for any company to maintain a lead against others by the excellence in their production system at that time. Rather, they exploited the economies of scale in the low cost operation areas.

SCM for Product Advancement

The entire supply chain was a powerful tool for the product development as well as the laboratory. Technologies and market needs are the drivers for the new products. New products driven by the latest certain technologies need the peripheral technologies to be realized. For commercial success, the coordination within the supply chain, particularly the upstream supply chain, is the key. Without it, it would take too much cost and time to succeed in the market. The downstream also plays a very important role in the middle and late product development phase. Companies set the target product specifications and price, and then start a full-dress investment in the facilities based on the information from the downstream supply chain. After the basic design is

Copyright © 2008, IGI Global. Copying or distributing in print or electronic forms without written permission of IGI Global is prohibited.

determined, companies should prepare for the differentiation. It is a critical factor for any company and supply chain to understand the market needs and to launch new products which realize those appropriately.

In the VCR case, companies put the color recording to practical use. It was possible for strong companies to realize a certain function. However, to satisfy the minimum requirement for the commercial success, VCRs should have been much smaller, cheaper, and more stable than the existing models. To do those efficiently and quickly, the headquarters companies had to coordinate with downstream suppliers. JVC gathered the suppliers in their VCR factory and asked them to supply the parts for the prototypes repeatedly. VCR products became smaller than before by trial and error. Therefore, suppliers also shared large risks with their headquarters in product development. Once the headquarters judged that their products met the minimum requirement for commercial success, some suppliers needed to make molds which were very expensive and burdensome to them. Many companies had started mass production and gambled to explore the initial home-VCR market. Some of them went bankrupt and others suffered losses because the downstream supply chains had had great losses. Miniaturization was the critical factor for the VCR industry, which needed the close coordination among downstream supply chain partners.

After VHS overtook Betamax, VHS group members started severe competition with each other. Their basic designs were the same because all of them had adopted the VHS format and made cross license agreements. Only JVC could upgrade the VHS format and others could only differentiate their products by minor devices. In this case, customer needs became the most powerful source for effective product differentiation. The upstream supply chain made it possible to add new functions without major changes. Although various minor functions were added to the VCR, most of them disappeared quickly. The product advancement came to be driven by the upstream supply chain rather than the downstream supply chain.

The price decline induced the global outsourcing. But, once companies utilized global outsourcing, they lost the flexibility to make major changes. The timing and the degree of the global outsourcing were different by company types. The leading companies held global sourcing to a low level even in the maturity phase, because if they succeeded in de-maturing, the market would be active and they could profit from the new products. To do so, they should restrain global outsourcing and retain flexibility as a whole supply chain. Sometimes a leading company might miss the next trend because of success

Copyright © 2008, IGI Global. Copying or distributing in print or electronic forms without written permission of IGI Global is prohibited.

in the past. JVC is a good example. JVC earned an honorable achievement in the analogue TV and VCR but they missed the trend toward the digital world. JVC continued to be in the red in the middle 2000s and Matsushita disposed of its JVC stocks in 2007. The followers utilized global sourcing effectively and, on the other hand, they tried to launch alternative products earlier than the leading companies. The cost cutters also utilized global sourcing and prepared to ride on the next trend. SCM related to global sourcing constrained the direction companies could take.

Conclusion

The development of the products continues to be a vital activity even after a dominant design emerges. A dominant design satisfies minimal requirements for commercial success but is still immature. Companies have much room for improving the product performance and adding new functions to a new product. The product can advance incrementally and cumulatively because the dominant design sets a standard design of the product and a framework of the competition. To spur the market sales, new generation products would appear in the market multiple times and some of them make the alternation of generations. However, the impact of the later generation becomes smaller and smaller in the performance/cost ratio because, although the price continues to go down by the establishment of the mass production, the experience-curve effect, and the economies of scale, companies have a smaller space for the improvement of the product performance than before.

In the VCR case, after the monaural VHS, the first generation of VHS, became a dominant design, many new functions were added to it. Furthermore, JVC launched new generation products a couple of times, HiFi, S-VHS, W-VHS, and D-VHS. Many consumers bought a monaural VHS machine and, then, replaced to a HiFi VHS machine. On the other hand, a limited number of consumers purchased S-VHS, D-VHS, and other advanced machines. Those were too expensive in comparison with their performance/cost ratio for the majority of people to purchase. In the VCR, the distinct alternation of generations occurred only once, from the monaural to the HiFi machine.

Copyright © 2008, IGI Global. Copying or distributing in print or electronic forms without written permission of IGI Global is prohibited.

References

Abernathy, W. J. (1978). *The productivity dilemma.* Baltimore: The John Hopkins University Press.

Abernathy, W. J, Clark, K.B., & Kantrow, A.M. (1983). *Industrial renaissance*. New York: Basic Books.

Christensen, C. M. (1997). *The innovator's dilemma.* Boston: Harvard Business School Press.

Economic and Social Research Institute, Cabinet Office, Government of Japan. Retrieved March 15, 2006 from http://www.cao.go.jp/index-e.html

Electronics Industries Association of Japan. Retrieved March 15, 2006 from http://www.jeita.or.jp/eiaj/english/index.htm

Funai. Retrieved March 15, 2006 from http://www.funaiworld.com/index.html

Higuchi, T, Troutt, M. D., & Polin, B. A. (2004). Life cycle considerations for supply chain strategy. In C.K. Chan & H.W.J. Lee (Eds.), *Successful strategies in supply chain management* (pp. 67-89). Hershey, PA: Idea Group.

Hitachi. Retrieved March 15, 2006 from http://www.hitachi.com/

Hitachi. (1985). *Hitachi seisakujo shi 4* (The history of Hitachi, Part 4), Tokyo: Hitachi.

Itami, H. (1989). *Nihon no VTR sanngyou naze sekai wo seiha dekitanoka.* Tokyo: NTT Publishing.

Japan Electronics and Information Technology Industries Association. Retrieved March 15, 2006 from http://www.jeita.or.jp/english/index.htm

Japan Victor. (1977). *Nippon Victor 50 nennsi.* Kanagawa, Japan: Japan Victor.

Japan Victor. (1987). *Nippon Victor no 60 nenn.* Kanagawa, Japan: Japan Victor.

Japan Video Software Association. Retrieved March 15, 2006 from http://www.jva-net.or.jp/en/index.html

JEITA (Japan Electronics and Information Technology Industries Association). (2005). *Minnseiyou Dennsi kiki Deta Shyu.* Tokyo: JEITA.

Matsushita. Retrieved March 15, 2006 from http://panasonic.net/

Copyright © 2008, IGI Global. Copying or distributing in print or electronic forms without written permission of IGI Global is prohibited.

Mitsubishi Electric. Retrieved March 15, 2006 from http://global.mitsubishi-electric.com/

Mitsubishi Electric. (1982). *Mitsubishi denki Kennkyuujo 60 nennsi*. Tokyo: Mitsubishi Electric.

Mitsubishi Electric. (1986). *Mitsubishi Denki Kennkyuujo 50 nennsi*. Tokyo: Mitsubishi Electric.

National Science Museum, Tokyo. Retrieved March 15, 2006 from http://sts.kahaku.go.jp/sts/

NHK (Japan Broadcasting Corporation.). (2000). *Project X: Challengers 1*. Tokyo: Japan Broadcast Publishing.

Rosenbloom, R. S. & Abernathy, W. J. (1982). The climate for innovation in industry. *Research Policy, 11*(4), 209-225.

Sanyo Electric. (2001). *Sanyo denki 50 nennsi*. Osaka, JPN: Sanyo Electric.

Sato, M. (1999). The story of a media industry. Tokyo: Nikkei Business Publications.

Schoenherr, S. Retrieved March 15, 2006 from http://home.sandiego.edu/~ses/

Sony Corporation. Retrieved March 15, 2006 from http://www.Sony.net/

Toshiba. Retrieved March 15, 2006 from http://www.toshiba.co.jp/index.htm

Utterback, J. M. (1994). Mastering the dynamics of innovation: How companies can seize opportunities in the face of technological change. Boston: Harvard Business School Press.

Victor Company of Japan. Retrieved March 15, 2006 from http://www.jvc-victor.co.jp/english/global-e.html

Copyright © 2008, IGI Global. Copying or distributing in print or electronic forms without written permission of IGI Global is prohibited.

Chapter VIII

Emergence of Destructive New Technologies

In this chapter, we discuss the emergence of alternative products. Under conditions of severe competition, companies launch alternative products to gain the initiative and to promote increased sales, although the R&D and manufacturing cause large cost increases. In the case of the VCR, many alternative products, such as EVR, TED, and the laser disk, have existed since the 1960s and were improved many times in attempts to beat VHS. However, VHS (a magnetic recording system) remained the de facto standard until the DVD and HDD recorders replaced the VHS around 2000. Here we explain the reasons why VHS products were able to become the de facto standard for a few decades and how the alternative products, DVD and HDD recorders, ultimately replaced the VHS.

Copyright © 2008, IGI Global. Copying or distributing in print or electronic forms without written permission of IGI Global is prohibited.

Source of Longevity

The life cycle of VHS, a de facto standard of the home-use VCR, is about three decades. Those of the automobile, the TV, the washing machine and the refrigerator are much longer. The demand for the automobile has continued with vigor for about a hundred years. However, that of the home-use VCR was very long considering the existence of alternative products, such as LD (Laser Disc), CD Video, and so on. The demand for the device which can record TV programs and play back prerecorded media would continue until we can access to the huge video storage through the Internet or cable TV at any time and for a reasonable price. Other factors of the longevity, except for the practicability of the product, were divided into two categories: the industry and market side.

From the industry side, the longer the life cycle, the better it is. The longer life cycle makes it easier for companies to recover the cost easily and earn a big cumulative profit. To enlarge the life cycle, the industry must control the diffusion rate and should exclude the alternative products. The home-use VCR market was a seller's market for the first decade because the number of manufacturers was limited. Japanese companies almost monopolized the industry from the beginning to end. In addition to the limited number of the companies, the industry had a core, VHS format, on which the product development advanced orderly based. The market was not confused by other formats after VHS became de facto standard. Those companies did not want to bring out new alternative products by themselves while they had the enough and stable profit in the market, in which various types of companies had the enough space to earn their living.

From the market side, the short life cycle was not preferable because users bought machines and owned recorded cassettes, which were their property. They did not want their property to become obsolete quickly. Most products gradually come up in quality and decline in price as time passes, so that many users repeated the purchase a few times over the years. In addition, some users who were weak in the manipulation did not want to make the change to an alternative product. On the other hand, others did not want the long life cycle because they wished for a new product after the long doldrums in the product development.

The home-use VCR enjoyed the longevity. The practicality of video storage and playback machines continues. Although the alternative products attacked the home use VCR many times, it pushed them aside. The industry

Copyright © 2008, IGI Global. Copying or distributing in print or electronic forms without written permission of IGI Global is prohibited.

Table 1. Sources of longevity

Item	Attribution	Effect on Longevity
Practicability (Fashion)	Product	Positive (Negative)
Expandability	Product	Positive
Orderly (Cut Throat) Competition	Industry	Positive (Negative)
Technological Revolution	Industry / Outside	Negative
High (Low) Switching Cost	Industry / Market	Positive (Negative)
Alternative Product	Industry / Market	Negative
Attachment (Capriciousness)	Market	Positive (Negative)
Quick Diffusion (Repeat Purchase)	Market	Negative (Positive)
Environmental Change	Outside	Positive / Negative

developed steadily and orderly because the number of companies was limited and the VHS format had been a pillar for three decades. The defense of the VHS format meant that the property the home-use VCR owner had also was protected. As a result, the switching cost increased to such a high level that other alternative products encountered the difficulty.

Alternative Products

The home-use VCR had been surrounded and attacked by many alternative products since its birth. In 1967, ABC experimented with the Ampex HS-100 color video magnetic disc for broadcasting. By the middle of the 1970s, EVRs had been introduced by CBS. These were quite different from VCRs because they used silver chloride film and plastic tape, rather than magnetic tape. AVCO introduced Cartrivision, a video cassette recorder, in the beginning of the 1970s. In addition, RCA and Telefunken developed SelectaVision and TED. TED was the extension of the record. At that time, these were practicable but none were practical until Betamax and VHS emerged in the middle of the 1970s (see details in Chapter V).

In the late 1970s, the battle between Betamax and VHS made the sparks fly, which were resemble so that other alternative products lurked behind. Both started mass-production and added new functions soon after they introduced the products. After VHS surpassed Betamax in 1980, it diffused very widely because the early adopters purchased and the early majority started purchas-

Copyright © 2008, IGI Global. Copying or distributing in print or electronic forms without written permission of IGI Global is prohibited.

ing. Under the prosperity of VHS, some companies who had optical disks technology watched for the chance.

LD (Laser Disk), an optical disk which was 30 cm in diameter, was developed by Philips in 1977. MCA (Music Corporation of America) first introduced DiscoVision, an optical disk, and its player to the market in 1978. The format was called Reflective Optical Videodisk System or Disco-Vision. For the distribution of it, MCA owned the rights of a great number of movies and manufactured the software under the MCA DiscoVision label. MCA manufactured disks for other software companies, such as Paramount and Disney. At almost the same time, Pioneer started selling an optical disk, called a Laser Videodisk, in Japan. They sold software specializing in movies and Karaoke. Laser Disks, however, did not diffuse very much. Pioneer continues to sell Laser Disk products, guaranteeing that they keep all parts for the product for eight years after they stop manufacturing.

Other optical video storage mediums were introduced in the 1980s. JVC introduced VHD (Video High Density), which was 25 cm in diameter, only to the Japanese market in 1983. The machines were used for Karaoke and anime video games. McDonnell-Douglas introduced Laserfilm, which was made out of photographic film, in 1984. CD Video was introduced, which was an analogue/digital hybrid type. Video CD was introduced by Sony, Philips, Matsushita, and JVC in 1993. It treated only the digital signals. Both sizes were same as that of the audio CD.

In 1996, DVD-ROM, standing for Digital Versatile Disk or Digital Video Disc, was introduced to the Japanese market first by Toshiba and Warner Home Video. DVDs can hold a huge size and wide variety of information, such as video, audio, and data. There were two different DVD formats, Multi Media Compact Disk (MMCD) and Super Density Disk (SD) in the early 1990s. Philips and Sony formed the MMCD group and Toshiba, Time-Warner, Matsushita, Hitachi, Mitsubishi, Pioneer, Thomson, and JVC formed the SD group. IBM arbitrated and prevented the battle between the formats again. DVD Consortium (changed to DVD Forum in 1997) was formed and recommended SD (see details in the DVD Forum's homepage). HP added a new function and DVDs became recordable and rewritable. Although various optical disks (see details in Wikipedia's homepage) have been developed since the middle of 1990s, none could surpass the DVD in the near future. However, the DVD is not a perfect product. For an example, on Feb.25, 2006, 86 DVD machines were available on the Internet at Bic Camera, one of the largest retailers of household appliances in Japan, which sold 433 billion Yen

Copyright © 2008, IGI Global. Copying or distributing in print or electronic forms without written permission of IGI Global is prohibited.

Table 2. Variety of VCR, DVD players and recorders on sale in 2006 and 2007 (data collected on February 25, 2006 and 2007)

	Only VCR	DVD Player (37)			DVD Player & Recorder (49)			
		Only DVD	With VCR	With HDD	Only DVD	With VCR	With VCR & HDD	With HDD
2006	28	26	11	0	1	7	8	33
2007	10	29	4	0	0	1	6	18

in the 2004-2005 fiscal year. Table 2 shows the variety of the VCR, DVD player, and recorders on February 25, 2006 and 2007. In 2006, 37 DVD player models, without the DVD recording function, were sold. Twenty-six of them were exclusively used as a player. The rest, 11, were with a VCR. Forty-nine DVD recorder models naturally incorporate the player function. Only one machine was exclusively for DVD player and recorder, others were combined with the HDD recorder (33), VCR (7), or both (8). In 2007, 33 DVD player models were sold and 29 of them were only DVD player. The number of DVD recorder with HDD models decreased to 18 almost half of the previous year. This is because the DVD recording function is used supplementary or the companies squeeze the variety of models as a whole. On the other hand, the VCR declines sharply. In 2006, 54 VCR models were sold. Twenty-eight of them were only VCRs and the remainder were combination products with DVD players, DVD recorders and HDD. However, in 2007, only 21 models were sold. All VCR models without DVD and HDD remained in 2007 were the surviving models from previous years.

PC has progressed steadily. In the 1990s, some machines could playback the CD Video, Video CD and DVD, and a few machines had a TV tuner could record TV programs on the hard disk, DVD, USB memory, and others. In the 2000s, a lot of PCs had these functions. PC replaced TV and VCR to some extent. In addition to PC, some video game machines could play DVDs. Although these machines are not for the exclusive use of video playback and storage, they supported the diffusion of the DVD software. Otherwise, it would take much longer time for DVD players and recorders to surpass the VCR.

The HDD (Hard Disk) recorder is the extension of the PC. It is used exclusively for the video storage and playback. Along with the DVD player and recorder, it has made the VCR obsolete. Although it incorporates a huge memory, a cassette and a disk are needed for the storage and transfer. Bic Camera sold

Copyright © 2008, IGI Global. Copying or distributing in print or electronic forms without written permission of IGI Global is prohibited.

42 machines incorporated with HDD recording function out of 104 including the VCR, DVD, and HDD by February 25, 2006. All of them except one combined with DVD recorders (33) and DVD recorders and VCR (8). HDD recorders had a serious problem because until recently, the recorded materials were too huge to store in media except the cassette and disk until now. It wasn't until the USB or SD memory could storage more than 10 gigabytes at reasonable cost that HDD recorders could stand on their own feet.

Table 3 is the list of the DVD and HDD products that were available for sale at Bic Camera in 2006 and 2007. Almost all DVD and HDD products on sale were new models except for only VCR models, only DVD player models, and the combination products of VCR and DVD player models. In addition, many companies came to and went out of the market within a year. Companies who launched only DVD player models, such as BOSE, Cyberhome, Momitsu, and Orion, withdrew from the market. On the other hand, Bluedot, IO Data, Kenwood, Rockridgesound, and Zox entered into the market. They have different product lines from each other. Although Bluedot, Kenwood, Rockridgesound, and Zox sold only DVD player models, their target customers were not same. IO data launched two HDD models. As a whole, the number of new models on February 25, 2007 became less than that of 2006.

Although Turin Olympic and Germany World Cup Soccer were held in 2006, the sales of the DVD recorder in 2006 were much less than expected. The number of domestic shipment decreased to 82 percent of the previous year. It is true that some customers waited for the next generation machine, Blu-ray Disk, but Blue-ray machines are so expensive and professional that it became popular. DVD recorders themselves are too difficult to operate because of the copyright and new functions. In Japan, the dubbing is strictly control. The dubbing of digital broadcasting programs is permitted only once. DVD recorders have various innovative new functions, such as recording by electric TV programming and reserving TV programs in the HDD. These functions make the operation of DVD recorders more difficult. For further distribution, DVD machines should be simplified so that most people can use them easily. The adoption rate of household is about 40 percent. That number is slightly smaller than the half of that of VCR. It can be concluded that there is a chasm between the early majority and late majority in the case of the DVD recorder. In other words, the DVD recorders did not satisfy the minimum requirements, such as the price, the easy manipulation, the size and weight, for the wide distribution yet.

Copyright © 2008, IGI Global. Copying or distributing in print or electronic forms without written permission of IGI Global is prohibited.

Table 3. DVD and HDD products for sale in 2006 and 2007 (Source:Bic Camera; http://www.biccamera.com/

	Year	Model	Sale	Actual Price (Yen)	Weight (Kg)	VCR	Only DVD Player	DVD Recorder	HDD	HDD Size (G)
AVOX	2006	AND-100S	05/6/21	7812.6	2	0	1	0	0	
	2007	ACP-500R		6282	1.5	0	1	0	0	
Bluedot	2006									
	2007	BDP-200W		12420	0.46	0	1	0	0	
BOSE	2006	PLS-1510	03/11/1	99876	8.9	0	1	0	0	
	2007									
Cyber-home	2006	CH-DVD300	04/1/1	8952.3	1.6	0	1	0	0	
	2007									
Denon	2006	DVD-M330(S)	04/10/21	20097	2.1	0	1	0	0	
	2007	DVD-M330		20790	2.1	0	1	0	0	
DIGIX	2006	DVP-M21	05/9/21	3462.6	1.4	0	1	0	0	
	2007	DVP-BL01		3582	1.3	0	1	0	0	
DX Broad Tech	2006	DVHR-V160	05/12/20	38976	5.6	1	0	1	1	160
	2007									
Hitachi	2006	DV-RV8500	05/6/21	43326	6.5	1	0	1	0	
		DV-DH161T	05/6/11	71730	5.4	0	0	1	1	160
		DV-DH160W	05/9/21	71840	5.2	0	0	1	1	160
		DV-DH250W	05/9/21	81836	5.2	0	0	1	1	250
		DV-DH500W	05/9/21	97280	5.2	0	0	1	1	500
		DV-DH1000W	05/10/21	138600	6	0	0	1	1	1000
	2007	DV-DH250S		71820	5.4	0	0	1	1	250
		DV-DH500S		87932	5.5	0	0	1	1	500
		DV-DH1000S		144480	6.2	0	0	1	1	1000
IO Data	2006									
	2007	HVR-HD500RRL		49320	3.8	0	0	0	1	500
		HVR-HD800R		62240	3.8	0	0	0	1	800
JVC	2006	XV-P313	05/5/1	8238.03	1.5	0	1	0	0	
		XV-A707	04/7/1	17226	1.9	0	1	0	0	
		HR-DV4	04/6/1	16456	4.1	1	1	0	0	
		HR-DF3	03/8/1	24186	4.4	1	1	0	0	
		DR-HD250	06/2/1	120060	5	0	0	1	1	250
		DR-HD400	06/2/1	137460	5.1	0	0	1	1	400

continued on following page

Copyright © 2008, IGI Global. Copying or distributing in print or electronic forms without written permission of IGI Global is prohibited.

Table 3. continued

	Year	Model	Sale	Actual Price (Yen)	Weight (Kg)	VCR	Only DVD Player	DVD Recorder	HDD	HDD Size (G)
JVC (continued)	2007	XV-P313*		8238.03	1.5	0	1	0	0	
		RX-DV31		35820	8.4	0	1	0	0	
		HR-DV4		16456	4.1	1	1	0	0	
		SR-DVM70		255150	6.2	0	0	1	1	160
		SR-DVN700		187200	5.8	0	0	1	1	250
Kenwood	2006									
	2007	DVF-8100		17820	2.3	0	1	0	0	
LG Elec-tronics	2006	KDR-C500	05/7/25	25926	5.5	1	0	1	0	
	2007	LDR-H50		22230	5.6	0	0	1	1	160
Matsushita	2006	DVD-S39	05/3/1	10005	2.14	0	1	0	0	
		DVD-S97-S	04/10/15	43326	2.5	0	1	0	0	
		NV-VP33-S	05/6/1	18357	4	1	1	0	0	
		NV-HV90B-S	03/4/20	21402	3.5	1	1	0	0	
		NV-VP51-S	03/10/10	36279	4	1	1	0	0	
		DMR-EH53-S	05/10/15	42330	4.5	0	0	1	1	200
		DMR-EH66	05/10/1	66300	4.7	0	0	1	1	200
		DMR-EX100	05/7/15	70044	5.6	0	0	1	1	200
		DMR-EX200V-S	05/11/10	95784	8.4	0	0	1	1	250
		DMR-EH73V-S	05/10/15	54444	4.5	1	0	1	1	200
	2007	DVD-S50-S		8722	2.2	0	1	0	0	
		NV-HV90B-S		21402	3.5	1	1	0	0	
		DMR-BW200		226809	5.8	0	0	1	1	500
		DMR-XP20V		69120	7	1	0	1	1	250
Mitsubishi	2006	DJ-P250	05/3/25	7812.6	1.3	0	1	0	0	
		DVR-T110	05/5/21	28730	2.3	0	0	1	0	
		DVR-S310	05/4/1	26550	4.3	1	0	1	0	
		DVR-HE50W	05/9/22	59330	5.4	0	0	1	1	250
		DVR-HS315	05/10/21	42330	5.6	1	0	1	1	160
	2007	DJ-P260		7832	1.3	0	1	0	0	
		DVR-DV635		79032	6.5	1	0	1	1	250
Momitsu	2006	MDP1500		4872	1.89	0	1	0	0	
	2007									
Onkyo	2006	DV-L55(S)	04/7/12	24186	3.6	0	1	0	0	
		DV-S155X(S)	04/3/20	27144	1.8	0	1	0	0	
		DPS-1	03/4/25	109620	5	0	1	0	0	
	2007									

continued on following page

Copyright © 2008, IGI Global. Copying or distributing in print or electronic forms without written permission of IGI Global is prohibited.

Table 3. continued

	Year	Model	Sale	Actual Price (Yen)	Weight (Kg)	VCR	Only DVD Player	DVD Recorder	HDD	HDD Size (G)
Orion	2006	DBF-8000	04/1/21	18087.3	4.5	1	1	0	0	
	2007									
Pioneer	2006	DV-484	05/5/1	9135	1.8	0	1	0	0	
		DV-585A	05/5/21	14616	1.8	0	1	0	0	
		CLD-R5	96/9/21	36279	6.4	0	1	0	0	
		DVL-919	98/10/21	68295	8.5	0	1	0	0	
		DVL-K88	98/2/1	100485	8.6	0	1	0	0	
		DV-AX5AVi	05/11/1	164430	10	0	1	0	0	
		DVK-900	98/10/21	166257	43.5	0	1	0	0	
		DVR-330H-S	05/10/11	38080	4.3	0	0	1	1	160
		DVR-331H-W	05/10/12	38976	4.3	0	0	1	1	160
		DVR-530H	05/5/1	40630	4	0	0	1	1	200
		DVR-555H	05/6/1	52026	4.1	0	0	1	1	250
		DVR-DT70	05/12/10	75840	6.2	0	0	1	1	250
		DVR-DT90	05/12/10	102400	6.9	0	0	1	1	500
	2007	DV-393		8973	1.8	0	1	0	0	
		DV-490V		11392	1.8	0	0	0	0	
		DV-696AV		14220	1.8	0	0	0	0	
		CLD-R5		36279	6.4	0	1	0	0	
		DVL-919		68295	8.5	0	1	0	0	
		DVL-K88		100485	8.6	0	1	0	0	
		DVD-V8000		103950	10	0	1	0	0	
		DV-AX5AVi		168210	10	0	1	0	0	
		DVK-900		171990	0	0	1	0	0	
		DVR-DT75		61462	6.2	0	0	1	1	250
		DVR-DT95		67387	6.2	0	0	1	1	400
		DVR-RT700D		67017	7.3	1	0	1	1	250
		DVD-RT900D		77964	7.3	1	0	1	1	400
		DVR-DT100		89820	6.9	0	0	1	1	800
		DMR-BR100		148520	5.5	0	0	1	1	200
Rockridge sound	2006									
	2007	DVP-3B		8982	0	0	1	0	0	
		DVP-3W		8982	0	0	1	0	0	

continued on following page

Copyright © 2008, IGI Global. Copying or distributing in print or electronic forms without written permission of IGI Global is prohibited.

Table 3. continued

	Year	Model	Sale	Actual Price (Yen)	Weight (Kg)	VCR	Only DVD Player	DVD Recorder	HDD	HDD Size (G)
Sharp	2006	DV-SF1	05/8/25	11136	1.3	0	1	0	0	
		DV-NC750	05/4/1	17226	3.1	1	1	0	0	
		DV-GH750	05/4/25	18096	3.1	1	1	0	0	
		DV-RW60	05/10/21	29580	4.3	1	0	1	0	
		DV-AR11	05/11/4	65436	5.2	0	0	1	1	160
		DV-AR12	05/11/4	70044	5.2	0	0	1	1	250
		DV-ARW12	05/12/1	101120	5.2	0	0	1	1	250
		DV-ARW15	05/12/1	115440	5.3	0	0	1	1	500
		BD-HD100	04/12/9	229460	8.5	0	0	1	1	160
		DV-TR11	05/12/1	41184	6.9	1	0	1	1	160
		DV-TR12	05/12/1	47242	6.9	1	0	1	1	250
		DV-TR14	05/12/1	81836	6.9	1	0	1	1	400
	2007	DV-NC750		17226	3.1	1	1	0	0	
		DV-GH750		18096	3.1	1	1	0	0	
		DV-AC52		80820	5	0	0	1	1	250
		DV-ACW60		178200	6.2	0	0	1	1	1000
		DV-ACV32		95040	0	1	0	1	1	250
		DV-ACW52		97200	5	0	0	1	1	500
		DV-ACW55		124200	5.8	0	0	1	1	500
Sony	2006	DVP-NS50P(N)	05/5/20	9396	1.9	0	1	0	0	
		DVP-NS50P(S)	05/5/20	9396	1.9	0	1	0	0	
		DVP-M20P	05/5/11	9396	1.7	0	1	0	0	
		SLV-D383P	05/5/20	17226	3.6	1	1	0	0	
		RDR-VX30	05/6/1	25330	4.9	1	0	1	0	
		VRP-T5	04/10/1	53550	2.5	0	0	0	1	160
		RDR-HX65	05/12/1	33830	4.2	0	0	1	1	200
		RDR-AX75	05/11/21	63840	6	0	0	1	1	250
		RDZ-D50	05/11/10	69146	5.5	0	0	1	1	250
		RDZ-D70	05/11/21	83160	5.6	0	0	1	1	250
	2007	DVP-NS53P		8722	1.7	0	1	0	0	
		DVP-M20P		9396	1.7	0	1	0	0	
		RDR-HX67		38520	4.2	0	0	1	1	250
		RDZ-D900A		124200	5.5	0	0	1	1	400
		BDZ-V7		205200	9.6	0	0	1	1	250
		BDZ-V9		268200	9.6	0	0	1	1	500
		RDZ-D60V		89820	7	1	0	1	1	250

continued on following page

Copyright © 2008, IGI Global. Copying or distributing in print or electronic forms without written permission of IGI Global is prohibited.

Table 3. continued

	Year	Model	Sale	Actual Price (Yen)	Weight (Kg)	VCR	Only DVD Player	DVD Recorder	HDD	HDD Size (G)
Toshiba	2006	SD-V600	05/10/1	18966	3.6	1	1	0	0	
		SD-B600	05/10/1	24186	3.6	1	1	0	0	
		D-VR3	05/12/1	33830	5.5	1	0	1	0	
		RD-XS38	05/12/1	41958	6	0	0	1	1	200
		RD-XD71	05/11/21	54444	6	0	0	1	1	200
		RD-XS48	06/2/1	76330	6	0	0	1	1	300
		RD-XD91	05/11/21	75848	6.2	0	0	1	1	400
		RD-X6	05/12/1	105144	7.2	0	0	1	1	600
		RD-T1	05/12/21	154860	7.2	0	0	1	1	1000
		AK-V100	05/11/1	44030	7.6	1	0	1	1	160
		AK-V200	05/10/1	57630	7.8	1	0	1	1	250
	2007	SD-580J		12282	1.62	0	1	0	0	
		SD-280J		7992.2	1.6	0	1	0	0	
		HD-XF2		49320	4.1	0	1	0	0	
		HD-XA2		115200	6.2	0	1	0	0	
		D-VR5		25920	5	1	0	1	0	
		RD-A1		313200	15.2	0	0	1	1	1000
		RD-E160		48772	5.6	0	0	1	1	160
		RD-E300		52777	5.6	0	0	1	1	300
		RD-W300		67830	9	1	0	1	1	300
Yamaha	2006	DVD-S657	05/7/1	18009	2.6	0	1	0	0	
	2007	DVD-S657		18423	2.6	0	1	0	0	
		DVD-S658		18423	2.6	0	1	0	0	
ZOX	2006									
	2007	ZTO-2101		4482	0	0	1	0	0	

**Models written in the bold were sold more than a year.*

Drivers for Obsolescence of VCR (VHS)

VCR built up a stronger defense against other products as time passed. While VCR dominated the market for about three decades, time increased the number of users and sales. These led to the learning curve effect and the price down. The price down induced users and non-users to purchase a product. Time also gave the VCR industry a chance to upgrade VCR products at the end of its performance. Users became accustomed to the operation of VCRs

Copyright © 2008, IGI Global. Copying or distributing in print or electronic forms without written permission of IGI Global is prohibited.

and came to store many recorded cassettes as time passed. The switching cost became higher and higher. However, the defense came to close the ceiling in the middle of 1990 because the VCR had improved very much in the function and been manufactured very efficiently. As result, the target of the alternative products became obvious.

The combination product of HDD and DVD recorder surpassed the VCR in the early 2000s. The background of the issue was the change from the analogue technology to a digital one. The VCR was an embodiment of the most sophisticated skills of analogue technologies. HDD and DVD recorders are representative of digital technologies. The users did not jump to DVD and HDD recorders suddenly. Other digital devices such as the PC and the video game machine played very important roles in the initial market.

The PC and the video game machines functioned as an intermediary. People who do not own a DVD player can watch the DVD software, which is high in quality and has many more functions than the VCR. They assisted the distribution of the DVD software to some extent. The growth of the DVD software brought about that of the hardware. DVD machines were on equal terms with VCRs in terms of software. In addition, PCs made us understand the convenience of the recording and the editing on the hard disk. The production of the optical disks grew rapidly in the early 2000s, while that of videotapes decreased sharply around 2000.

The VCR lost its advantage gradually against HDD and DVD recorders. The replacement proceeded though many steps. First, the DVD software rose first by the aid of other digital devices. Without their assistance, it would have taken longer. Second, the PC demonstrated the recording TV programs on the hard disk. It saved spaces for the cassette and time to find a program. Third, the combination of HDD and DVD recorders demonstrated better performance than that of VCR. The early majority of people started replacing HDD and DVD recorders instead of VCR.

From the industry side, they need an innovation for the break through doldrums. Cost cutters eroded the low end and ordinary products. The rest, the high end markets, was not so big that leaders and followers could live together. S-VHS could not have a major impact on the market. JVC developed D-VHS to change from the analogue product to the digital one. However, VCRs had strong limits inside because they were designed to make most of the analogue technologies. The cassette tape became out of date in the digital age. The industry should have gotten down to the search for the post-VCR products and shifted DVD and HDD recorders.

Copyright © 2008, IGI Global. Copying or distributing in print or electronic forms without written permission of IGI Global is prohibited.

Consumers can be divided into two groups based upon whether they welcome the new products or not. The innovators tried other digital devices first to check the performance in the late 1990s. Around 2000, the early adopters started shifting DVD and HDD recorders after they saw the great possibility in the near future. At that time, the DVD software industry stood on its feet. In the early 2000s, the early majority followed the early adopters after they had full assurance of the DVD and HDD recorders' success. In the middle of 2000s, the late majority started shifting after they saw the trend. On the other hand, laggards hesitate to switch because of the attachment, the recorded cassettes, and the anxiety in the operation.

End of the VCR

The volume of domestic shipments of the VCR has steadily decreased since its peak in 1998 although for a few years it had maintained a higher level than that of the trough between the first and second peaks. However, the volume fell below the trough between the first and second peak in 2002 and decreased sharply after that. This indicates that plenty of customers stopped replacing VCRs and instead started purchasing alternative products such as DVD and HDD recorders.

According to the decline of the VCR industry, the number of companies that sold VCRs decreased by half to just 10. As a result, the share of the top three, JVC, Matsushita, and Sony, increased to about 60 percent in 2004. Most surviving manufacturers began to sell VCR products in combination with DVD player & recorder, HDD recorder, or both. Table 4 lists up the VCR products on sale at Bic Camera in 2006 and 2007. During the period, Funai, Hitachi, and Orion stopped launching VCR products. JVC, LG Electronics, and Matsushita continued to sell the previous models. On the other hand, Pioneer, Sharp Sony, and Toshiba introduced new VCR models with DVD or HDD recorders. These types of products bridged the gap between these machines for the alternation of generations.

The volume of the domestic shipment of DVD players and recorders caught up with that of the VCR in 2002. It reached 7,240,000 units, thereby exceeding the peak of the VCR of 7,156,000 in 1998. Although the DVD and HDD recorders are presently much more expensive than the VCR, they are capable of greater performance. The DVD is much smaller and more durable than

Copyright © 2008, IGI Global. Copying or distributing in print or electronic forms without written permission of IGI Global is prohibited.

Table 4. VCR products on sale in 2006 and 2007 (Source: Bic Camera; http://www.biccamera.com/)

	Year	Model	Sale	Actual Price (Yen)	Weight (Kg)	Only DVD Player	DVD Recorder	HDD	HDD Size (G)
Funai	2006	FV-N70R	-	7289.73	2	0	0	0	
	2007								
Hitachi	2006	DV-RV8500	2005/6/21	43326	6.5	0	1	0	
	2007								
JVC	2006	HR-B13	-	8682.6	3.1	0	0	0	
		HR-G13	2005/7/11	11049	3.1	0	0	0	
		HR-F13	2003/2/10	15486	3.2	0	0	0	
		HR-DV4	2004/6/1	16456	4.1	1	0	0	
		HR-DF3	2003/8/1	24186	4.4	1	0	0	
		HR-S700	2003/6/1	16356	3.2	0	0	0	
		HR-V700	2003/6/1	17226	3.2	0	0	0	
		HR-ST700	2003/6/1	19836	3.1	0	0	0	
		HR-VT700	2003/6/1	20706	3.2	0	0	0	
		DR-MV5	2005/6/1	40320	5.7	0	1	0	
	2007	HR-B13		8682.6	3.1	0	0	0	
		HR-G13		11049	3.1	0	0	0	
		HR-F13		15486	3.2	0	0	0	
		HR-DV4		16456	4.1	1	0	0	
LG Electronics	2006	GV-HIA5	-	7203.6	2.87	0	0	0	
		KDR-C500	2005/7/25	25926	5.5	0	1	0	
	2007	GV-HIA5		7203.6	2.87	0	0	0	
Matsushita	2006	NV-HV62	2005/3/1	12006	2.5	0	0	0	
		NV-HV72G	2005/3/1	12876	2.5	0	0	0	
		NV-SV120-S	2003/4/20	21402	3.5	0	0	0	
		NV-SV150B-S	2003/4/20	25317	3.6	0	0	0	
		NV-VP33-S	2005/6/1	18357	4	1	0	0	
		NV-HV90B-S	2003/4/20	21402	3.5	1	0	0	
		NV-VP51-S	2003/10/10	36279	4	1	0	0	
		DMR-EH73V-S	2005/10/15	54444	4.5	0	1	1	200
	2007	NV-HV62		12006	2.5	0	0	0	
		NV-HV72G		12876	2.5	0	0	0	
		NV-HV90B-S		21402	3.5	1	0	0	

continued on following page

Copyright © 2008, IGI Global. Copying or distributing in print or electronic forms without written permission of IGI Global is prohibited.

Table 4. continued

	Year	Model	Sale	Actual Price	Weight	Only DVD Player	DVD Recorder	HDD	HDD Size
				(Yen)	(Kg)				(G)
Mitsubishi	2006	DVR-S310	2005/4/1	26550	4.3	0	1	0	
		DVR-HS315	2005/10/21	42330	5.6	0	1	1	160
	2007	DVR-DV635		79032	6.5	0	1	1	250
Orion	2006	DBF-8000	2004/1/21	18087.3	4.5	1	0	0	
	2007								
Pioneer	2006								
	2007	DVR-RT700D		67017	7.3	0	1	1	250
		DVD-RT900D		77964	7.3	0	1	1	400
Sharp	2006	VC-H220	2002/4/1	9396	2.6	0	0	0	
		VC-GH20	2002/4/1	11658	2.6	0	0	0	
		DV-NC750	2005/4/1	17226	3.1	1	0	0	
		DV-GH750	2005/4/25	18096	3.1	1	0	0	
		DV-RW60	2005/10/21	29580	4.3	0	1	0	
		DV-TR11	2005/12/1	41184	6.9	0	1	1	160
		DV-TR12	2005/12/1	47242	6.9	0	1	1	250
		DV-TR14	2005/12/1	81836	6.9	0	1	1	400
	2007	DV-NC750		17226	3.1	1	0	0	
		DV-GH750		18096	3.1	1	0	0	
		DV-ACV32		95040	-	0	1	1	250
Sony	2006	SLV-NX15	2004/5/21	10701	2.6	0	0	0	
		SLV-NX35	2004/5/21	12528	2.8	0	0	0	
		SLV-D383P	2005/5/20	17226	3.6	1	0	0	
		RDR-VX30	2005/6/1	25330	4.9	0	1	0	
	2007	RDZ-D60V		89820	7	0	1	1	250
Toshiba	2006	SD-V600	2005/10/1	18966	3.6	1	0	0	
		SD-B600	2005/10/1	24186	3.6	1	0	0	
		D-VR3	2005/12/1	33830	5.5	0	1	0	
		AK-V100	2005/11/1	44030	7.6	0	1	1	160
		AK-V200	2005/10/1	57630	7.8	0	1	1	250
	2007	D-VR5		25920	5	0	1	0	
		RD-W300		67830	9	0	1	1	300

Copyright © 2008, IGI Global. Copying or distributing in print or electronic forms without written permission of IGI Global is prohibited.

the VHS cassette and it can record much longer and at higher quality than the VCR. In Japan, the sales volume of DVD software exceeded that of the VHS in 2001. In 2004, 100 million DVD units were sold; VHS software sold only 10 million units. The video disk which was equal to the video cassette in 1990 was regarded as a suitable medium for the prerecorded programs because of the high quality images, the sounds, and easy manipulation. However, it requires a special equipment to replay the video disks whose size was much bigger than that of the DVDs. The production of the video disks continues to decrease slowly because the adopters can not switch easily and quickly. The production of video cassettes met its peak in 1998. After that, it also declined steadily. Instead, the production of DVDs grew dramatically in the 2000s. DVDs whose size is much smaller than that of the video disks demonstrate the high quality of sound and images at reasonable price. In addition, we can watch the DVDs on the various equipments. As a result, the production of the DVDs surpassed the maximum production amount of the video cassette in 1998 easily. The alternation of players and recorders became obvious after 2002.

Strategies for Alternative Products

Yamada (1997) demonstrated the strategies against alternative products from the view of de facto standard. Companies face two tradeoffs in the maturity. The first one is how long they keep the format or how fast they switch to others. It is a tough decision when they abandon the profitable products and promote the unexpectable products. They should balance present and future profits. The second one is the choice of compatibility or innovativeness. Compatible products are acceptable from the users. Leaders have much possibility to maintain the lead against others than innovative products. On the other hand, innovative products may have a big impact on the market and industry. These products explore the new users and give a company a chance to overtake others. Each company has the appropriate timing, which is differ from others based on the market position and technological advantage.

There are three types of strategies for an alternative product: passive, neutral, and active. Passive strategies are the protection of an existing product by the sales promotion and the development of products, and the interruption of consensus among companies for a new product by complaints about the

Copyright © 2008, IGI Global. Copying or distributing in print or electronic forms without written permission of IGI Global is prohibited.

Table 4. Altitude for alternative products

Type	Introduction	Early Growth	Late Growth	Maturity	Decline
Leaders	Conflict	Passive	Passive	Neutral	Active
Followers	-	Passive	Neutral	Active	Active
Cost cutters	-	-	Neutral	Neutral	Neutral

technical and commercial problems. Active strategies are the extension of an existing product to absorb the new technology, the introduction of a bridge product for the smooth transaction, the partial preoccupation of the new technology for the trial, and the abandonment of existing products. The neutral strategies incorporate and balance the passive and active strategies.

Table 4 shows the attitude against the alternative products by the periods and company type. Leaders explore the market by competing alternative products and accustom the experience and the publicity. They have a big lead against followers and cost cutters in the technology and the market. On the other hand, they have a responsibility for users to protect the product for a certain time. Therefore, they are passive against the alternative product while the market grows. In the maturity, they start searching the possibility of the alternative products with protecting existing products. In the decline, they should switch to the alternative products. Followers fall behind the leaders. They enter into the market later with their unique technology and contribute to the market growth. Just after they enter into the market, they are passive against the alternative products because they incurred the big sunk cost. However, they start researching alternative products and develop alternative products earlier than leaders to get a better position in the next products. Cost cutters are unique. Their competence is in the manufacturing system, which is very flexible in the variance of quantity and the variety of products because most of them have the experience as an OEM maker. They earn the some money in the existing product and prepare for the next products simultaneously.

SCM Against Alternative Products

SCM on the alternative products can be divided into three, the passive, neutral, and active as same as an individual company. The passive attitude of the supply chain required the upstream supply chain partners to evolve the product collaboratively and the downstream partners to do the sales

Copyright © 2008, IGI Global. Copying or distributing in print or electronic forms without written permission of IGI Global is prohibited.

promotion. The active attitude all partners to prepare for the switch to the alternative products. It means that upstream partners should obsolete their facility for the parts to existing products and take the risk whether alternative products become popular or not. Some small size partners do not have enough resources to maintain multiple lines for existing and alternative products. Others face the rise of manufacturing costs because of the preparation for the switch and the weakness of the economies of scale. On the other hand, the downstream partners are easy to accept the alternative products because they can respond to the switch quickly by changing the selection and the layout of the products. In addition, new products contribute to an increase in sales in the short run. The leaders of the supply chain should decide the altitude to the alternative products concerning the fact that although upstream partners hope that the existing products have a longer life due to the decrease of the manufacturing, downstream partners want the alternative products for the increase of sales

Each supply chain has the best timing to change the attitude toward the alternative products. A supply chain which a leader of the industry created tends to have a strong relation with suppliers and a good position in the market due to the technological advantage and the brand royalty. This type of supply chain might miss the timing to switch. Sometimes, they continue to produce the industry after they lost the profitability in the industry. JVC was a good example, who ran the great deficit, 31 and 45 billion yen in 2002 and 2006. A supply chain whose leading company is a follower makes importance on the profitability and the future. They have a weaker relationship with supplier and position in the market because they took a short cut to the commercialization. They could refer to the previous products and purchase some key parts. This enabled them to switch to alternative products quickly. A supply chain of a cost cutter focused on cost reduction and flexibility. They purchase a lot of standard parts from outside sources and prepare for the quick response to the present market demand with flexibility. Their flexibility causes them to take a neutral position.

Table 4. Attitude toward alternative products

	Attitude toward Alternative Products	**Orient**	**Strength**
Leader	Passive	Past	R&D / Development
Follower	Active	Future	R&D / Speed
Cost Cutter	Neutral	Present	Flexible Production

Copyright © 2008, IGI Global. Copying or distributing in print or electronic forms without written permission of IGI Global is prohibited.

Emergence of New Technology and Life Cycle

The emergence of DVD technology was an extreme case of a disruptive technology (DT). DVDs essentially made VHS obsolete very quickly. Recent work on DT has revealed that more generally, a new technology may have various impacts other than complete obsolescence of the incumbent technology (Weisenbach & Keller, 2005; Weisenbach, Keller & Shanklin, in press). Their work has looked at three categories or patterns that have been exhibited. The first of these is the *zero-sum* case in which one technology obsoletes the incumbent one essentially completely. That is, one wins and the other loses. Of course DVD has been an example of this case.

Second is what is called *competitive disruption* and is said to occur when the existing technology and the new entrant eventually coexist and serve differing segments of the market. The interactions between television and movie theatres provide one example. Mainframe and personal computers provides another. In each case both technologies have continued to coexist.

Positive-sum disruption expands the size of the whole industry market. This is in the sense that consumers way want or need both the new and old one. Cellular telephones and landline telephones provide an example. While some consumers may use only one of these, many used both of them. Also, VCRs initially and DVDs later and can be thought to bear this relationship with movie theatres. The Hollywood movie product can be sold first in theatres and then later on DVDs to the same consumer. Thus the movie product get a larger market than before as a huge aftermarket was developed for movies premiered in theatres. Weisenbach et al. (in press) noted that by the turn of the 21st century, video store revenues and box office gross sales were virtually identical, at $8 billion each, and that sales of both were on the increase.

Zero-sum, positive-sum and competitive disruption all share common characteristics in the beginning of their life cycles. Incumbent competitors are prone to see an emerging technology as inferior to the existing technology. That is generally true at its inception but appears to lead to underestimation of its threat. Determining which pattern of disruption may be developing involves careful monitoring of the market's reaction to the new technology. Weisenbach et al. (in press) suggest that the changing sentiments and buying behavior of marginal customers and non-customers are particularly important to observe. These are likely low-priority or fringe customers as far as incumbents are concerned, but are precisely those that need to be carefully monitored. In particular, they suggest to watch i) what inroads are being

Copyright © 2008, IGI Global. Copying or distributing in print or electronic forms without written permission of IGI Global is prohibited.

made by the entrant technology, ii) what changes there are in the sales rate of sale of the incumbent technology, and iii) what is the interest level of all customer segments, not just among mainstream buyers.

The points raised in this research (Weisenbach & Keller, 2005; Weisenbach et al., in press) should be also helpful in making judgments about which product design will emerge as dominant and whether coexistence, replacement, or market expansion will likely occur. Similarly, our work on the industry life cycle should help inform the future studies of the interplay of competing technologies.

Conclusion

We discussed the emergence of new technologies and the alternative products, which is not as severe a change as the alternation of generation. Companies launch alternative products whose base technology is different from existing products to obtain the initiative in the market. On the other hand, consumers such as the innovators and the early adopters want brand-new products.

In the VCR case, many alternative products existed from the beginning. In 1960s, the VCR competed with EVR, TED, and LD. The laser disk continued to be a competitor in 1970s and 1980s. In the 1990s, DVD and HDD recorders emerged in the market. The VCR was deposed as the de facto standard around 2000 because of the big wave of digital technologies. DVD software started spreading dramatically in the middle of 1990s because they can be played back by PCs and video game machines, which demonstrated the new functions and use. The diffusion of DVD software promoted the sales of the DVD machines. As a result, VCRs lost their stronghold. DVD and HDD recorders forced the VCR into a corner gradually in the late 1990s. Although JVC launched digital machines, those could not compete with DVD and HDD recorders because the VHS was designed to fit analog technologies.

References

Abernathy, W. J. (1978). *The productivity dilemma.* Baltimore: The John Hopkins University Press.

Copyright © 2008, IGI Global. Copying or distributing in print or electronic forms without written permission of IGI Global is prohibited.

Abernathy, W. J, Clark, K.B., & Kantrow, A.M. (1983). *Industrial renaissance*. New York: Basic Books.

Bic Camera. Retrieved March 15, 2006 from http://www.biccamera.com/

Christensen, C. M. (1997). *The innovator's dilemma*. Boston: Harvard Business School Press.

DVD Forum. Retrieved March 15, 2006 from http://www.dvdforum.org/forum.shtml

Economic and Social Research Institute, Cabinet Office, Government of Japan. Retrieved March 15, 2006 from http://www.cao.go.jp/index-e.html

Electronics Industries Association of Japan. Retrieved March 15, 2006 from http://www.jeita.or.jp/eiaj/english/index.htm

Funai. Retrieved March 15, 2006 from http://www.funaiworld.com/index.html

Higuchi, T, Troutt, M. D., & Polin, B. A. (2004). Life cycle considerations for supply chain strategy. In C.K. Chan & H.W.J. Lee (Eds.), *Successful strategies in supply chain management* (pp. 67-89). Hershey, PA: Idea Group.

Hitachi. Retrieved March 15, 2006 from http://www.hitachi.com/

Itami, H. (1989). Nihon no VTR sanngyou naze sekai wo seiha dekitanoka. Tokyo: NTT Publishing.

Japan Electronics and Information Technology Industries Association. Retrieved March 15, 2006 from http://www.jeita.or.jp/english/index.htm

Japan Video Software Association. Retrieved March 15, 2006 from http://www.jva-net.or.jp/en/index.html

JEITA (Japan Electronics and Information Technology Industries Association). (2005). Minnseiyou Dennsi kiki Deta Shyu. Tokyo: JEITA.

Matsushita. Retrieved March 15, 2006 from http://panasonic.net/

Mitsubishi Electric. Retrieved March 15, 2006 from http://global.mitsubishi-electric.com/

National Science Museum, Tokyo. Retrieved March 15, 2006 from http://sts.kahaku.go.jp/sts/

NHK (Japan Broadcasting Corporation.). (2000). Project X: Challengers 1. Tokyo: Japan Broadcast Publishing.

Rosenbloom, R. S. & Abernathy, W. J. (1982). The climate for innovation in industry. *Research Policy*, *11*(4), 209-225.

Copyright © 2008, IGI Global. Copying or distributing in print or electronic forms without written permission of IGI Global is prohibited.

Sato, M. (1999). The story of a media industry. Tokyo: Nikkei Business Publications.

Schoenherr, S. Retrieved March 15, 2006 from http://home.sandiego.edu/~ses/

Sony Corporation. Retrieved March 15, 2006 from http://www.Sony.net/

Toshiba. Retrieved March 15, 2006 from http://www.toshiba.co.jp/index.htm

Utterback, J. M. (1994). *Mastering the dynamics of innovation: How companies can seize opportunities in the face of technological change*. Boston: Harvard Business School Press.

Victor Company of Japan. Retrieved March 15, 2006 from http://www.jvc-victor.co.jp/english/global-e.html

Wikipedia. Retrieved March 15, 2006 from http://en.wikipedia.org/wiki/Main_Page

Copyright © 2008, IGI Global. Copying or distributing in print or electronic forms without written permission of IGI Global is prohibited.

Section III

Consumer Aspects

Based on Moore (2002, 2005), we discuss the transition of consumer behavior according to the life cycle of the VCR. From the technology adoption life cycle which models the acceptance of the new model by the consumers, the consumers can be divided into five groups: the innovators (techies or technology enthusiasts), the early adopters (visionaries), the early majority (pragmatists), the late majority (conservatives) and the laggards (skeptics). The model explains the way to develop a high-tech market by moving smoothly through the target consumer types. The market is viewed as two distinct periods—the early market and the mainstream market. The innovators and the early adopters lead the early market and the early majority and late majority play a major role in the mainstream. Laggards are almost out of the picture in the model because sales promotion to them is not considered to be effective. In the first chapter in this part, the extreme innovators and innovators are discussed. These consumers are those who purchased an incomplete product at a high price. We also place emphasis on the supply chain aspects rather than the consumer aspects. It is true that the goal of the supply chain is consumer satisfaction, but it also must be taken into consideration that their behavior is the result of the supply chain activity or management to some extent.

Chapter IX

Extreme Innovators and Innovators

In this chapter, the innovators and extreme innovators are discussed. These types of consumers are very important because they grow the infant market. The extreme innovators purchase an incomplete product at high price and contribute to the product development. Then the innovators purchase an immature product at a relatively high price and their reviews have great effects on the future diffusion. We also follow their second and later purchases because their behavior in repeat purchases has a strong relation with the alternation of product generations. In other words, they are the motive power for the alternation of products or the change to the alternative products.

Copyright © 2008, IGI Global. Copying or distributing in print or electronic forms without written permission of IGI Global is prohibited.

Characteristics of Extreme Innovators and Innovators

Moore (2002) has observed that:

> *Innovators pursue new technology products aggressively. They sometimes seek them out even before a formal marketing program has been launched. This is because technology is a central interest in their life, regardless of what function it is performing. They are inherently intrigued with any fundamental advance and often make a technology purchase simply for the pleasure of exploring the new device's properties. There are not very many innovators in any given market segment, but winning them over at the outset of a marketing campaign is key nonetheless, because their endorsement reassures the other players in the marketplace that the product does in fact work.* (p. 12)

Innovators have four main characteristics. First, they adopt new products and dominate the new market at the very beginning of the product life cycle. Roger (1995) and Kotler (2005) called the first 2.3 percent of adopters the innovators. Thus, they constitute a small number outside two standard deviations from the mean. Second, they are good with technology; most of them are able to operate new products by themselves. Third, they spend much more money for new products than others. The price and performance tend to improve over time. Although the product performance and the price have much room for later improvement in such new products, innovators nevertheless are willing to buy them now. Finally, they enjoy evaluating new products. They also express their opinions about new products by word of mouth or especially through the internet.

The extreme innovators are considered to be the first 0.13 percent of the total consumers, and amount to a fraction outside three standard deviations. They buy the product while the industry is still involved in trial and error. They often are very affluent—for example medical doctors, movie stars, and executives—and they prize the novelty of the products very much. They do not have to be techies because the manufacturers themselves can support them with great care or they can hire others to operate the devices. It is natural that the number of consumers be very small at this stage in accordance with the low product performance, the product performance/price ratio, and the inefficient mass-manufacturability.

Copyright © 2008, IGI Global. Copying or distributing in print or electronic forms without written permission of IGI Global is prohibited.

Figure 1. Customer types and their distribution

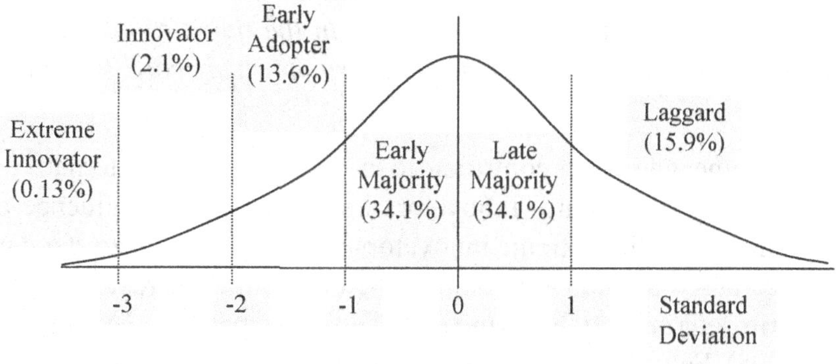

Innovators' Role in Industry

The extreme innovators are very rare, being just 0.13 percent of the total consumers. They likely are jaded people willing and able to indulge themselves with gadgetry and experimental or incomplete products well before the commercial success of such products. However, the manufacturers can benefit from them and develop these products step by step through transactions and communications with this group. They also tend to give manufacturers feedback about the products. The complaints from them are a precious driver for the growth of the product and its further development for success in the market.

Innovators follow the extreme innovators but have a weak relationship with them because the products purchased by these groups are different in both performance and price range. However, they also are critical at the beginning of developing markets. Although the number of them is still very small, namely about 2.3 percent of the total consumers, they test and evaluate the new immature products, which are nearing the commercial success level. Obviously, those products most admired by the consumers have the greatest chance of becoming widely popular. Moore (2002) observes the following about this group:

They are the ones who first appreciate the architecture of your product and why it therefore has a competitive advantage over the current crop of products established in the marketplace. They are the ones who will spend hours trying to get products to work that, in all conscience, never should have been shipped

Copyright © 2008, IGI Global. Copying or distributing in print or electronic forms without written permission of IGI Global is prohibited.

in the first place. They will forgive ghastly documentation, horrendously slow performance, ludicrous in functionality, and bizarrely obtuse methods of invoking some needed function—all in the name of moving technology forward. They make great critics because they truly care. (p. 31).

The extreme innovators contribute to the development of alternative products of the de facto standard. Innovators impart a stronger influence on future consumers than the extreme innovators because the de facto standard can be decided based on their evaluation. Their evaluation is regarded as a critical factor in determining the future diffusion of the product (Moore, 1991; Rogers, 1995). Future consumers buy the extensions of the product the innovators admire highly and essentially co-design.

Failures due to Prematurity in Technology and Market

Untimely sales, that is, sales too early in the life cycle, sometimes cause great financial damage to companies. This is due to technology factors, market factors, or both. From the point of view of technology, the products have not matured sufficiently in the areas of specifications and price. Companies can not launch products that satisfy the minimum requirements for commercial success at this point in time. The specifications include the available functions, the size, weight, quality (reliability), and the ease of the operation. If the products are missing one or more of these specifications, it follows that they would be too far from commercial success readiness. In addition to the specifications, the price also has a great effect on the market. The experience curve effect illustrates the decline of the manufacturing cost. In the beginning, the products are manufactured experimentally. This means the product design is changeable and far from the sophisticated ultimate design. Companies can not start mass production at this time. Therefore, generally, these products can not satisfy the price and market requirements. These products need much more product and process innovations to reach the stage of commercial success.

Much depends on the balance between the product specifications, including the price and customer needs, as to whether the product becomes popular or not. In the case where a great number of customer needs clearly exist,

Copyright © 2008, IGI Global. Copying or distributing in print or electronic forms without written permission of IGI Global is prohibited.

the product advancement and the price decline of the product determine the level of diffusion. The stronger the needs someone has, the more he or she pays. On the other hand, in the case where most customer needs are latent, companies should explore the peripheral areas that constrain the diffusion. For example, in the case of the VCR, the scarcity and costliness of the video software packages were the main bottlenecks in the beginning. The recording function was one of the most critical factors for commercial success.

Ampex was a leading producer of magnetic recording equipment for broadcasting and business use in the 1960s and 1970s. They were heavily oriented toward R&D and the expansion of the state-of-the-art frontiers in magnetic recording. William E. Roberts, Ampex's President, predicted that home audio and video recorders, plus prerecorded entertainment tapes, would create a huge market by 1970. However, they could not make much of a ripple in the growth of the home-use VCR after the late 1970s. Ampex's products were adapted mainly for broadcasting and business use. The adopters for these products had a strong need and huge budget for the VCR. They also had enough space and expertise for the VCR. On the other hand, the early adopters and the majority of people did not have these elements. Even though some extreme innovators and innovators adopted these products, they could not become popular. These products could not cross the chasm between the innovators and early adopters by any means. Although several companies made attempts to commercialize the VCR market in the early 1970s, very few people adopted them. Most companies, with the exception of Sony and JVC, failed to successfully commercialize the VCR.

Price Decline and Diffusion of the VCR

The price of VCR products decreased rapidly according to the accumulated number of the VCR production. The shape looks like a fractional function. At the beginning, although there was great potential for price decreases, the number produced increased very slowly. There was a strong interrelation between the price and the accumulated number produced. It was after the price became reasonable that the majority of people purchased the product. Although the price continued to decrease even after the majority of people adopted the product, the rate of further reductions became smaller and smaller as shown by Figure 2.

Copyright © 2008, IGI Global. Copying or distributing in print or electronic forms without written permission of IGI Global is prohibited.

Figure 2. Production cost and accumulated number of production units (Source: JEITA, 2005)

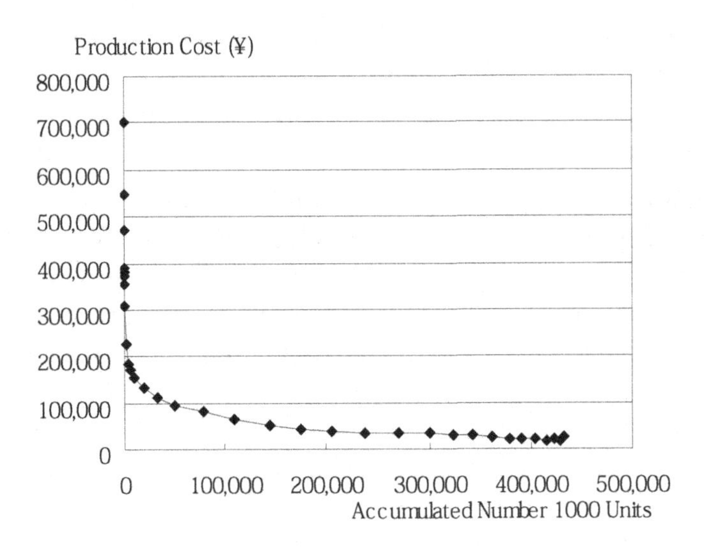

The decrease of the production cost per unit can be expressed by the experience curve. The semiconductor is a good example of the experience curve, also known as the learning curve. Its effect demonstrates that the production cost per unit decreases 20 to 30 percent each time the accumulated number of the production doubles. In addition, the semiconductor is so minute that the material cost and logistical costs per unit are essentially negligible. However, in reality, the production cost per unit will never decrease exactly to zero because of the material costs and other indispensable costs, however small they may be. We applied the experience curve to the VCR case. The production cost decreases, as Figure 3 shows, according to the accumulated number of products.

Figure 4 demonstrates the estimated cost from the experience curve effect and the actual production cost per unit, in which the horizontal axis is time. Figure 4 is similar to Figure 2 in a sense, but there are two differences. First, the horizontal axis is time rather than the accumulated number. However, we checked the number of the accumulated production per year and plotted the corresponding costs to the number of the accumulated production in each year. Second, we added the estimated costs from the experience curve effect for the comparison. Table 1 shows the estimated costs. We made these estimates by the following method:

Copyright © 2008, IGI Global. Copying or distributing in print or electronic forms without written permission of IGI Global is prohibited.

Figure 3. Experience curve of the VCR (Source: JEITA, 2005)

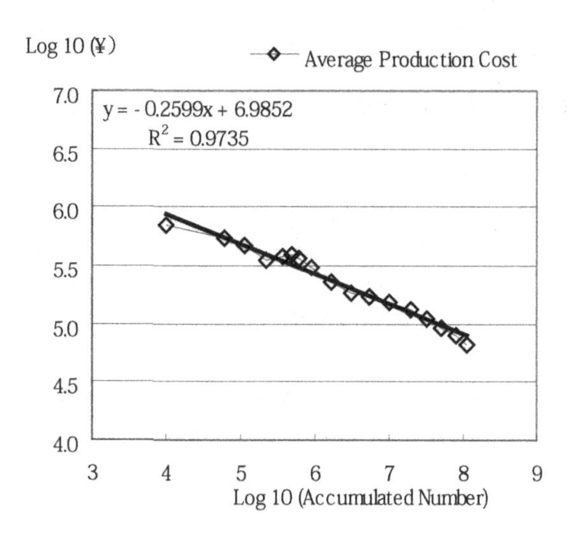

Figure 4. Estimated cost and actual cost

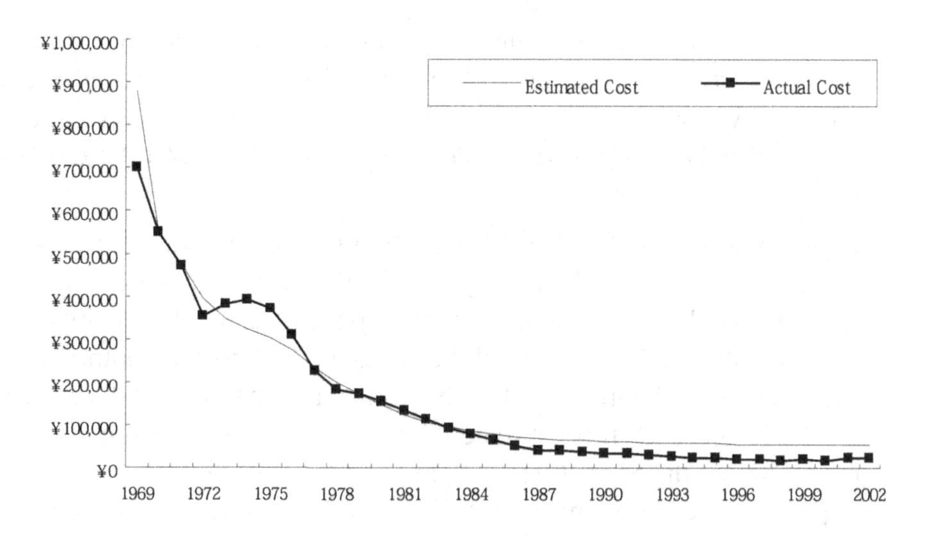

Copyright © 2008, IGI Global. Copying or distributing in print or electronic forms without written permission of IGI Global is prohibited.

(Step 1) Accumulate the number of production at the end of year i, Xi

(Step 2) Identify, Log Xi

(Step 3) Apply the result of the regression in Figure 9-3 and identify Yi

(Step 4) Calculate 10 to the power of Yi and plot them

We deduced that the trend of company strategies created the gap between the estimated and actual costs. At the beginning, the actual production cost per unit was lower than the estimated cost partly because companies made a great effort to lower the production cost to a reasonable level. During 1973-1976, the actual costs were above the estimated costs because companies upgraded the target product specifications according to customer needs. After they achieved this upgrade, they reestablished the price decline.

Extreme Innovators and Innovators in the Home-use VCR Market

In March 1979, VCR ownership by percentage of Japanese households reached 2.0 percent. This number is equal to 2.4 (=2.0/82.6) percent of the maximum 82.6 percent reached in 2004. The first 2.3 percent of adopters (beyond two standard deviations) are called the innovators (Kotler, 2005; Rogers, 1995). Thus, those people who purchased the VCR before 1979 can be regarded as innovators.

Figure 5 shows the trend of the adoption rate and the number of units possessed per household adopted. By March 1978, 1.3 percent of houses owned at least one VCR and their average number possessed was 1.077 units. Roughly 7.7 percent (0.154 percent of all households) had owned another VCR. The percentage 0.154 percent is slightly larger than the three standard deviations' gap (0.13 percent). It is estimated that most of them had purchased another model before the SL-6300 (Betamax) in March 1976. It was not reasonable to purchase both Betamax and VHS at the beginning because they were incompatible. Furthermore, very few owners bought two Betamax machines simultaneously at that time. Hence, we call the first consumers outside the three standard deviation range extreme innovators here.

The extreme innovators were very wealthy and prized innovative products very much. They were medical doctors, movie stars, executives, and the like,

Copyright © 2008, IGI Global. Copying or distributing in print or electronic forms without written permission of IGI Global is prohibited.

Table 1. Calculation of the estimated cost

		Step 1: Xi	Step 2	Step 3: Yi	Step 4
	Production (K Units)	Accumulated Number (K Units)	Log Xi	Yi=-0.2599Xi +6.9852	10 to the power of Yi
1969	10	10	4.00	5.94	880,737
1970	50	60	4.78	5.74	554,510
1971	49	109	5.04	5.68	474,387
1972	114	223	5.35	5.60	393,827
1973	137	360	5.56	5.54	347,692
1974	124	484	5.68	5.51	321,932
1975	119	603	5.78	5.48	304,026
1976	288	891	5.95	5.44	274,684
1977	762	1,653	6.22	5.37	233,905
1978	1,470	3,124	6.49	5.30	198,257
1979	2,199	5,323	6.73	5.24	172,613
1980	4,441	9,764	6.99	5.17	147,433
1981	9,480	19,244	7.28	5.09	123,598
1982	13,134	32,378	7.51	5.03	107,966
1983	18,217	50,595	7.70	4.98	96,140
1984	28,611	79,206	7.90	4.93	85,568
1985	30,581	109,787	8.04	4.90	78,607
1986	33,879	143,667	8.16	4.87	73,300
1987	30,563	174,230	8.24	4.84	69,716
1988	31,660	205,891	8.31	4.82	66,755
1989	32,015	237,905	8.38	4.81	64,294
1990	31,640	269,545	8.43	4.79	62,241
1991	30,699	300,244	8.48	4.78	60,521
1992	23,366	323,610	8.51	4.77	59,353
1993	19,986	343,596	8.54	4.77	58,436
1994	19,202	362,798	8.56	4.76	57,616
1995	16,115	378,913	8.58	4.76	56,969
1996	12,725	391,638	8.59	4.75	56,482
1997	12,615	404,253	8.61	4.75	56,018
1998	12,051	416,304	8.62	4.75	55,592
1999	7,933	424,237	8.63	4.74	55,320
2000	5,513	429,750	8.63	4.74	55,135
2001	2,309	432,059	8.64	4.74	55,058
2002	1,563	433,622	8.64	4.74	55,007

continued on following page

Copyright © 2008, IGI Global. Copying or distributing in print or electronic forms without written permission of IGI Global is prohibited.

Table 1. continued

		Step 1: Xi	Step 2	Step 3: Yi	Step 4
	Production (K Units)	Accumulated Number (K Units)	Log Xi	Yi=-0.2599Xi +6.9852	10 to the power of Yi
2003	334	433,956	8.64	4.74	54,996
2004	0	433,956	8.64	4.74	54,996

and their home managers. VCR makers needed to offer the video software by itself by the early 1970s. JVC collaborated with an expert in Japanese flower arrangements and a publisher to make video software for flower arrangement to sell to school members along with VCR machines. They also made video software for computer education with a computer maker. The number of consumers was very low in accordance with the product performance, the product performance/price, and the mass-manufacturability. However, JVC's engineers had many useful opportunities to interact with their customers because they did the sales and the repairs by themselves. They figured out the prerequisite conditions, the basic functions, the quality of image and sound, the desired recording time, the compatibility, the right price range, the simplicity of operation, and other requirements for success in the home-use VCR market. This resulted in the enormous success of VHS.

Customers who purchased the Betamax or VHS VCR in the late 1970s were regarded as the innovators. They were very sophisticated and they hoped that these products were the basis of future machines. The majority of people recognized these new products because of mass media advertising. However, the home-use VCR was diffused very slowly because there was still a lot of room for improvement in the performance, the price and the size. Innovators had a much higher regard for the innovativeness or had a greater necessity for the product than the early adopters. Most of them were capable with high technology products and they tended to spread their evaluations. Innovators contributed to an accelerated recognition rate and had a great effect on the decision making process of the early adopters. Although Betamax and VHS were almost equal in performance, the innovative home-use VCR owners put a higher value on the VHS than the Betamax because of the longer recording time. In 1977, Sony introduced Beta II which could record two hours but which was not compatible with the original Betamax. In that way, Sony threw away their lead that the Betamax had accumulated since 1975.

Copyright © 2008, IGI Global. Copying or distributing in print or electronic forms without written permission of IGI Global is prohibited.

Figure 5. Trend of the adoption rate and number of possessions per household (Source: JEITA, 2005)

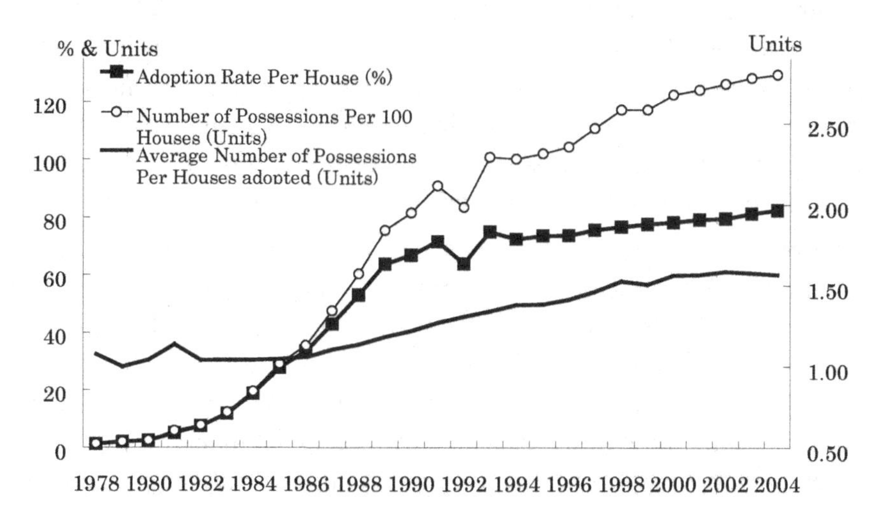

In 1981, the average number of VCR units possessed per household adopted temporarily increased to 1.14 units. VHS exceeded Betamax in units sold by 1980. Some innovators who owned Betamax were expected to purchase a VHS product, the de facto standard in the home-use VCR market and to dispose of the older device because there were very few merits for owning and holding onto incompatible machines in the same home.

VCR Products for Innovators

In our arguments above, the existence of the extreme innovators has been supported in addition to that of the innovators. It may be said that they bought different products at different times. The extreme innovators bought the incomplete innovative products very early on and at highly inflated prices. On the other hand, the innovators bought the still immature new products at a similarly expensive but slightly lower price.

The products for the extreme innovators had fatal deficiencies. Most extreme innovators bought the open-reel machines and U-format VCRs before 1975.

Copyright © 2008, IGI Global. Copying or distributing in print or electronic forms without written permission of IGI Global is prohibited.

Compared with Betamax and VHS (see details in Chapter IV), they were incomplete because of the lack of some critical functions, low performance, cumbersome size and weight, and the expensive price. The VCR ideally should have had basic functions similar to the TV tuner and recording functions. At that time, the selection of video software was very limited and carried a high price tag. Furthermore, early models had low performance in recording time and quality of image and sound. The extreme innovators contributed to the product development and commercialization to a great degree.

The products for the innovators were immature products that satisfied the minimum requirements for commercial success, but which still left much opportunity for improvement. Most innovators bought the SL-6300 or the SL-7300 (Betamax) or the HR-3300 (VHS) during 1975-1980. In the case of the VCR, unfortunately, two different formats, Betamax and VHS, existed at the same time. Innovators used and evaluated the machines simultaneously. They played a very important role in deciding the de facto standard.

Repeat Purchases of Extreme Innovators and Innovators

In addition to the first purchase of the extreme innovators and innovators, their repeat purchases also had a great effect on product development and market growth. These consumers are very affluent and value innovativeness so highly that they are willing to pay big money for an incomplete product. The interval of the repeat purchases may be shorter than that of others because they tend to immediately consider a repeat purchase when new series appear. On the other hand, the majority of consumers only consider a repeat purchase when they have had some trouble with their existing machine or get a desire for another usage, such as dubbing or other special uses. The repeat purchases of the extreme innovators and innovators help decide the popularity of a new series to some extent.

The extreme innovators had a long purchase history. They purchased a first unit when the product was terribly expensive and had low performance. After that, the product became cheaper and its performance improved. Compared to the first unit, the extreme innovators could buy the better ones at a lower price. In the VCR case, although they were very few, they were the most valued customers when a new product series emerged. Thus, they played

Copyright © 2008, IGI Global. Copying or distributing in print or electronic forms without written permission of IGI Global is prohibited.

important roles at the introductions of each of the U-Format, Betamax, VHS, HiFi VHS, and S-VHS innovations.

The innovators also had a long purchase history. They first purchased an immature and incomplete product. The price was still expensive but the product satisfied the basic requirements. The product matured gradually though the addition of new functions with each newly emerging series. Some of them enjoyed collecting the products and making comparisons with various products. Along with the extreme innovators, the innovators played a very important role in the popularity of new products.

Supply Chains to Innovators

The supply chain can be divided into two parts: the upstream (suppliers) and the downstream (distribution). The goal of the upstream supply chain is considered to be developing and manufacturing new products. Similarly that of the downstream supply chain is considered to be informing the upstream of the art of the market and maximizing sales. Both should change according to the types of customers.

In the beginning, the extreme innovators are the primary customer targets. At that time, the R&D activity is prior to the manufacturing activity because the products are experimental and far from completely practical use. Upstream suppliers should locate closely to each other and near the extreme customers for the quickest response to design changes and improvements. The downstream supply chain need not be systemized at the beginning because the extreme innovators are very scarce and widely distributed. Also, the products might still be improved dramatically. Thus, direct sales may be the most appropriate distribution channel to them at this time.

Innovators are still rare but are sufficient in number to start mass production. The upstream supply chain should be systemized for mass production. The upstream supply chain members, the suppliers, still should locate near the integrator because the products have much room for improvement in product design. They also should increase their manufacturing capability to respond to market expansion subject to time, cost, and quality. Close communication between the integrator and upstream supply chain members is highly desirable in order to promote and cope with product and process innovations. The economy of scope is very important because the number

Copyright © 2008, IGI Global. Copying or distributing in print or electronic forms without written permission of IGI Global is prohibited.

of suppliers with whom the integrator can communicate directly and whom they can trust is limited.

The downstream supply chain, the distribution channel, should also be systemized for mass marketing. Direct sales will likely have achieved its limit of efficient operation because the number of potential customers begins to increase rapidly. In addition, the innovators are very sensitive to the latest information and decide to purchase voluntarily. Hence, sales through retail shops and the Internet become very effective with mass advertising.

Conclusion

We discussed the extreme innovators and innovators in this chapter. These types of consumers are very important in the infant market. The extreme innovators contribute to the product development before the emergence of the dominant design because they purchase an incomplete product at an extremely high price. On the other hand, innovators purchase an immature product at a relatively high price. Although the innovators are 2.3 percent of all adopters, the number is enough large to trigger mass-production. Their reviews also have great effects on the future diffusion because their sense is much closer to the majority than the extreme innovators. Furthermore, their second and later purchases have a strong relationship with the alternation of product generations and the end of life cycle. They promote the alternation of products or the change to the alternative products.

In the VCR case, the extreme innovators purchased relatively defective or incomplete products at an extremely high price before the emergence of the dominant design. Their reviews oriented the direction of the product development and led companies to figure out the minimum requirements for commercial success. Most innovators adopted the first Betamax or VHS at their first purchase. These satisfied minimum requirements for commercial success and were no longer incomplete products. In the VCR case, the innovator's behaviors and reviews decided the de facto standard.

Copyright © 2008, IGI Global. Copying or distributing in print or electronic forms without written permission of IGI Global is prohibited.

References

Bass, F. M. (1969). A new product growth model for consumer durables. *Management Science, 15*, 215-227.

Buzzell, R. D. (1966). Competitive behavior and product life cycles. In J.S. Wright & J. L. Goldstucker (Eds.), *New ideas for successful marketing* (pp. 46-68). Chicago: American Marketing Association.

Economic and Social Research Institute, Cabinet Office, Government of Japan. Retrieved March 15, 2006 from http://www.cao.go.jp/index-e.html

Electronics Industries Association of Japan. Retrieved March 15, 2006 from http://www.jeita.or.jp/eiaj/english/index.htm

Funai. Retrieved March 15, 2006 from http://www.funaiworld.com/index.html

Higuchi, T. (2006). Industrial life cycle in the VCR industry. *Sakushin Policy Studies,* 6, 19-34.

Higuchi, T., & Troutt, M.D. (2004). Dynamic simulation of the supply chain for a short life cycle product - Lessons from the *Tamagotchi* case. *Computers & Operations Research, 31*(7), 1097-1114.

Higuchi, T, Troutt, M. D., & Polin, B. A. (2004). Life cycle considerations for supply chain strategy. In C.K. Chan & H.W.J. Lee (Eds.), *Successful strategies in supply chain management* (pp. 67-89). Hershey, PA: Idea Group.

Hitachi. Retrieved March 15, 2006 from http://www.hitachi.com/

Itami, H. (1989). *Nihon no VTR sanngyou naze sekai wo seiha dekitanoka.* Tokyo: NTT Publishing.

Japan Electronics and Information Technology Industries Association. Retrieved March 15, 2006 from http://www.jeita.or.jp/english/index.htm

Japan Video Software Association. Retrieved March 15, 2006 from http://www.jva-net.or.jp/en/index.html

JEITA (Japan Electronics and Information Technology Industries Association). (2005). *Minnseiyou dennsi kiki deta shyu.* Tokyo: JEITA.

Kohli, R., Lehman, D. R, & Pae, J. (1999). Extent and impact of incubation time in new product diffusion. *Journal of Product Innovation Management, 16*(2), 134-144.

Kotler, P. (1999). *Marketing management.* New Jersey: Prentice Hall.

Copyright © 2008, IGI Global. Copying or distributing in print or electronic forms without written permission of IGI Global is prohibited.

Lilien, G., Kotler, P., & Moorthy, K. (1992), *Marketing models*. Englewood Cliffs, NJ: Prentice-Hall.

Matsushita. Retrieved March 15, 2006 from http://panasonic.net/

McIntyre, S. H. (1988). Market adaptation as a process in the product life cycle of radical innovations and high technology products. *Journal of Product Innovation Management, 5*(2), 140-149.

Mitsubishi Electric. Retrieved March 15, 2006 from http://global.mitsubishi-electric.com/

Moore G. A. (1991). *Crossing the chasm*. New York: Harper Business.

Moore G. A. (2005). *Dealing with Darwin*. New York: Portfolio.

National Science Museum, Tokyo. Retrieved March 15, 2006 from http://sts.kahaku.go.jp/sts/

NHK (Japan Broadcasting Corporation.). (2000). *Project X: Challengers 1*. Tokyo: Japan Broadcast Publishing.

Rogers, E. M. (1995). *Diffusion of innovations* (4th ed.). New York: The Free Press.

Sato, M. (1999). *The story of a media industry*. Tokyo: Nikkei Business Publications.

Schoenherr, S. Retrieved March 15, 2006 from http://home.sandiego.edu/~ses/

Sheth, J. N. (1971). Word of mouth in low-risk innovations. *Journal of Advertising Research, 11*(3), 15-18.

Sony Corporation. Retrieved March 15, 2006 from http://www.Sony.net/

Toshiba. Retrieved March 15, 2006 from http://www.toshiba.co.jp/index.htm

Victor Company of Japan. Retrieved March 15, 2006 from http://www.jvc-victor.co.jp/english/global-e.html

Copyright © 2008, IGI Global. Copying or distributing in print or electronic forms without written permission of IGI Global is prohibited.

Chapter X

Early Adopters and Early Majority

In this chapter, the characteristics and the role of the early adopters and early majority are reviewed from the life cycle perspective following the VCR case study. Both of these groups of consumers purchase a mature or more advanced product at a reasonable price. Their adoption demonstrates that the product performance has enough future possibilities. On the other hand, manufacturers should expand their manufacturing facilities quickly because the demand starts growing dramatically. The early adopters (13.6 percent) are almost six times as many as innovators (2.3 percent) and the early majority (34.1 percent) is about 2.5 times as many as early adopters. Once a diffusion process starts, the spread is very rapid in the first group.

Copyright © 2008, IGI Global. Copying or distributing in print or electronic forms without written permission of IGI Global is prohibited.

Characteristics of Early Adopters and Early Majority

Early adopters, like innovators, buy into new product concepts very early in their life cycle, but unlike innovators, they are not technologists. Rather they are people who find it easy to imagine, understand, and appreciate the benefit of new technology, and to relate these potential benefits to their other concerns. Whenever they find a strong match, early adopters are willing to base their buying decision upon it. Because early adopters do not rely on well-established references in marketing these buying decisions, preferring instead to rely on their own intuition and vision, they are key to opening up any high-tech market segment. (Moore, 2002, p. 12)

The next 13.6 percent of the adopters following innovators (2.3 percent) are called the early adopters. They are the people who are between the double and the single standard deviation ranges in the first half. Their number is sufficiently large to draw other companies to the market and to decide the future of the industry.

The early adopters are very thoughtful and take some risk in adopting new products. They understand the meaning of the new product in the practical sense. They collect ample information from the manufacturers and are keenly aware of the risk involved. Unlike the innovators, they put emphasis on the potential benefit of the new product. It is natural for them to consider the benefit/cost ratio under some risk. In the event two different products compete for the de facto standard, they try to foresee the winner. They are very rational and future-oriented, and as a result, they are also called visionaries. Moore (2002) gives the following description:

The early majorities share some of the early adopter's ability to relate to technology, but immediately they are driven by a strong sense of practicability. They know that many of these newfangled inventions end up as passing fads, so they are content to wait and see how other people are making out before they buy in themselves. They want to see well-established references before investing substantially. Because there are some so many people in this segment—roughly one-third of the whole adoption life cycle—winning their business is key to any substantial profits and growth. (pp. 12-13)

Copyright © 2008, IGI Global. Copying or distributing in print or electronic forms without written permission of IGI Global is prohibited.

The early majority is the next 34.1 percent of the adopters following the innovators (2.3 percent) and early adopters (13.6 percent). They are the people who are between the lower single standard deviation and the average, and they emphasize the newness and practical use. They have an immense impact on the industry because of their number, which is so huge that many companies can not only survive in the industry, but can even expand their manufacturing ability as well with additional product differentiation.

The early majority is sensitive to trends and is made up of pragmatists that make decisions based on practical considerations. The diffusion process of this 34.1 percent of consumers occurs very quickly. In addition, they prefer additional functions and features for their convenience and for product or brand differentiation. Excellent instruction manuals are essential for them to use the most attractive products because of the level of their technical skills and the sophistication of the products. On the other hand, they place special emphasis on confidence in the product, which is proven to them by the previous users.

Early Adopter's Effect on Future Adopters

Decisions made by the early adopters have a significant effect on future adopters. Early adopters judge the potential benefits and product performance under a certain cost and decide to adopt it. In effect, future adopters, such as the early majority, late majority, and laggards, inherit the adopter's decisions because the future products are extensions of the products admired by them.

There exist four boundaries that separate these groups of consumers: the innovators, the early adopters, the early majority, the late majority, and the laggards. That between the early innovators and the early majority is called *the chasm*, and is considered the most important among them (Moore, 2002). However, we conclude that the partitions between the earlier stages are bigger than those of the later stages, from the industrial life cycle viewpoint. Consumers adopt the promised products and contribute to the development of those products by their input and requests. As a result, the possibility of the discontinuance of the early adopter's decision is much less than that of the innovators, especially the extreme innovators. This differs from the view of Moore (2002) because we put more emphasis on the growth of the industry, the progress of the product and the supply chain.

Copyright © 2008, IGI Global. Copying or distributing in print or electronic forms without written permission of IGI Global is prohibited.

Role of Early Adopters in Industry

Visionaries are that rare breed of people who have the insight to match an emerging technology to a strategic opportunity, the temperament to translate that insight into a high-visibility, high-risk project, and the charisma to get the rest of their organization to buy into that project. They are the early adopters of high-tech products. Often working with budgets in the multiple millions of dollars, they represent a hidden source of venture capital that funds high-technology business. (Moore, 2002, pp. 33-34)

Early adopters play a very important role in industry growth. From the product innovation viewpoint, their adoption demonstrates that the product performance satisfies the minimum requirements for commercial success and that the product has sufficient future possibilities. In effect, they order the industry to mature the products by adding new functions and improving the basic design.

From the perspective of process innovation, the industry can boost mass production after a good response, so that rapid market growth can occur. The industry can confirm that they are on the right track. The number of these adopters is large enough to permit the start of mass production. The industry should therefore expand manufacturing capability and improve product performance while maintaining high product quality.

From the marketing perspective, they set a de facto standard from the alternatives by judging future possibilities favorably and by informing the potential later customers of their evaluation. If companies want to make their product a de facto standard, they should introduce products by the end of the time when the early adopters finish making their first purchases. Otherwise, it is very difficult for any product that is superior to the existing products in performance and price to later prevail because of the trend and the externalities.

Early Majority's Impact on the Industry

Throughout the 1980s, the early majority, or pragmatists, have represented the bulk of the market volume for any technology product. You can succeed with the visionaries, and you can thereby get a reputation for being a high

Copyright © 2008, IGI Global. Copying or distributing in print or electronic forms without written permission of IGI Global is prohibited.

flyer with a hot product, but that is not ultimately where the dollars are. Instead, those funds are in the hands of more prudent souls, who do not want to be pioneers ("Pioneers are people with arrows in their backs"), who never volunteer to be an early test site ("Let somebody else debug your product"), and who have learned the hard way that the "leading edge" of technology is all too often the "bleeding edge." (Moore, 2002, pp. 41-42)

The early majority has a great impact on the industry in the following ways. First, they boost the expansion of the market. Although the expansion rate is decreasing, the absolute number of sales increases tremendously. At the beginning, the expansion rate tends to be exaggerated because the base, the number of innovators or the early adopters, is very minute. The early majority forces the manufacturers to expand their manufacturing ability and gives others the chance to enter the market.

Secondly, they widen the variety of the product. A lot of new functions have been added to the products. Some of them become essential to all products and others are optional. Manufacturers make combinations among the various functions to respond to customer needs. The price of the products diversifies from the low end to the high end. In our analysis based on price, the products are divided into three categories: the low end, the middle class (standard), and the high end. The range of companies promotes the diversification of the products for differentiation.

Finally, they stabilize the framework of the competition among manufacturers under a certain level of risk. The manufacturers accumulate experience and reputation during the rapid market growth. Based on their strengths, they focus on the target customers. It would be very difficult to overcome a leader in the low-growth phase after the rapid growth. On the other hand, manufacturers run a risk. They expect the future sales and expand their manufacturing capability accordingly. They might misjudge the future demand, the future market size, and their popularity, the self-evaluation. Sometimes, manufacturers miss the great opportunity and suffer overproduction.

Early Adopters and Early Majority of the VCR

In March 1983, VCR ownership by percentage of Japanese households, reached 11.8 percent. This number is equal to 14.3 (=11.8/82.6) percent of

Copyright © 2008, IGI Global. Copying or distributing in print or electronic forms without written permission of IGI Global is prohibited.

the maximum 82.6 percent reached in 2004, which is slightly smaller than that of a single standard deviation, 15.9 percent. The early adopters (between the double and single standard deviation, 2.3-15.9 percent) are called the early adopters. Those people who purchased the VCR during 1980-1983 for the first time can be regarded as early adopters.

During the early 1980s, the average number of VCRs possessed per household remained almost at the same level, about 1.4 units, except for a bulge in 1981, 1.137 units. This is because the innovators, especially Betamax users, bought the VHS machines. Then, some abandoned or gave up the Betamax machine while others kept both Betamax and VHS machines. Most early adopters did not have the demand for another machine in their home for dubbing or other personal uses.

The early adopters contributed to the maturation of the VCR products. In 1983, Sony and JVC introduced HiFi sound models. HiFi models then became the mainstream, replacing the standard (monaural) VHS, S-VHS (Super-VHS), W-VHS (World-VHS) and D-VHS (Digital-VHS). These latter models needed more practical functions rather than more extraneous functions. Hence, companies improved the basic performance of the VCR products.

Furthermore, the VCR industry should have advanced the process innovation to respond to the rapid market growth. The rate of growth accelerated greatly. The extreme innovators (the first 0.13 percent adopters) purchased incomplete VCR products for about a decade, until 1975, and the innovators (the next 2.14 percent) adopted immature products for about four years, during 1975-1979, as first purchases. On the other hand, the early adopters (the next 13.59 percent) purchased for three years. The diffusion speed to the early adopters per year was much faster than that of previous consumers. The domestic production in units increased to 28,611 from 2,199 thousands units during 1980-1984 and the peak was 33,879 thousands units in 1986.

In March 1987, VCR ownership by percentage of Japanese households, reached 43.0 percent. This number is equal to 52.1 percent of the maximum 82.6 percent reached in 2004. Those people who bought a VCR for the first time during 1983-1987 were called the early adopters and 37.8 percent of the total users, adopted during this four year period. The average penetration rate was 9.4 percent.

VCRs were the symbol of rapid growth. While the unit number of VCR production and the domestic shipment were increasing, the yield met its peak in 1983. This was mainly because the competition promoted discounts even during the rapid market growth and partly because the offshore production

Copyright © 2008, IGI Global. Copying or distributing in print or electronic forms without written permission of IGI Global is prohibited.

Figure 1. Transition of the number and cost of domestic shipment (Source: JEITA, 2005)

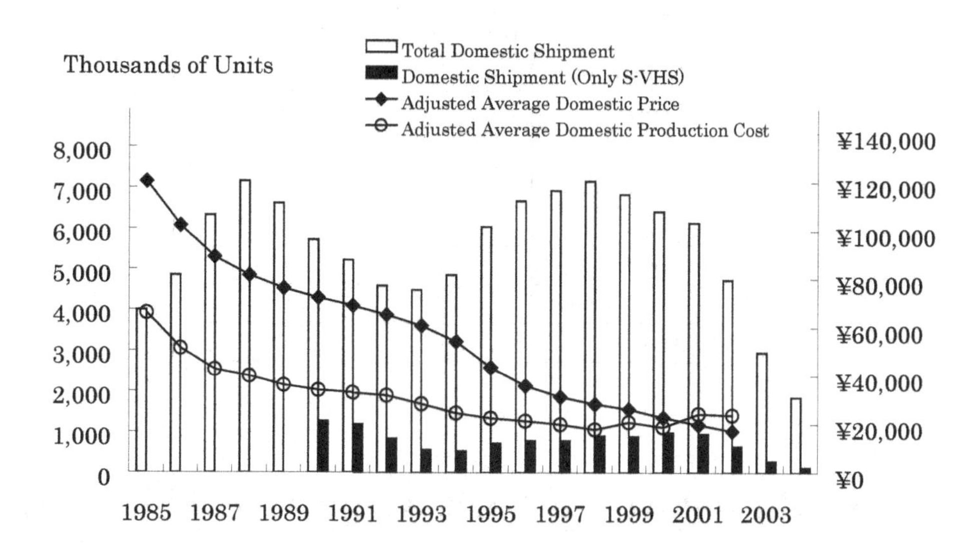

had begun with lower costs. Figure 10-1 shows the transition of the number of units and the adjusted average domestic price of domestic shipments and the production cost. The first peak of the domestic shipment in units was in 1988. The average domestic shipment price and production cost decreased sharply until 1987. Then those decreasing rates slowed down. The early majority also showed a sharp decline just before or after they bought for the first time.

By 1986, the VHS had diffused worldwide. The accumulated number of units shipped is shown in Table 1. It shows that North America had a great purchasing power, in which the majority of people purchased the product earlier than others. However, JVC set up only one facility in North America while having seven plants in Europe. This was partly because Matsushita explored the North America markets and JVC focused on the European market. In addition, European countries adopted protection against other areas. It was said that more than 150 companies produced or received OEM VHS products throughout the world (JVC 1987).

Copyright © 2008, IGI Global. Copying or distributing in print or electronic forms without written permission of IGI Global is prohibited.

Table 1. Accumulated number of units shipped and location of JVC's plants in 1986

	Units (million)	No. of JVC Plants
North America	58.2	1
Europe	32.5	7
Japan	22.9	1
Asia less Japan	8	2
Middle East	4.1	0
Oceania	3.6	0
Latin America	2.1	0
Africa	1.6	0

VCR Products for Early Adopters and Majority

In the early 1980s, VHS products became popular. The VCR was diffused widely beyond the early VCR fans and extreme techies. JVC called the HR-7100 and the HR-D120, which had been designed by a woman, the *Fashionable Video Cassette*. These products were so convenient for everyone because of the large buttons and pictographs that they came to be regarded as the popular version. JVC advertised and reinforced this user-friendliness at their retail dealers. They explained to the customers the excellence of their product in detail and gently guided the customers in their choice of a product. The VCR products therefore penetrated to daily life.

Most early majority consumers purchased the monaural or HiFi VCR. There were two trends influencing the product. First, the prices of the monaural and HiFi VCRs continued to decrease. The price of monaural and HiFi VCRs dropped to about a half and one-third, respectively, of their previous price, during 1983-1987. As a result, the price difference between them shrank. In addition, the price difference of the monaural VCR among the leaders and cost cutters diminished. It was not profitable for the leaders to manufacture and sell the monaural VCR because their cost had become higher than that of the cost cutters.

Secondly, the mainstream product changed from monaural to HiFi VCRs during this period. Both the leaders and followers offered HiFI VCRs. HiFi VCRs continued to decline steadily in price after it was introduced in 1983 because of the development of the production process that enabled localizing

Copyright © 2008, IGI Global. Copying or distributing in print or electronic forms without written permission of IGI Global is prohibited.

the impact of changes. This meant that manufacturers could make the most of previous parts and experience. Some leaders decided to receive OEM products from the cost cutters to focus on the HiFi VCRs and keep the product lineup. The cost cutters waited for the best timing to manufacture the HiFi VCRs while manufacturing the monaural products for the OEM and themselves.

In the middle of the 1980s, in addition to the sound, the quality of images of VHS products improved. JVC developed the HQ (High Quality) mode by which the images of the products were very much improved while maintaining compatibility with previous machines. They launched the HR-D565 and the HR-D470, which were called the HQ & HiFi VHS.

In 1987, JVC launched the S-VHS VCRs. However, most purchasers were the innovators and the early adopters, while very few of them were early majorities. In addition to the high cost, these VCRs were unproven as far as performance, practicability, and future utility. The early majority scrutinized the developing market. Hence, some of them selected the S-VHS VCR for their first purchase. On the other hand, the innovators and early adopters purchased these products as replacements or repeat purchases. Typical service lives of VCRs are five to 10 years. In addition, the innovators and the early majority value innovativeness highly. As a result, they were willing to pay more for the VCRs and take bigger risks than others.

Figure 2(a) and Figure 2(b) show the price ranges of JVC, Matsushita, Sanyo, Funai, and Aiwa, during 1985-1986 and 1987-1988. In 2(a), Funai and Aiwa had not introduced the products to the customers yet. The ranges between the

Figure 2(a). Price range during 1985-1986

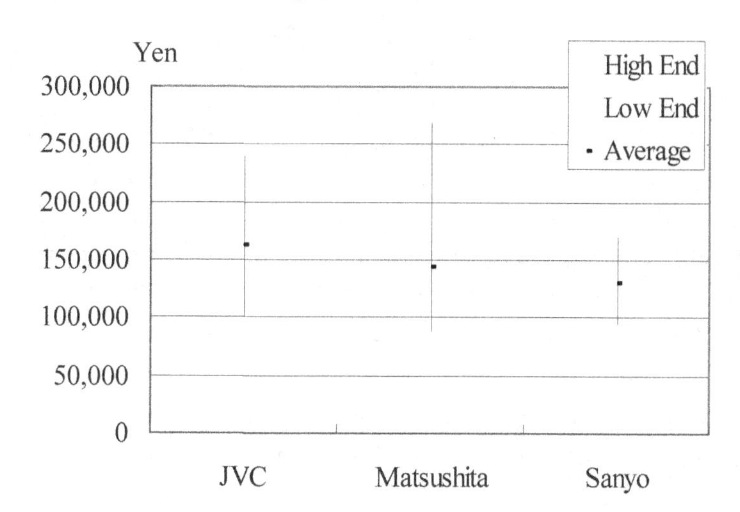

Copyright © 2008, IGI Global. Copying or distributing in print or electronic forms without written permission of IGI Global is prohibited.

Figure 2(b). Price range during 1987-1988

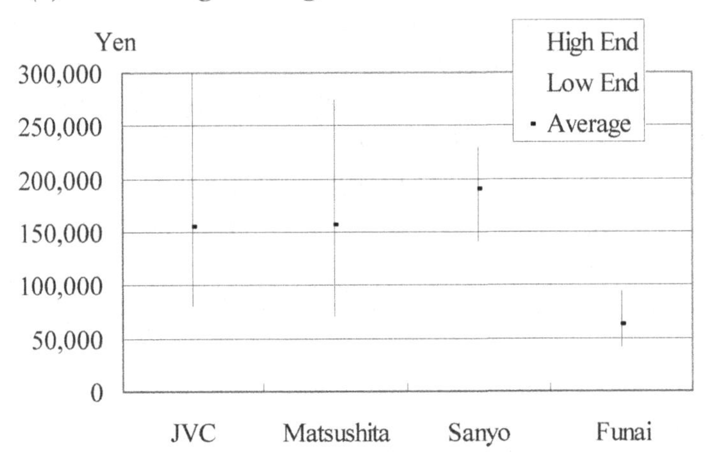

high end and the low end were relatively small because the impact of HiFi VHS became smaller than before. Figure 2(b) shows two things. First, the ranges of price became bigger than before because of the emergence of the S-VHS. As a result, the average price of products of JVC, Matsushita and Sanyo increased. The average price was calculated based on the models, not the shipment base. The best selling models were HiFi VHS. Second, Funai entered the market with a low-end target. To avoid price competition, the others would have upgraded their products.

Repeat Purchases of Early Adopters and Early Majority

We regarded the people who purchased the first VCR during 1980-1983 as the early adopters. The adjusted average production cost per unit deflated by the GDP deflator was 153,238 and 93,278 Yen in 1980 and 1983, respectively. The shipment price would be about twice the production cost. Compared to the production cost for the early majority during 1984-1987, from 79,385 to 42,710 Yen, the early adopters were lavish in purchasing expensive products.

There were several reasons for the early adopters to make repeat purchases including the replacement of an inadequate unit and the need for multiple

Copyright © 2008, IGI Global. Copying or distributing in print or electronic forms without written permission of IGI Global is prohibited.

units. Early adopters would have replaced an old unit with an improved new one because of wear and tear and age problems, as well as for other reasons such as inadequate features. Generally, the approximate lifetime of the machine was from five to 10 years. Additionally, the price decline of the VCR encouraged them to purchase another unit. The early adopters could buy another one at almost half the previous price within a few years after the first purchase. Some of them purchased another one for personal use, dubbing, and multiple uses. Lastly, the product advancement motivated their repeat purchase. The early adopters were influenced by the emergence of the new series, S-VHS, and new functions. Some of them purchased an additional unit while others replaced old ones.

Their repeat purchases had two major effects on the development of the VCR industry. First, they positively evaluated the usefulness of the S-VHS VCR. They spent a lot of money on the VCR but the prices were deemed very reasonable. Their evaluation was the key to further adoption. They did not evaluate the S-VHS machine highly compared to the price. As a result, the adoption rate of the S-VHS remained at a low level, and was mostly adopted by professionals and techies. Regarding company strategy, it is very important to determine the most advantageous timing for introducing a new series and to get positive feedback about it from the early adopters. Second, they heightened the peak of the domestic shipment in 1988 to some extent. To cope with the peak demand, companies should ideally expand their facilities but there is a risk that demand might decrease abruptly after the peak. The behavior of the early adopters serves as a good indicator of how the early majority moves. To prepare for future demand, companies should analyze adopter behavior and feedback about the product.

The repeat purchases of the early majority mainly occurred about 10 years after their first purchase. The second peak of domestic shipments in the units was slightly bigger than the first one and occurred 10 years later. The total domestic shipment in units during 1979-1992 was 57 million units and was fewer than that during 1993-2004, namely 60 million units. The repeat purchases of the early majority proceeded at a slightly faster rate than the first purchases because the time to the peak was shorter. In addition to replacements, they also purchased extra units for dubbing and personal use. This raised the number of domestic shipments to some extent.

Copyright © 2008, IGI Global. Copying or distributing in print or electronic forms without written permission of IGI Global is prohibited.

Supply Chains to Early Adopters and Early Majority

Upstream supply chains should finish localizing the impact of changes because they have been formulated. Otherwise, they should revise themselves from the start to respond to product development. They are required to increase manufacturing capability rapidly under the condition that the basic design might be changed to some extent. If they can not localize the impact of changes, they should change the manufacturing process. Without localization, it would be very difficult for them to expand their capability incrementally after they launch mass production. To be flexible for product development and to cope with rapid demand growth simultaneously, they need to localize the impacts.

Rapid market growth attracts followers to the market. The differences between the followers and the leaders are technology and brand image. Although the followers often have unique peripheral technologies, they are inferior to the leaders in the core technologies of the product. In addition to patents, the followers sometimes procure the core parts from the leaders or their suppliers. The followers launch new products by adding new functions.

Rapid market growth also incentivizes the OEM (Original Equipment Manufacturers) of the standard products. In this stage, cost cutters are the OEM, unlike the introduction stage where the leaders are the original equipment manufacturers because of manufacturability. Product development creates many products and series of products. It is very costly and risky to maintain all lines in a company and it is reasonable for any company to focus on emerging new products rather than the old or standard ones.

The major way to sell products to early adopters is retailing. The downstream supply chains expand the number of sales items. This increases the inventory cost of products so much so that they need to become concerned with logistics, which is the part of the supply chain process that plans, implements, and controls the efficient, effective flow and storage of goods, services, and related information from the point of origin to the point of consumption in order to meet the customers' requirements (Council of Logistics Management). Otherwise, they can not balance cost and sales (service level) to maximize benefits. The Internet or mail orders are as yet unsuitable for the early adopters because the reputation of the products has not been established since the product is still under development. The variety of products is so wide that it

Copyright © 2008, IGI Global. Copying or distributing in print or electronic forms without written permission of IGI Global is prohibited.

is very risky to purchase a product without checking directly. Hence, an effective and efficient distribution channel including the retailers, wholesalers and integrators is required.

The downstream supply chains gain initiatives in the industry. The early majority has great purchasing power at a certain time and their preferences are diversified. A wide variety of products are offered to the early majority. Although production size is very important to obtain the economies of scale, it entails great risk including unsold products or overproduction. Hence, the whole supply chain should become very sensitive to demand that is very trendy. Otherwise, they lose their sales opportunity and suffer from huge inventories.

To be sensitive to demand, the distribution channel is the key to efficient sales and the latest market information. Retail shops have the latest customer information through direct contact with them. The quantitative information, such as orders, is processed easily within the supply chain as well as mail order or Internet sales. However, it is very difficult to transact qualitative information, such as the desired specifications, because it is transmitted mainly by word of mouth. The qualitative information is distorted and transmitted with lags. The downstream supply chains play the lead in the supply chain.

The downstream supply chains are required to be geared to the upstream supply chains to increase profitability by reducing lead time and inventory. They make the most of standardization and postponement. They standardize the parts to increase the number of common parts and delay the timing to specify certain products. As a result, they can correspond to a wide variety of products with speed and efficiency.

The normal direction of OEM is reversed. In the introduction stage, the leaders become the OEM for the followers because of the technological difficulties. On the other hand, the leaders receive the OEM products, the low-end products, in order to maintain the product lineup while focusing on the high-end and middle-class products. This means that the parts market has developed and the technological difficulties are decreasing.

Conclusion

We discussed the early adopters and early majority. Their characteristics and role are reviewed from the life cycle viewpoint. In the first half of the diffu-

Copyright © 2008, IGI Global. Copying or distributing in print or electronic forms without written permission of IGI Global is prohibited.

sion process, the diffusion speed is very fast because the numbers of these types of customers grow rapidly. The early adopters are almost six times as many as innovators and the early majority is about 2.5 times as many as early adopters. Their adoption forces companies to expand their production facilities and gives new entrants a chance to compete in the market. On the other hand, their adoption is the proof that the product performance has been improved enough for a wide diffusion.

In the VCR case study, most early adopters purchased a monaural VHS at first in the early 1980s. In the middle of 1980s, the early majority purchased for the first time. Although HiFi and monaural VHS machines existed at that time, most early majority buyers were thought to select a monaural VHS machine because monaural VHS exceeded HiFi VHS in availability at that time. The early adopters and the early majority were likely to purchase a mature or completed product at a reasonable price.

References

Bass, F. M. (1969). A new product growth model for consumer durables. *Management Science, 15*, 215-227.

Buzzell, R. D. (1966). Competitive behavior and product life cycles. In J.S. Wright & J.L. Goldstucker (Eds.), *New ideas for successful marketing* (pp. 46-68). Chicago: American Marketing Association.

Economic and Social Research Institute, Cabinet Office, Government of Japan. Retrieved March 15, 2006 from http://www.cao.go.jp/index-e.html

Electronics Industries Association of Japan. Retrieved March 15, 2006 from http://www.jeita.or.jp/eiaj/english/index.htm

Funai. Retrieved March 15, 2006 from http://www.funaiworld.com/index.html

Higuchi, T. (2006). Industrial life cycle in the VCR industry. *Sakushin Policy Studies,* 6, 19-34.

Higuchi, T., & Troutt, M.D. (2004). Dynamic simulation of the supply chain for a short life cycle product - Lessons from the *Tamagotchi* case. *Computers & Operations Research, 31*(7), 1097-1114.

Copyright © 2008, IGI Global. Copying or distributing in print or electronic forms without written permission of IGI Global is prohibited.

Higuchi, T, Troutt, M. D., & Polin, B. A. (2004). Life cycle considerations for supply chain strategy. In C.K. Chan & H.W.J. Lee (Eds.), *Successful strategies in supply chain management* (pp. 67-89). Hershey, PA: Idea Group.

Hitachi. Retrieved March 15, 2006 from http://www.hitachi.com/

Itami, H. (1989). *Nihon no VTR sanngyou naze sekai wo seiha dekitanoka.* Tokyo: NTT Publishing.

Japan Electronics and Information Technology Industries Association. Retrieved March 15, 2006 from http://www.jeita.or.jp/english/index.htm

Japan Video Software Association. Retrieved March 15, 2006 from http://www.jva-net.or.jp/en/index.html

JEITA (Japan Electronics and Information Technology Industries Association). (2005). *Minnseiyou dennsi kiki deta shyu.* Tokyo: JEITA.

Kohli, R., Lehman, D. R, & Pae, J. (1999). Extent and impact of incubation time in new product diffusion. *Journal of Product Innovation Management, 16*(2), 134-144.

Kotler, P. (1999). *Marketing management.* New Jersey: Prentice Hall.

Lilien, G., Kotler, P., & Moorthy, K. (1992), *Marketing models.* Englewood Cliffs, NJ: Prentice-Hall.

Matsushita. Retrieved March 15, 2006 from http://panasonic.net/

McIntyre, S. H. (1988). Market adaptation as a process in the product life cycle of radical innovations and high technology products. *Journal of Product Innovation Management, 5*(2), 140-149.

Mitsubishi Electric. Retrieved March 15, 2006 from http://global.mitsubishi-electric.com/

Moore G. A. (1991). *Crossing the chasm.* New York: Harper Business.

Moore G. A. (2005). *Dealing with Darwin.* New York: Portfolio.

National Science Museum, Tokyo. Retrieved March 15, 2006 from http://sts.kahaku.go.jp/sts/

NHK (Japan Broadcasting Corporation.). (2000). *Project X: Challengers 1.* Tokyo: Japan Broadcast Publishing.

Rogers, E. M. (1995). *Diffusion of innovations* (4th ed.). New York: The Free Press.

Sato, M. (1999). *The story of a media industry.* Tokyo: Nikkei Business Publications.

Copyright © 2008, IGI Global. Copying or distributing in print or electronic forms without written permission of IGI Global is prohibited.

Schoenherr, S. Retrieved March 15, 2006 from http://home.sandiego.edu/~ses/

Sheth, J. N. (1971). Word of mouth in low-risk innovations. *Journal of Advertising Research, 11*(3): 15-18.

Sony Corporation. Retrieved March 15, 2006 from http://www.Sony.net/

Toshiba. Retrieved March 15, 2006 from http://www.toshiba.co.jp/index.htm

Victor Company of Japan. Retrieved March 15, 2006 from http://www.jvc-victor.co.jp/english/global-e.html

Copyright © 2008, IGI Global. Copying or distributing in print or electronic forms without written permission of IGI Global is prohibited.

Chapter XI

Late Majority and Laggards

The characteristics and the role of the late majority and laggards are discussed in this chapter. Both of them adopt the product later than others and purchase a practical product at a lower price. Although their population is huge, their impact on the market is not big because various types of customers exist in the market. Manufacturers offer a wide selection of products for various customers. The late majority and the laggards are a part of all customers. In addition, most manufacturers finish enlarging their manufacturing facility before the late majority purchases occur. However, from the view of the life cycle, they play an important role as a last purchaser including the repeat purchases.

Copyright © 2008, IGI Global. Copying or distributing in print or electronic forms without written permission of IGI Global is prohibited.

Characteristics of Late Majority and Laggards

The late majority shares all the concerns of the early majority, plus one major additional one: Whereas people in the early majority are comfortable with their ability to handle a technology product, should they finally decide to purchase it, members of the late majority are not. As a result, they wait until something has become an established standard, and even then they want to see lots of support and tend to buy, therefore, from large, well-established companies. Like the early majority, this group comprises about one-third of the total buying population in any given segment. Courting its favor is highly profitable indeed; for a while profit margins decrease as the products mature, so do the selling costs, and virtually all the R&D costs have been amortized. (Moore, 2005, p. 13)

The late majority is the next 34.1 percent of the adopters following the early majority (34.1 percent). They are the huge group of people who are between the average and a single standard deviation and who put emphasis on practical use and price. Their adoption rate per year decreases as time passes.

The late majority is conservative toward new technology. They are hesitant to adopt a new product until it has been proven to be very practical and the price goes down to a reasonable level for them. In addition, they are not good at the operation of new technology products. Hence, some of them want models with limited functions for simplification and a correspondingly lower price. Others rely on well-established brand names or cling to a specific function. Customization is necessary for them.

Laggards simply don't want anything to do with new technology, for any of a variety of reasons, some personal and some economic. The only time they ever buy a technological product—the way, say, that a microprocessor is designed into the braking system of a new car—that they don't even know it is there. Laggards are generally regarded as not worth pursuing on any other basis. (Moore, 2002, p. 13)

The last 15.9 percent of adopters are called laggards and they follow the late majority (34.1 percent). They are a large group of people who are between one and two standard deviations above the mean. They are also called skeptics because most of them adopt cautiously or passively. Some purchase the

Copyright © 2008, IGI Global. Copying or distributing in print or electronic forms without written permission of IGI Global is prohibited.

product because of unavoidable reasons, and others receive the product from others who purchase one as a gift for instance. The diffusion process for them is very slow because they do not purchase the product on their own initiative. However, the products gradually penetrate to laggards as time passes.

Laggards want an easy-to-manipulate product because most of them are unskilled in high-tech products. They also do not have sufficient knowledge about the product and do care for high-tech products in general. Hence, they will ask others to purchase a simple one for them or take the recommendation of a sales person.

Effect of Late Majority and Laggards on the Industry

Conservatives (late majority) have enormous value to high-tech industry in that they greatly extend the market for high-tech components that are no longer state-of-art. The fact that the United States has all but conceded great hunks of this market to the Far East is testimony not so much to the cost advantages of offshore manufacturing as to the failure of onshore product planning and marketing imagination. Many Far East solutions today still bring only one value to the table---low cost. That is, they are nowhere near the goal of being a "whole product solution." Thus, they typically have to go through VAR channel in order to be upgraded to the kind of complete system that a conservative can purchase. The difficulty in this distribution strategy is that few VARs are large enough to achieve the volume needed to leverage a conservative market. Far more dollars could be mined from this segment of the high-tech market place if American leading-edge manufacturers and marketers, with their high-volume channels and vast purchasing sources, simply paid more attention to it. (Moore, 2002, p. 48)

The industry does not focus on the late majority in sales promotion, unlike the innovators, the early adopters and the early majority. In the market, the innovators, the early adopters, and the early majority make a good number of repeat purchases and prefer the standard and high-end products with new functions and technologies. The industry reinforces the high-end and the low-end products simultaneously. Some of the late majority follows the previous adopters and others seek the low-end products.

Copyright © 2008, IGI Global. Copying or distributing in print or electronic forms without written permission of IGI Global is prohibited.

However, they can bring a great amount of profit to the industry. At this stage, most companies will have recovered their initial costs, such as R&D expenditures, and will have finished studying and improving the manufacturing system. The demand including the repeat purchases remains at a high level. In cases where the manufacturing facilities maintain a high rate of operations linked with the market demand, they become the source of profitability.

Skeptics (Laggards)—the group that makes up the last one-sixth of the Technology Adoption Life Cycle—do not participate in the high-tech marketplace, except to block purchases. Thus, the primary function of high-tech marketing in relation to skeptics is to neutralize their influence. In a sense, this is a pity because skeptics can teach us a lot about what we are doing wrong—hence this postscript. (Moore, 2002, p. 54)

Laggards have the weakest impact on the industry among all types of adopters. At that time, the previous adopters start repeat purchases and that accounts for most of the demand in the market. The laggards adopt a product passively. It means that it takes a longer time to spread to them than others and that they are influenced by others, namely the previous adopters. The presence of the laggards does not have much effect on the industry, when maturity continues.

However, the late majority and the laggards are still very important because they might determine the end or duration of the life cycle. If they do not adopt a product, diffusion would stop and the market would become smaller. In addition, this situation might make companies withdraw from the market before the repeat purchases of the early adopters and the early majority. Companies should care for them in order to expand the market and to manage the life cycle of the product.

Late Majority and Laggards of the VCR

In March 1987, VCR ownership by percentage of Japanese households reached 43.0 percent. This number is equal to 52.1 percent of the maximum 82.6 percent reached in 2004. VCR ownership in 1991 reached 71.5 percent, which equals 86.6 percent of the 2004 percentage. Those people who bought the VCR for the first time during 1987-1991 are regarded as the late majority.

Copyright © 2008, IGI Global. Copying or distributing in print or electronic forms without written permission of IGI Global is prohibited.

The average number of units possessed per household increased from 1.06 to 1.27 units during 1986-1991. Some people decided to purchase another unit before the end of the useful life of existing units because of the excellent new products, dubbing, or personal use. It is believed that mainly the innovators and the early majority purchased a second machine during this period. The average number continued to increase to 1.59 units in 2002. The late majority was considered to have mainly contributed to the increase after 1992 because it would not have been reasonable to purchase two VCRs at the time of the first purchase.

In March 1991, VCR ownership by percentage of Japanese households reached 71.5 percent, which was equal to 86.6 percent of that of 2004. After that it took almost a decade to increase to 79 percent, which was equal to 96 percent of that of 2004. It was obvious that the diffusion speed to the laggards had slowed down. Those people who bought the VCR for the first time after 1991 are regarded as the laggards.

They were conservative in adopting the DVD and HDD recorders as well as the VCR. The adoption rate of the DVD recorders reached 40 percent in 2006 but the domestic shipment in units decreased by 18 percent from the previous year. In the VCR case, VCR products became sufficiently cheap, functional, and convenient for the late majority and the laggards to adopt at a time when most of the early majority had finished adopting the VCR. DVD and HDD recorders have significant room for improvement in order to be adopted by the late majority and the laggards. In addition, they mull over whether to purchase a DVD and HDD recorder or a VCR because VCR products are sufficiently inexpensive, functional and convenient for them.

There is a sizeable chasm between the early majority and the late majority, which is indicative of the management difficulties in product advancement at that stage. If they missed the timing to reduce the price and the simplification of the new products, they choose the old products. This interval might impede the advancement of the products when the interval of the repeat purchase of the adopters is lengthy. The diffusion speed of the DVD and HDD recorders was too fast in the early 2000s to prepare for the late majority and the laggards. Furthermore, the DVD and HDD recorders were much more stable and had longer lives than that of the VCR because DVD and HDD recorders are mainly composed of electronic parts rather than mechanical parts. People do not have to own multiple DVD and HDD recorders in a single household because these machines can dub by themselves and it is very reasonable to choose only the DVD player machine as a second unit. The decline of the

Copyright © 2008, IGI Global. Copying or distributing in print or electronic forms without written permission of IGI Global is prohibited.

demand in the mid-2000s might damage the members of the DVD and HDD recorders' supply chain. The industry might seek out new products with a large amount of investment. Some of them might shift to other products and stop manufacturing certain parts.

The late majority and the laggards are the key to the longevity of the products. It is a critical matter for the industry whether these groups adopt a product or not as mentioned above. The continuous and stable movement of the demand helps to even out the advancement. Once they have adopted, they become loyal users. When new products become popular, they do not care for them. However, their existence also delays the switch to other products as JVC did in the case of the VCR. The time management or the correspondence to the late majority and the laggards is still very important in the second half of the life cycle.

VCR Products for the Late Majority

The late majority had ample options because many companies existed in the industry and they offered customers various types of products, from the monaural VCR to S-VHS. They actively launched numerous models into the market. For example, JVC launched 40 new models during 1987-1991 in the domestic market, which was much more than the 15 models launched during 1983-1986, and the 27 models during 1992-1995.

Various types of customers also existed in the market, the innovators, the early adopters, the early majority, and the late majority. Compared to the others, the late majority has a passive nature in the adoption of products because they value the innovativeness less than the others. They are mindful of the price and the performance (including the quality) of the products and the difficulty of operation or use. They wait for the price to go down to a reasonable level for them, after previous users have validated the excellence of the products, and when they are offered the easier-to-use versions of the products.

In the case of the VCR, few of the late majority purchased the S-VHS launched in 1987 because of the reasons stated above. Most of them chose monaural and HiFi VCRs. The reduction rate of the average domestic shipment price per product slowed down during 1987-1994. It shows that the price went down enough for the late majority. After 1994, imports from Asian countries contributed toward the further reduction of the price. The technical differ-

Copyright © 2008, IGI Global. Copying or distributing in print or electronic forms without written permission of IGI Global is prohibited.

ence among the companies in those VCRs had diminished. As a result, the share of leaders including Matsushita, Sony and JVC, decreased to about 40 percent in 1991. It decreased by 10 percent during 1987-1991.

Figures 1(a) and 1(b) show the price range during 1991-1992 and 1993-1994. From these figures we can conclude the following things. First, as a whole,

Figure 1(a). Price range in 1991-1992

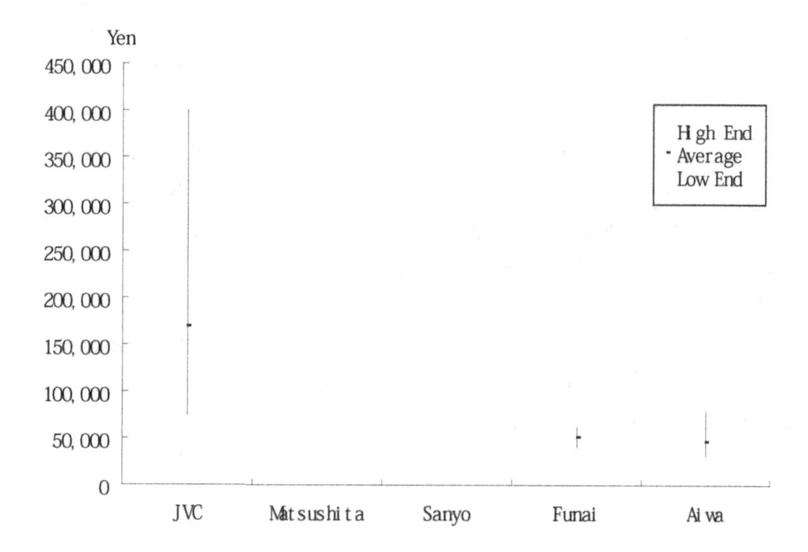

Figure 1(b). Price range in 1993-1994

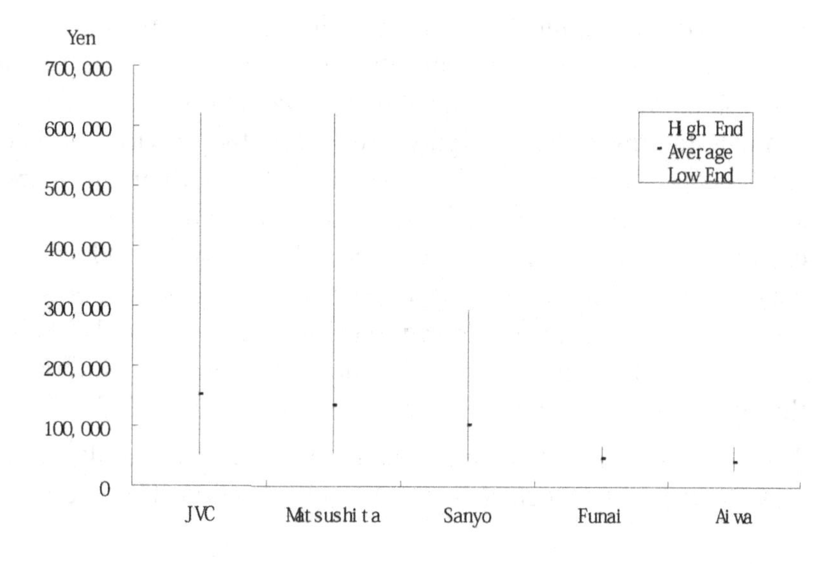

Copyright © 2008, IGI Global. Copying or distributing in print or electronic forms without written permission of IGI Global is prohibited.

the price range during 1991-1992 was narrower than that during 1993-1994. The high end of products became cheaper and cheaper because the main customers of this period were the late majority who were not willing to pay as much money on the VCR as the innovators, the early adopters and the early majority. From 1993 on, the main customers changed to repeat purchasers. The repeat buyers of this period were previous adopters who were willing to pay much more money than the late majority and the laggards. On the other hand, the low end of products continued to decrease. Further price declines contributed to spread the VCR to the laggards. Various types of companies existed in the market, and they established their own territories. With the market in confusion, the variety of products became more extensive, while at the same time the number of new models decreased as shown in Table 1.

The late majority and the laggards eased the sharp decline of the demand at the end of the life cycle. In the case of the VCR, the demand decreased sharply after 2001 because of the emergence of the DVD and HDD recorders. However, at that time, the late majority and laggards were accustomed to the VCR and they were very passive toward the new technology. Although the mainstream changed to digital technology, they continued to use analog technology longer than the others. They preferred simple products and made the most of their recorded programs on the VCR.

Supply Chains to the Late Majority and Laggards

Retailers ideally should have a wide variety of products from various companies. In the market, the late majority can be a mainstay, but laggards can not because of the size of the demand in a period and their characteristics. Some evaluate the track record highly and follow the previous adopters. Others seek cheaper products. However, they do not overwhelm the market because some innovators and early adopters start repeat purchases. Various types of customers want the industry to offer a wide variety of products. On the other hand, the climactic importance of the late majority is smaller than that of the innovators and the early majority. The downstream supply chains continue the framework of the existing distribution channel and improve efficiency through logistical management based on better use of information systems.

The upstream supply chain members are forced to simultaneously offer a wide variety of products and an efficient operation. They need to change to

Copyright © 2008, IGI Global. Copying or distributing in print or electronic forms without written permission of IGI Global is prohibited.

Table 1. Number of new models

	Year	VHS	Mono	HiFi	SVHS	BS	BS-HiFi	BS-SVHS
JVC	82	1	1	-	-	-	-	-
	83	4	3	1	-	-	-	-
	84	3	2	1	-	-	-	-
	85	5	3	2	-	-	-	-
	86	7	2	5	-	-	-	-
	87	8	4	2	2	-	-	-
	88	10	1	3	5	1	-	1
	89	11	3	4	4	-	-	-
	90	11	1	6	3	1	-	1
	91	6	-	1	1	4	1	3
	92	8	-	1	-	7	2	5
	93	5	1	1	-	3	-	3
	94	6	-	3	-	3	1	2
Matsushita	77	1	1	-	-	-	-	-
	78	4	4	-	-	-	-	-
	79	2	2	-	-	-	-	-
	80	4	4	-	-	-	-	-
	81	4	4	-	-	-	-	-
	82	6	6	-	-	-	-	-
	83	7	5	2	-	-	-	-
	84	9	8	1	-	-	-	-
	85	7	5	2	-	-	-	-
	86	10	5	5	-	-	-	-
	87	13	3	8	2	-	-	-
	88	10	2	2	6	-	-	-
	89	12	3	5	4	-	-	-
	90	9	-	3	3	3	1	2
	91	9	1	2	4	2	-	2
	92	7		1	1	5	2	3
	93	7	1	3	1	2	1	1
	94	9	-	4	3	2	2	-
Sanyo	85	3	3	-	-	-	-	-
	86	3	1	2	-	-	-	-
	87	4	-	2	1	1	1	-
	88	1	-	-	-	1	-	1
	89	4	-	1	3	-	-	-

continued on following page

Copyright © 2008, IGI Global. Copying or distributing in print or electronic forms without written permission of IGI Global is prohibited.

	Year	VHS	Mono	HiFi	SVHS	BS	BS-HiFi	BS-SVHS
Sanyo (continued)	90	2	-	-	2	-	-	-
	91	2	-	-	-	2	-	2
	92	2	-	-	1	1	1	-
	93	3	1	1	-	1	-	1
	94	10	2	3	1	4	3	1
Funai	87	3	2	1	-	-	-	-
	88	10	6	4	-	-	-	-
	89	12	6	6	-	-	-	-
	90	4	2	-	-	2	2	-
	91	3	2	1	-	-	-	-
	92	3	1	2	-	-	-	-
	93	3	2	1	-	-	-	-
	94	7	2	3	-	2	2	-
Aiwa	91	3	3	-	-	-	-	-
	92	4	3	1	-	-	-	-
	93	7	4	3	-	-	-	-
	94	3	2	1	-	-	-	-

flexible manufacturing to be able to respond quickly to capricious market demand. To reduce inventory and lead time, they reinforce the postponement through modular manufacturing and the standardization of parts.

On the other hand, they come to recognize the limits of further cost cutting. They make the most of the economies of scale under a wide variety of products through the standardization of most processes and parts. However, room for further cost cutting is diminishing. The upstream supply chain, the manufacturing system, has become so efficient that the only way to reduce cost further is to source less costly resources. First, they increase the ratio of the parts on the market. In addition to the low technology products, some outsource the high technology products. Second, to compete in foreign markets, global sourcing is the key to penetrating and competing in foreign markets and to avoiding trade frictions.

Copyright © 2008, IGI Global. Copying or distributing in print or electronic forms without written permission of IGI Global is prohibited.

Conclusion

We discussed the characteristics and the role of the late majority and laggards in this chapter. Both of them adopt the product in the later half of the cycle. They prefer to choose a practical product at a lower price from a wide selection of products. Their impact on the market is not big while their population is huge. This is because companies finish expanding their production capacity and various types of customers are still in the market.

In the VCR case, the late majority adopted mainly a HiFi VHS for the first time around 1990. Although three generations of products, the monaural VHS, HiFi VHS and S-VHS, existed in the market around 1990, they choose the HiFi VHS because it demonstrated a high performance at reasonable cost. While the monaural VHS declined in popularity, S-VHS could spread to limited types of customers, the extreme innovators and the innovators, because they were too expensive and sophisticated for most people to use. In addition, the alternation of generations from the monaural to HiFi VHS occurred in the late 1980s. Although JVC launched many generations of products, the laggards still purchased HiFi VHS after 1991 because HiFi VHS demonstrated an excellent performance/cost ratio.

References

Bass, F. M. (1969). A new product growth model for consumer durables. *Management Science, 15*, 215-227.

Buzzell, R. D. (1966). Competitive behavior and product life cycles. In J.S. Wright & J.L. Goldstucker (Eds.), *New ideas for successful marketing* (pp. 46-68). Chicago: American Marketing Association.

Economic and Social Research Institute, Cabinet Office, Government of Japan. Retrieved March 15, 2006 from http://www.cao.go.jp/index-e.html

Electronics Industries Association of Japan. Retrieved March 15, 2006 from http://www.jeita.or.jp/eiaj/english/index.htm

Funai. Retrieved March 15, 2006 from http://www.funaiworld.com/index.html

Higuchi, T. (2006). Industrial life cycle in the VCR industry. *Sakushin Policy Studies,* 6, 19-34.

Copyright © 2008, IGI Global. Copying or distributing in print or electronic forms without written permission of IGI Global is prohibited.

Higuchi, T., & Troutt, M.D. (2004). Dynamic simulation of the supply chain for a short life cycle product - Lessons from the *Tamagotchi* case. *Computers & Operations Research, 31*(7), 1097-1114.

Higuchi, T, Troutt, M. D., & Polin, B. A. (2004). Life cycle considerations for supply chain strategy. In C.K. Chan & H.W.J. Lee (Eds.), *Successful strategies in supply chain management* (pp. 67-89). Hershey, PA: Idea Group.

Hitachi. Retrieved March 15, 2006 from http://www.hitachi.com/

Itami, H. (1989). *Nihon no VTR sanngyou naze sekai wo seiha dekitanoka*. Tokyo: NTT Publishing.

Japan Electronics and Information Technology Industries Association. Retrieved March 15, 2006 from http://www.jeita.or.jp/english/index.htm

Japan Video Software Association. Retrieved March 15, 2006 from http://www.jva-net.or.jp/en/index.html

JEITA (Japan Electronics and Information Technology Industries Association). (2005). *Minnseiyou dennsi kiki deta shyu*. Tokyo: JEITA.

Kohli, R., Lehman, D. R, & Pae, J. (1999). Extent and impact of incubation time in new product diffusion. *Journal of Product Innovation Management, 16*(2), 134-144.

Kotler, P. (1999). *Marketing management*. New Jersey: Prentice Hall.

Lilien, G., Kotler, P., & Moorthy, K. (1992), *Marketing models*. Englewood Cliffs, NJ: Prentice-Hall.

Matsushita. Retrieved March 15, 2006 from http://panasonic.net/

McIntyre, S. H. (1988). Market adaptation as a process in the product life cycle of radical innovations and high technology products. *Journal of Product Innovation Management, 5*(2), 140-149.

Mitsubishi Electric. Retrieved March 15, 2006 from http://global.mitsubishi-electric.com/

Moore G. A. (1991). *Crossing the chasm*. New York: Harper Business.

Moore G. A. (2005). *Dealing with Darwin*. New York: Portfolio.

National Science Museum, Tokyo. Retrieved March 15, 2006 from http://sts.kahaku.go.jp/sts/

NHK (Japan Broadcasting Corporation.). (2000). *Project X: Challengers 1*. Tokyo: Japan Broadcast Publishing.

Rogers, E. M. (1995). *Diffusion of innovations* (4th ed.). New York: The Free Press.

Copyright © 2008, IGI Global. Copying or distributing in print or electronic forms without written permission of IGI Global is prohibited.

Sato, M. (1999). *The story of a media industry*. Tokyo: Nikkei Business Publications.

Schoenherr, S. Retrieved March 15, 2006 from http://home.sandiego.edu/~ses/

Sheth, J. N. (1971). Word of mouth in low-risk innovations. *Journal of Advertising Research, 11*(3): 15-18.

Sony Corporation. Retrieved March 15, 2006 from http://www.Sony.net/

Toshiba. Retrieved March 15, 2006 from http://www.toshiba.co.jp/index.htm

Victor Company of Japan. Retrieved March 15, 2006 from http://www.jvc-victor.co.jp/english/global-e.html

Copyright © 2008, IGI Global. Copying or distributing in print or electronic forms without written permission of IGI Global is prohibited.

Section IV

Location Aspects

The allocation of facilities is one of the most effective means for fulfilling strategy. The location of supply chain partners has a direct impact on the competitiveness of the supply chain. From the life cycle viewpoint, location plays a very important role. For example, if a company wants to succeed in developing a brand-new product, they need the latest information from the innovators and a strong linkage among suppliers. Hence, most supply chain partners are better served by locating near the extreme innovators, and in the most advanced countries. On the other hand, if a company decided to do sales promotion in a certain country, they would introduce a production facility in that country for quick response and to gain publicity for that local market. Furthermore, if a company decided to adopt the cost leadership strategy, the offshore production or global sourcing in areas of low cost operation is the key. In a declining industry, companies decide to withdraw from the market. At that time, they would close production facilities in the most advanced countries first, and finally, downsize to one in a low cost operation area because of economies of scale, or close production facilities altogether.

Chapter XII

Physical Location

Vernon's product cycle theory is reexamined and extended in the first half of this chapter. It did not originally address the recent environmental changes, the global economy, supply chain management, and the rise of developing countries. The geographic location of manufacturing facilities is discussed from the viewpoint of the innovations. The physical location at the beginning of the VCR industry is explained in the latter half of this chapter. Although there was a great global demand for the VCR from the beginning, Japanese companies exclusively manufactured almost all VCRs in Japan and exported them all over the world. Manufacturability is the critical factor at the beginning as Vernon's product cycle theory demonstrated.

Copyright © 2008, IGI Global. Copying or distributing in print or electronic forms without written permission of IGI Global is prohibited.

Vernon's Product Cycle Theory

The Product Life Cycle was developed by Vernon (1966, 1977, 1998) from the standpoint of location. This theory, now called the product cycle, is oriented toward explaining the production process and the role of the physical location of facilities. He described the geographical location of manufacturing facilities according to the state of the technology (the product and process innovations), international trade, and economies of scale, unlike PLC (Product Life Cycle), which focuses on the market and consumer behavior. Vernon (1966) classified the time periods of this life cycle into three stages, *New*, *Maturing,* and *Standardized.* In this section, we review the product cycle from the individual process and product innovation stages in the terms of the global supply chain under the condition that national borders are becoming less of a problem and that the uniqueness of the countries and regions is becoming more important than before.

However, Vernon's product cycle theory did not cover the following recent environmental changes. Therefore, we need to discuss the relationship between the product and process innovations and the geographic location of the manufacturing facilities in view of the following modern developments. They are: the global economy, supply chain management, and the rise of developing countries.

1. **Global economy:** Technical innovations in telecommunications and international transportation have blurred national borders. The global division of labor is taken as a matter of course. The demand for new products spreads worldwide much quicker than before. On the other hand, facilities in the low cost operation areas can start manufacturing high-tech products much earlier in the life cycle than before because they have gained experience through practice. Any company should make the most of the global market and global sourcing simultaneously.
2. **Supply chain management:** Supply chains have become the basic unit of competition. No single company can do everything efficiently. Various suppliers have unique expertise or advantages in one or more ways. It is a waste of resources for a single company to attempt to cover all aspects of development, production, and logistics. Sometimes, supply chain members are very flexible for sound strategic reasons. This view can shorten the time to market and at the same time achieve more

Copyright © 2008, IGI Global. Copying or distributing in print or electronic forms without written permission of IGI Global is prohibited.

reasonable costs. Nowadays, most major companies have established an efficient and effective supply chain network beyond national borders.

3. **Rise of developing countries:** Many developing countries have accumulated various technologies and have become tough rivals for advanced countries. In some areas, they had a considerable lead against advanced countries. On the other hand, they have become a serious purchasing power. Companies must determine their strategies vis a vis the rise of the developing countries at the beginning of the life cycle.

The new product stage includes the period before the category of products is launched, and the period just after the first products are launched and recognized by the majority of consumers. Vernon assumed that enterprises in any one of the advanced countries are not appreciably different from those in any other advanced country before the decisions on where the new product is born and manufacturing facilities are to be located. However, it has been stated that equal access to the latest technology does not mean equal probability of the application of it in the generation of new products. This is because of the large gaps between the knowledge of such technologies and their application to the product. In this stage, R&D departments play a vital role in commercializing and enhancing a product. It is therefore most effective for R&D departments to be located in the most advanced countries in which sophisticated and affluent consumers live and advanced technologies are concentrated. The close communication among concerns, latent consumers, and suppliers, is an essential factor in identifying the latent market needs and in commercializing a new category of products. Therefore, manufacturing facilities, as well as R&D departments, tend to be located in highly developed countries.

The propensity of multinational enterprises to use their home markets to develop and introduce new products stems from a series of powerful forces. It has been confirmed again and again in empirical studies of various sorts that successful innovations tend to be those that respond to the market conditions surrounding the innovation. The original idea may be developed almost anywhere, but successful innovation depends strongly on the compelling character of the demand. (Vernon, 1977, pp. 40-41)

Copyright © 2008, IGI Global. Copying or distributing in print or electronic forms without written permission of IGI Global is prohibited.

The next stage is maturing. In this stage, the basic product design has become so stable that the manufacturing process has progressed significantly and competing newcomers enter the market. The manufacturers expand mass-production via capital-intensive facilities, in order to achieve economies of scale. Reduction in cost then begins to predominate. The domestic and international demand for the particular category of products expands dramatically because of price decline and the increased variety of products. As a consequence, the barriers to maintaining manufacturing facilities in the highly developed countries begin to decrease.

The final stage is the standardized stage. In this stage, the product and the process designs have become very stable or fixed. As a result, the performance, quality, and price of the product achieve an excellent level. On the other hand, little room is left for improvements. Additional product and process improvements have little effect on the product. At this stage, the remaining viable way to reduce cost is offshore or foreign production. Finally, the origin country itself may become a large importer because of the gap in manufacturing costs. Figure 1 summarizes the Product Cycle Theory, in which the vertical axis is the distance between customers and the production facilities. Generally, new products are developed and manufactured near the extreme innovators and innovators at the beginning. Then, the range of customers widens and the demand for the product grows globally. Some companies decide to position an assembly factory near the offshore customers. As competition becomes severe, the companies change their manufacturing system in order to obtain much greater efficiency. Finally, they locate tightly together

Figure 1. Change of distance between customers and factories

Distance
Far
Close
Time
NEW
Customers and Factories Are in Highly Developed Countries
MATURING
Exportation to Other Developed Countries
STANDARDIZE
Local Production
Exportation from Developing Countries
Exclusive Production in a Developin g Country

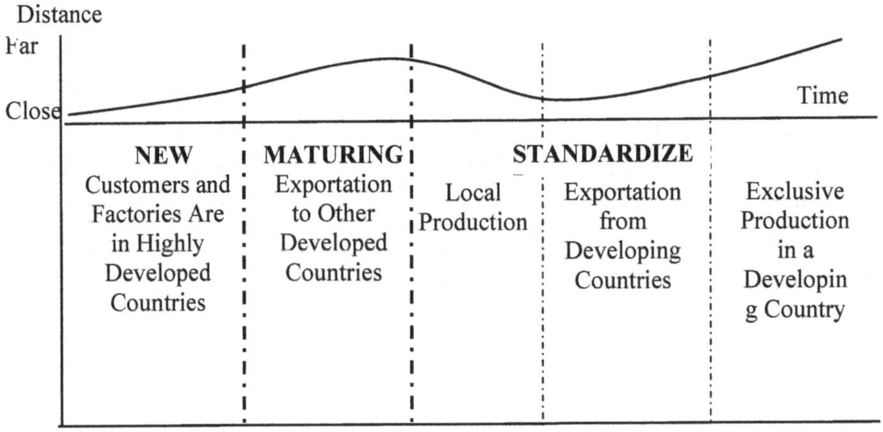

Copyright © 2008, IGI Global. Copying or distributing in print or electronic forms without written permission of IGI Global is prohibited.

the number of factories in a certain country or area and export throughout the whole world.

New Stage: Innovative but Incomplete Product

Innovative products are brand new products that embody a novel technology or usage, or both. Generally, these products are initially very expensive and experience problems in their functions and quality. Most products in this stage are therefore incomplete. Some of them need improvements in the product design and the production process to achieve commercial success, while others must upgrade in the very near future to achieve further success.

In this stage, supply chains should focus on R&D activity for the first mover advantage and emphasize effective and efficient communication among the customers, R&D departments, and manufacturers (including suppliers). R&D departments should generally locate in the most advanced countries because of the availability of more innovative or sophisticated customers and the merits of aggregation of technologies. Not only R&D departments, but also the manufacturing facilities of the high-tech products are better located in the most advanced countries in the beginning. In this stage, the supply chain concentrates on the linkages within the supply chain and rapid response to customer needs. Leading companies formulate strong partnerships with key suppliers and promote product development. Without the collaboration with the key suppliers, they might overlook fatal problems, which can delay R&D activity or decrease feasibility because products, especially durable goods, are composed of a great number of parts and their product performances are decided by weakest parts or link. It is impossible for any single leading company, such as Toyota, Matsushita, Sony, and others, to excel over all suppliers in all related technologies. The physical location of partners is a good way to speed up R&D activities because they can more easily share information and execute parts transactions quickly.

The launch of an innovative product carries great uncertainty. All companies may naturally hesitate in making huge investments on facilities because the product design and manufacturing process might change dramatically in the near future. Therefore, in addition to product performance, there exists considerable potential for improvement in quality and cost. The production processes are far from being standardized such that mass production can be

Copyright © 2008, IGI Global. Copying or distributing in print or electronic forms without written permission of IGI Global is prohibited.

achieved. This means that thorough coordination among suppliers and leading companies is required for production planning and execution. Only after the dominant design emerges in the market, can supply chains start standardizing the production processes and begin mass manufacturing.

The extreme innovators are essential for the incubation stage of innovative products. They embrace and adopt these incomplete products first at an extremely high price. Their experiences and critiques guide an industry to the right place for commercial success. They contribute very much to the development of the product version which is called the dominant design or de facto standard later. They have a great influence on the other innovators who adopt the dominant design. Most extreme innovators and innovators live in the most advanced countries.

When a product is new (innovative but incomplete), the physical location of key members in the most advanced countries is very effective. Most of these products succeed only after several attempts. The leading companies need rapid feedback from suppliers and the extreme innovators to boost the R&D for the de facto standard, and to establish the mass production system with their suppliers. The location of supply chain members reflects the necessity of close communication among them and the extreme innovators. Given that the location of the manufacturing facilities is fixed, the number of adopters increases to the extreme innovators and the innovators. As a result, the distance between customers and the production facilities steadily increases.

Maturing Stage: Immature Product

In this stage, product performance matures and the manufacturing process also progresses significantly. The demand for the product expands because of the growth of global sales and penetration of the domestic market. The variety of products proliferates due to competitors. As the product becomes popular in other advanced countries and the manufacturing process is standardized, the barriers to maintaining manufacturing facilities in the most advanced country abate and the pressure to reduce costs begins to predominate at the end of this stage.

After a dominant design emerges, the product will generally still have much room for improvement in product performance and production process. The product performance advances through the addition of minor functions and

Copyright © 2008, IGI Global. Copying or distributing in print or electronic forms without written permission of IGI Global is prohibited.

the extension or enhancement of existing ones. The stabilization of the product design makes it possible to facilitate mass production, which increases quality and decreases the cost of the products. As a result, the number of adopters increases rapidly and the product becomes popular in other countries.

Although a dominant design is so widely adopted that it becomes a de facto standard, it is far from the matured product. For example, the Model T that Ford launched in 1905 was a dominant design in the automobile industry and it demonstrated the framework of the modern automobile. During this century, the automobile industry continued to add new functions and upgrade existing functions. Companies make efforts to mature their products by adding new and upgrading existing functions. As a result, product performance might mature to some extent by the end of this stage.

The distance between the customers and the manufacturing facilities is increased in this stage. In addition to the maturity of the product performance, the price of the products decreases sharply because of mass production. As a result, the number of customers grows and global diffusion of the product progresses rapidly. However, it is natural for any company to make the most of existing facilities as long as possible, and to postpone offshore production in a situation where there is some capacity for further improvement in product performance and production process. Customers are far removed from the manufacturing facility and products are shipped very long distances to them.

Supply chains need to be systematized in order to respond to a growing market domestically and globally while product performance is modified. Although each supply chain has a different core competence or strategy, most of them may allocate or keep most of their facilities in the most advanced countries in the beginning. Supply chains should establish mass production systems collaboratively so as to localize the impact of changes of the product design. Otherwise, the changes of the design might impair the existing production facilities. Close communication with those concerned, that is, the latent customers and suppliers, is an essential factor in speeding up the development of a new category of products. Therefore, one of the best guarantees of success is to locate production facilities near the most innovative and wealthiest customers.

Copyright © 2008, IGI Global. Copying or distributing in print or electronic forms without written permission of IGI Global is prohibited.

Physical Location at the Beginning of the VCR Industry

The 1950s and 1960s constituted the embryonic period for the home-use VCR. It took almost two decades to succeed in commercializing the home-use VCR after the first VCR for broadcasting was born in the US. Ampex and RCA developed most of the essential technologies for the VCR in the 1950s. Ampex monopolized the VCR for broadcasting in the 1950s. However, RCA invaded that province in the 1960s. On the other hand, Japanese companies that were excellent in the mass production of products related to the home-use VCR, namely the cassette recorder and the television, also tried to put the home-use VCR to practical use. One of the major differences between the VCR for broadcasting and that for home-use is the size (weight) and the price. The initial VCRs for broadcasting were so huge and expensive that only broadcasting stations had a sufficiently strong need and enough space for the device. On the other hand, the home-use VCR needed to be reasonably priced, and suitably smaller and lighter, so that an individual consumer could handle it.

Many companies attempted to develop the home-use VCR market. Ampex and RCA played very important roles. Although Ampex had a great advantage in the broadcasting VCR, they could not adapt it to the home-use VCR. Broadcasting stations had adequate space and budgets for the high-end VCRs. In addition, these models were somewhat specialized for broadcasting uses. They were also poorly adapted to mass production. When Ampex launched the VCR market in business and education, they had two departments that specialized in the VCR. One in Chicago was for broadcasting and the other in California was for consumer electronics. Both developed the home-use VCR models, the VR-303 and the VR-7000, separately. Ampex chose to deliver to the market the VR-7000, which was more expensive and lower in quality than their competitors' models. Technically, Ampex had a few weaknesses in the semiconductor design, which was necessary for downsizing, and in the magnetic head which was a core component of the VCR. They made strategic alliances with Sony and Toshiba to make up for the weaknesses. In the Ampex case, they could not absorb essential technologies because of the long distances and the gap with the leading products. On the other hand, RCA had already achieved excellence in consumer electronics and also made large investments in R&D. However, they had four research institutes for the VCR at different places in the US. In addition, the research activity was

Copyright © 2008, IGI Global. Copying or distributing in print or electronic forms without written permission of IGI Global is prohibited.

isolated from the production functions. RCA also could not capitalize on its ability because of the physical distance among key members.

Japanese companies fell behind US companies in the 1950s and 1960s. However, Japanese companies had some advantages to be able to catch up with them. First, many Japanese companies produced TVs and cassette recorders at that time. It is reasonable to say that the VCR is a complex product made from both a TV and a cassette recorder. In other words, the VCR may be called the first *mechatronics* product, a combination of both electronics and mechanics. Both technologies were essential to develop the home-use VCR market. Secondly, most R&D divisions were located within or near the factory in Japan. This facilitated communication between R&D and the manufacturing departments and contributed to an accelerated cycle time from design to prototype. Finally, most Japanese companies focused on magnetic tape recorders, the U format, while others tried various other recorders, such as an optical disk and other tapes. Too many heterogeneous devices placed a disadvantage on the users and the VCR industry in the late 1970s. In 1967, Ampex introduced the HS-100, a color magnetic disk recorder that provided the rapid-playback feature. CBS developed EVR (Electric Video Recording) which used a film in a cassette and which was invented by Goldmark who succeeded in the practical use of the TV and the Long-Play record. EVR was superior in vividness of image to the U Format VCR introduced later, controllability of the still frame, and the mass-productive capacity of the software. In 1969, RCA demonstrated SV (SelectaVision) which used a holographic tape and which could not record. Sony, JVC and Matsushita made a cross-license agreement in 1970. It accelerated the development of the home-use VCR so fast that US and European companies could not follow.

Sony had the largest R&D force for the VCR among Japanese companies in the 1960s and 1970s. In addition, they had the most advanced technology in semiconductors in the world, which was necessary for VCR product development. However, JVC with a lot less people and a smaller budget than Sony nevertheless caught up with them. At JVC, the R&D of the VCR was done in the VCR department, near the production facilities. Some R&D members did sales while others repaired the products on site. This communication with (potential) customers gave them a chance to identify a clear image of the goal for home-use VCR products. It led them to launch a lighter and longer-recording VCR than Sony in 1976.

By the middle of the 1970s, the basic design of the VCR changed dramatically and repeatedly. These changes had great impacts on the suppliers because

Copyright © 2008, IGI Global. Copying or distributing in print or electronic forms without written permission of IGI Global is prohibited.

the specifications and the size of parts changed frequently. The history of the development of the VCR is that of downsizing and the stabilization of the mechanical parts. This required very close coordination among the integrator and the suppliers. At that time in Japan, *keiretsu*, a company alliance system based on capital relationship and long-term transactions worked very well. *Keiretsu* relied heavily on face-to-face communication. The physical location of key suppliers was very effective in accelerating product development by finding problems very quickly and sharing that information.

Table 1 demonstrates amount of the investment in the VCR, TV, and radio facilities. It is interesting that the amounts of VCR facilities in 1975 and 1976 were very low. Although some companies, such as Matsushita, made a big investment for the VCR, the demand did not grow at all. As a result, they were seriously injured financially. Others learned the lesson and hesitated to place a big investment until the demand started growing. Companies increased production manually by the middle of the 1970s because of the lack

Table 1. Investments in billions of Yen in VCR, TV and radio facilities during 1971-1981 (Source: The Ministry of Economy, Trade and Industry)

	VCR (9)*	TV (11)**	Radio (2)***
1968	0.1	6.8	2.3
1969	0.1	14.6	2.7
1970	0.8	14.8	3.4
1971	0.4	5.0	1.9
1972	0.4	4.8	1.7
1973	2.4	7.1	1.9
1974	0.7	8.7	2.4
1975	0.4	4.8	1.4
1976	0.2	10.6	3.1
1977	1.9	16.8	5.3
1978	8.8	10.2	4.9
1979	10.8	10.1	5.0
1980	42.9	24.5	1.2
1981	87.3	26.0	2.8
1982	60.0	25.6	3.2

* 9 companies' declared amount of investment in VCR facilities.

** 11 companies' declared amount of investment in TV facilities.

*** 2 companies' declared amount of investment in radio facilities.

Copyright © 2008, IGI Global. Copying or distributing in print or electronic forms without written permission of IGI Global is prohibited.

of and/or the imperfection of their mass production systems. That is, while preparing for mass production, companies manufactured the VCR manually exclusively in a factory, which was close to the key suppliers.

At the beginning of the VCR industry, during the 1960s to 1970s, the physical location of key supply chain members was very effective for the acceleration of product development because of the following reasons. First, as noted above, the VCR is a combination product of the television and the cassette recorder. This means that a company having both technologies, electronics, and mechanics enjoyed a great advantage, namely, the mechanical parts needed for development. Mechanical parts especially required coordination among concerns. Second, the size and weight of VCR products decreased dramatically along with upgrading of basic functions and the addition of new functions. The supplier system, or the upstream supply chain, was therefore required to respond to changes quickly and correctly. Third, at that time, information technologies had not advanced sufficiently to tele-collaborate. It was after the 1990s that companies could exchange vast information, such as CAD data, promptly and stimulate product development on the computer. Finally, management methods using tele-collaboration were not yet established at that time. More recently, companies have accumulated experience with tele-collaboration and product standardization has advanced for a long time. Nowadays, most companies regard tele-collaboration in R&D and manufacturing activities as much easier than before.

Initial International Demand for the VCR

As time passed, global sales of the VCR expanded. By the beginning of 1980, VCRs were manufactured primarily and expensively in Japan. At that time only Japanese companies could produce VCRs because of the difficulty of the manufacturing process. To manufacture the VCR, in addition to the technologies of mechanics and electronics, they needed to have the technology of precision engineering because the magnetic heads write and read merely a few microns space of the videotape. Even though the technologies of the VCR were open, very few companies could manufacture at that time. It was not effective to locate the production facilities in other countries before the product design and manufacturing process were stabilized. Even after that, it was natural for any company to seek maximal utilization of its exiting

Copyright © 2008, IGI Global. Copying or distributing in print or electronic forms without written permission of IGI Global is prohibited.

facilities. As a result, the distance between the factories and the market were greatly increased because of the growth of global sales, as evidenced by an increased excessive imbalance of exports. Then, almost all VCRs were manufactured in and exported from Japan.

The European market played a very important role in the initial market. The initial demand in Europe was much more than that of North America. The European market was very significant. It was expected that the struggle for the de facto standard would happen in the near future in Europe because there were some European electronic companies developing their own format. If multiple formats, for example, VHS and Betamax, existed in the market, consumers and video package makers would suffer serious damages. JVC made alliances with European companies to sell VHS products in Europe. European countries adopted PAL (Phase Alternating Line), while Japan and the United States adopted the NTSC (National Television Standard Committee) standards. JVC prepared for accommodation to this difference. In addition, Europe established trade barriers against other areas. To increase the rate of the local component content, Japanese companies not only set up assembly plants, but they also manufactured some parts aggressively.

The North American market is the biggest market in the world. In the case of the VCR, it showed great purchasing power from the beginning as well as did Europe. A large number of the extreme innovators and innovators existed in North America. Matsushita explored the US market at first with the four-hour recordable VHS at less than US$1,000. This success of Matsushita in the United States gave VHS group members a great advantage in Japan and Europe.

Table 2. Global demand (exportation) in millions of Yen at the beginning of the 1980s (Source: The Ministry of Finance)

	Exportation from Japan to						
	Europe	**North America**	**Asia**	**Latin America**	**Africa**	**Oceania**	**Total**
1980	179,021	135,538	88,018	17,262	13,975	9,812	443,627
1981	359,835	285,534	128,390	23,714	27,528	28,504	853,505

Copyright © 2008, IGI Global. Copying or distributing in print or electronic forms without written permission of IGI Global is prohibited.

Conclusion

We reexamined Vernon's product cycle theory and tried to improve it. At the present time, companies operate environmentally friendly supply chains. In addition, many world economic powers are rising from the developing countries. However, the physical location of manufacturing facilities has had great practical significance in the real life. This is because countries have different positions in purchasing power, manufacturing ability, operating costs, and access to the large markets.

The VCR industry demonstrated a typical transition pattern of physical location. There are several reasons. First, the manufacturability had expanded globally and slowly. Second, multiple generations of products existed in the market simultaneously. Third, VCR machines are so light that companies could establish the global division of labor with little concern for transportation costs.

A global demand for the VCR was great from the beginning. But Japanese companies exclusively manufactured almost all VCRs domestically until the middle of the 1980s as Vernon's product cycle theory suggested. The manufacturability is the critical factor at the beginning. No companies could manufacture VCR machines and compete with Japanese companies until the 1990s. Japanese companies started the overseas production by themselves in 1980s.

References

Bowersox, J. & Closs, D. J. (1996). *Logistical management.* New York: McGraw-Hill.

Dunning, J. H. (1988). *Explaining international production.* London: Unwin Hyman.

Higuchi, T. (2006). Industrial life cycle in the VCR industry. *Sakushin Policy Studies,* 6, 19-34.

Higuchi, T, Troutt, M. D., & Polin, B. A. (2004). Life cycle considerations for supply chain strategy. In C.K. Chan & H.W.J. Lee (Eds.), *Successful strategies in supply chain management* (pp. 67-89). Hershey, PA: Idea Group.

Copyright © 2008, IGI Global. Copying or distributing in print or electronic forms without written permission of IGI Global is prohibited.

Itami, H. (1989). *Nihon no VTR sanngyou naze sekai wo seiha dekitanoka*. Tokyo: NTT Publishing.

JEITA (Japan Electronics and Information Technology Industries Association). (2005). *Minnseiyou dennsi kiki deta shyu*. Tokyo: JEITA.

NHK (Japan Broadcasting Corporation.). (2000). *Project X: Challengers 1*. Tokyo: Japan Broadcast Publishing.

Sato, M. (1999). *The story of a media industry*. Tokyo: Nikkei Business Publications.

Vernon, R. (1966). International investment and international trade in the product cycle. *Quarterly Journal of Economics*, *80*(1), 190-207.

Vernon, R. (1977). *Storm over the multinationals*. Boston: Harvard University Press.

Vernon, R. (1998). *In the hurricane's eye*. Boston: Harvard University Press.

Copyright © 2008, IGI Global. Copying or distributing in print or electronic forms without written permission of IGI Global is prohibited.

Chapter XIII

Partial Dispersion

In this chapter, we discuss the partial dispersion of manufacturing facilities and offshore production. It is better to locate most suppliers and manufacturers in close proximity at the beginning of the life cycle. However, as time passes, companies start locating assembly facilities in the other advanced countries for sales promotion and, sometimes, cost cutting. It becomes desirable for the manufacturers to avoid trade friction and to penetrate foreign markets quickly. The standardization of the product including the modularity of the parts makes it possible to do so. The partial dispersion at the standardized stage of the VCR industry is demonstrated in the latter part of this chapter.

Copyright © 2008, IGI Global. Copying or distributing in print or electronic forms without written permission of IGI Global is prohibited.

Beginning of the Standardized Stage

The concepts of standardization and modularity are important to the present chapter so some general background discussion is appropriate. Standardization may be regarded as the agreement within an industry on uniform and consistent design parameters of parts, components, and even systems. While the advantages of standardization might seem intuitive in many respects, Vollman, Berry, Whybark, and Jacobs (2005) describe several concrete instances of the value of standardization in what they call integrated manufacturing planning and control. Evidently, such integrated planning and control is especially critical for supply chain management. Some authors have argued that the primary advantage of standardization stems from economies of scale (Botschen & Hemetsberger, 1998; Levitt, 1983; Polin, Troutt, & Acar, 2005; Porter, 1980, 1985; Shoham, 1995). Economies of scale considerations extend to production, logistics, distribution and research and development (Hout, Porter, & Rudden, 1982; Porter, 1980, 1985; Shoham, 1995). While a firm producing standardized products for all consumers worldwide will incur a lower per-unit cost, this strategy addresses only the cost side of the equation. Thus, there is a caveat that strategies that succeed in reducing costs may not always be necessarily associated with greater profitability. Modularization is an important type of product design standardization in which several related functions are combined into one module.

The standardized stage begins after a product design becomes stable or standardized. Standardization includes product design, the production process and the parts. While a product specification is still in flux, the product design also is correspondingly very changeable. After a product specification reaches a certain level, companies can progress toward modularity. Modularity minimizes the impact of subsequent design changes while keeping open wide options for the future and also permits process innovations to proceed in earnest. Standardization also makes it possible to locate manufacturing facilities in other areas including foreign countries.

Some manufacturing facilities are shifted to other highly developed countries because of the rapid market growth or potential in those countries, and also to avoid various kinds of trade friction. By trade friction we refer mainly to barriers such as tariffs on imports. However, another barrier can be loyalty of the population to home country brands. By opening an assembly plant in a new country and using locally made parts and components (product content), these barriers can be reduced or eliminated.

Copyright © 2008, IGI Global. Copying or distributing in print or electronic forms without written permission of IGI Global is prohibited.

From the viewpoint of the manufacturing cost, companies wish to utilize existing facilities to the utmost degree. However, facility location in highly advanced countries becomes effective due to the logistic costs and sales promotion. In addition, trade friction might bring about protective tariffs. To avoid this, companies build mainly assembly plants in the highly advanced countries. The highly standardized parts made in the highly advanced country or the low cost operation areas and other parts made in the country of origin, are sent to the assembly plants in the highly advanced country.

Leading companies build their assembly plants in other highly advanced countries earlier than followers and niches. They exploit the standardization of product design and product process, as well as the global market. When leading companies standardize the products and processes to some extent, others are accumulating the experience and know-how. Their product design and process are less mature than those of the leaders. In addition, companies can not obtain most parts through the market at that time. The standardization process advances at each company level, not at the industry level. The earlier the stage of development, the bigger is the impact of the maturity of the product design and process. Therefore, only leading companies have the ability and necessity to operate assembly plants in other highly advanced countries in the beginning.

Assembly Plant in Highly Advanced Countries

At the beginning of the standardized stage, the price of the product is so expensive that it is only diffused to a limited number of affluent countries. In these countries, the customer can easily obtain product information through the Internet and by word of mouth. In addition, an effective and efficient distribution channel has already been established. From the standpoint of manufacturers or shippers, these wealthy countries constitute a large and reliable market. The manufacturers start considering overseas production in highly advanced countries because of the following reasons.

First, the total logistic costs increase with the sales. These are mainly composed of the freight and handling and inventory costs. It is true that the logistic cost per product decreases as sales increase, but there is a critical mass over which the local production becomes more efficient than exportation. The level depends on the working rate of their facilities, product characteristics,

Copyright © 2008, IGI Global. Copying or distributing in print or electronic forms without written permission of IGI Global is prohibited.

and protective tariffs if any. In a situation where the working rate is high and the demand is growing, companies must make decisions on investments in facilities. Economies of scale in manufacturing are not an obstacle in this case. The product characteristics related to the logistic costs are weight, size, shelf life, and care. As the products become smaller, lighter, and more robust, the logistical costs per product become smaller. As the desired service level becomes higher, the reasonable level of inventory at the retail shops and wholesalers increases. For further logistic cost cutting, local production near the customers becomes a good option for the leading companies.

Second, the leading companies decide to start overseas production to promote sales in the highly advanced countries. Such local production is a good means to penetrate the market in the country and contributes to improved and faster response by shortening the lead time and customization of the products to the area. In addition, the service level of the product in the area increases tremendously because they can repair the products and exchange the parts quickly at a reasonable cost. Thus, there is considerable merit to locating facilities in the highly advanced countries.

Third, as time passes, the obstacles for overseas production, such as the manufacturability and the existing facilities, become smaller than before. Companies promote the modularity of the product design and the production process for efficient and flexible production and the localization of the impact of changes, so the assembly plants can operate in other highly advanced countries.

Finally, trade friction encourages the companies to locate facilities in the new country. If a product whose demand is growing rapidly is manufactured exclusively in an exporting country, importing countries may consider protective tariffs to promote local production for job opportunities and technology transfer. To avoid such trade friction, leading companies increase the rate of the local product content by locating their facilities, particularly assembly plants, in the highly advanced countries.

Standardized Parts at the Beginning of the Standardized Stage

Companies use many standardized parts which they can get through the market, such as screws and bolts, from the beginning. The application of

Copyright © 2008, IGI Global. Copying or distributing in print or electronic forms without written permission of IGI Global is prohibited.

standardized parts saves time and resources for product development and reduces the manufacturing cost. However, not all parts can be standardized and gotten through the market. Some non-standardized parts are the source of the differentiation. In any case, the utilization rate of standardized parts gradually increases over time.

Companies use a lot of common parts that have been standardized by ISO (International Standards Organization) and other organizations. It is very useful for most companies to use standardized parts unless the non-standard parts are critical to product performance. Companies source inexpensive, quality parts globally. Nowadays, the low cost operation areas, ASEAN (Association of South-East Asian Nations) and NIEs (Newly Industrializing Economies) have increased their manufacturability so that they can produce quality parts at relatively low cost.

On the other hand, companies should develop and improve unique, special, or core parts of a product. The leading companies have standardized the basic design and manufacturing process at the beginning of the standardized stage. Suppliers of the leading companies also have accumulated experience and know-how. The fundamentals of the new entry companies therefore will have improved a great deal. Some new entrants may source most parts from the market and make a contract with other companies to get the core parts. They also can concentrate on cost cutting by refining the basic design to eliminate extra parts and functions.

The rates of utilization of the standardized parts are different according to the type of company, such as the leader, the follower, the niche, and the cost cutter, which influences the type and timing of the overseas production decision. The cost cutters make the most of standardized parts. Figure 1 shows the lowest price of Funai's products by VHS format. The differences among them were very small. They introduced new formats after they succeeded in localizing the impact of changes. While the leaders contribute to standardize the product design, the parts, and the production process, they exploit differentiation by advancing the core parts and adding new functions. Although they clear the technical fundamentals, such as modular manufacturing for overseas production first, they then try to induce innovations. As a result, they mainly locate assembly plants near the big buying power areas. Others will want to locate manufacturing facilities in the highly advanced countries for further cost cutting and sales promotion. But they have not accumulated enough technology and expertise for overseas production. From the viewpoint of standardization, the leading companies locate assembly plants in the

Copyright © 2008, IGI Global. Copying or distributing in print or electronic forms without written permission of IGI Global is prohibited.

Figure 1. Lowest price of Funai's products by VHS format

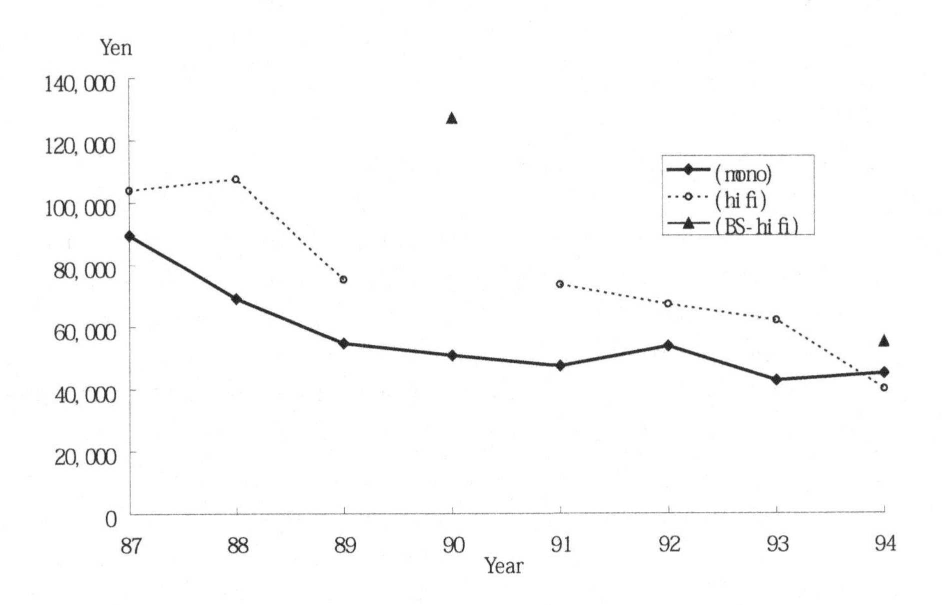

highly advanced countries at first. The followers chase the leaders with a lag because their target customers are almost the same as those of the leading companies. On the other hand, some niche and cost cutters follow a different strategy in that they do not set up facilities in the highly advanced countries because their customers are sparse there or the price sensitive.

Location of Supply Chain at the Beginning of the Standardized Stage

At the beginning of this stage the location of the supply chain partners starts diverging worldwide. When they make decisions on the overseas location, the order of priority is the feasibility, effectiveness, and efficiency. Feasibility is related to various factors that can be divided into two main parts: the product factors and the local factors. The product factors are composed of the degree of standardization in design and process and the product characteristics, such

Copyright © 2008, IGI Global. Copying or distributing in print or electronic forms without written permission of IGI Global is prohibited.

as the size, weight, and care. The local factors include the industry level and the importance of the transportation hub. The product factors decide the difficulty of local production. On the other hand, the local factors show the manufacturability of the local area. Feasibility is the result of the balance of difficulty and manufacturability.

Effectiveness is an indicator of future success. Supply chains make preparations for it by allocating their facilities. At the beginning of the standardized stage, supply chains aim to penetrate the worldwide market and to expand global sales. Market size at the time is one of the most important items. Supply chains allocate their facilities in the highly advanced countries because they have most of the extreme innovators and innovators who play a very important role in the early global diffusion process in an area. At the beginning of the standardized stage, the minimum requirements for the tele-collaboration within the supply chains are satisfied by the establishment of the modular manufacturing system, which makes the assembly process easy and enables them to upgrade the product with minor changes in the product design and manufacturing lines.

From the viewpoint of the life cycle, efficiency is delayed until feasibility and effectiveness are secured. Once efficiency becomes the most critical element in the market, it maintains that position for a very long time. However, at the beginning of the standardized stage, efficiency is not the most critical factor because the market is dominated by a limited number of companies. The price war does not start yet. Hence, when supply chains decide to set up a new overseas facility, they prefer the highly advanced countries to low-cost operations. Nowadays, China is an exception. China is regarded as a low cost operation area. On the other hand, they demonstrate great buying power. High-tech industries set up their facilities in China at the beginning of the standardized stage considering both their buying power and low cost operations.

Supply chains diverge by locating new assembly factories near the big markets. As a result, the distance between the market and the production facilities is shortened as Figure 1 shows. Unlike exportation, the partial dispersion of assembly plants worldwide changes the structure of the value added process.

Copyright © 2008, IGI Global. Copying or distributing in print or electronic forms without written permission of IGI Global is prohibited.

Partial Dispersion at the Beginning of Standardized Stage in the VCR Industry

The home-use VCR was developed and produced exclusively in Japan in the 1970s. Some foreign companies tried to manufacture the VCR, but they could not catch up with the Japanese companies because of the communication lag. When they appeared to be catching up, Japanese companies had proceeded even further because they had collaboratively sophisticated the VCR and, at the same time, had competed very effectively. However, in the early 1980s the leading Japanese companies JVC and Matsushita, started overseas production.

As seen in Table 1, large investments in the VCR facilities were made in 1980s and 1990s. It took a few years to standardize the production process so they could guarantee the product quality of overseas production. At least, companies *should* have established a standardized way of production before the overseas production. Otherwise, all companies or facilities should have started from scratch.

JVC set up an assembly plant, J2T, in West Germany in 1982. It was a joint venture with Thorne EMI (a British company) and Telefunken (a German company). The name of J2T came from the initials of JVC, Thorne EMI and Telefunken. However, the original plan was J3T, but Thomson (a French Company) had meanwhile dropped out. Later, Thomson acquired Telefunken. Therefore, in a sense the original plan was achieved. J2T had a great impact on the European market because three companies whose mother countries were different succeeded in a joint venture. It increased the brand image of their products. They also established a subsidiary to produce magnetic videocassette tapes in West Germany. As a joint venture with Robert Bosch (a German Company), Matsushita also set up an assembly plant in West Germany in 1983. To increase the local product contents in Europe, within a few years JVC manufactured the print boards in the UK, and drums and cylinders in France. Otherwise, JVC might suffer from protective tariffs. However, they continued to import the magnetic heads from Japan because these were high-precision parts and had room for further improvement. By 1988, JVC produced more than one million VCR units in Europe.

In the early 1980s, the leading companies partially started overseas production. Generally, they initially set up assembly plants in the highly advanced countries for sales promotion advantages. The European market was much

Copyright © 2008, IGI Global. Copying or distributing in print or electronic forms without written permission of IGI Global is prohibited.

Table 1. Number of VCR plants in Europe and Asia

	JVC		Sony		Matsushita		Sanyo		Funai		Aiwa	
	EU	Asia	EU	Asia	EU	Asia	EU	Asia	EU	Asia	EU	Asia
82	1	0	0	0	1	0	0	0	1	0	0	0
83	1	0	0	0	1	0	0	0	2	0	0	0
84	1	0	0	1	2	0	1	0	2	0	0	0
85	1	0	1	1	2	0	1	0	2	0	0	0
86	1	0	1	1	4	0	1	1	2	0	0	0
87	2	0	1	2	4	0	1	2	2	0	0	0
88	2	1	2	2	4	0	1	2	2	0	0	0
89	2	1	2	3	6	2	1	2	2	1	0	0
90	1	1	2	3	6	3	1	2	0	1	0	0
91	1	1	2	3	5	3	1	2	0	1	1	0
92	1	2	2	3	5	3	1	4	0	2	1	2
93	1	2	2	3	5	3	1	4	0	3	1	2
94	1	2	2	3	5	4	1	4	0	2	1	2
95	1	2	2	3	5	4	1	4	0	1	1	2

bigger than that of Asia and it enacted protective tariffs against others. The leading companies started making other parts in the area and increased the rate of local product components to avoid the protective tariffs. Table 1 shows the number of VCR factories of JVC, Sony, Matsushita, Sanyo, Funai, and Aiwa in Europe and Asia. Most companies established factories in Europe first. In the early 1980s, not only JVC and Matsushita, but also Funai, a typical cost cutter, had factories in Europe. Although Funai entered the market very late, they made their products so simple that they could easily transfer to new assembly plants. In the early 1980s, none of them, except Sony, had a factory in Asia. This is because of the manufacturability. At that time, much more technology and expertise were accumulated in Europe than in Asia. Companies decided to both assemble and source the standardized parts in Europe.

As Table 1 shows, Sony took a different strategy than JVC and Matsushita. At that time, the VHS group caught up to Sony. In 1984, Sony set up an assembly factory for the standard Betamax machines in Taiwan before Europe. They might turn the tables in the struggle for the de facto standard by cost cutting. Possibly, they realized that they were losing the game and kept open the possibility for other promising products. However, the strategy did not work.

Copyright © 2008, IGI Global. Copying or distributing in print or electronic forms without written permission of IGI Global is prohibited.

Table 2 shows the number of domestic shipments, exports and overseas production during 1982-1987. In this period, the number of the overseas production was much less than that of the exportation from Japan. The peak of the exportation from Japan was 27,689 thousand units in 1987 and that of overseas production was 27,048 thousand units in 1999. The overseas production increased very slowly, step by step. The overseas production before 1987 was the initial and experimental step. Companies demonstrated the feasibility of overseas production in other highly industrialized areas near big markets. In addition to the protective tariffs, many European electronic companies requested Japanese companies to enter into joint ventures for the production of the VCR in Europe. Although the number of units of overseas production was distributed unevenly in Europe, the exportation from Japan to Europe exceeded that of overseas production until 1987. The overseas production in Europe could not cover the demand in the area. However, JVC and Matsushita succeeded in penetrating European markets through the strategic alliance with European companies and local production in Europe. The next stage started with the late 1980s when some companies set up manufacturing facilities in Asia, the low cost operation area. After that, the number of units of overseas production increased rapidly.

The physical distance of supply chain members increased and its structure became more complex than before as the overseas production increased. When an assembly plant was set up in other countries, some suppliers were requested to cooperate to improve local product content and to obtain quicker response to market needs. However, every company has a minimum number

Table 2. Domestic shipments, exports, and overseas production in thousands of units after 1982 (Source: JEITA, 2005)

	Domestic Shipment	Exportation from Japan	Overseas Production
1982	2,343	7,355	
1983	3,658	10,652	309
1984	4,271	15,237	823
1985	4,006	22,071	2,063
1986	4,853	25,475	2,727
1987	6,331	27,689	3,826
1988	7,155	21,863	5,297
1989	6,617	23,130	5,259
1990	5,712	25,839	7,298

Copyright © 2008, IGI Global. Copying or distributing in print or electronic forms without written permission of IGI Global is prohibited.

of production units to achieve the break-even point. Such companies have a constraint on the economic number of production facilities. This constraint, together with the need for global sales promotion, induced the global formation of supply chains. The standardization of the production process, whose structure was thus controlled by economic reasons, and the collaboration among supply chain members, further enabled the growth of supply chains.

Country of Origin

Globalization and Country of Origin (COO) issues present a number of concerns for supply chain management and strategy. A recent comprehensive discussion has been given by Polin et al. (2005), who detail motivations and cautions for global integration of the supply chain. Some potential advantages may be at odds with cost-minimizing strategies. COO aspects can be impacted by trade legislation and may have marketing implications in meeting disparate demands of differing regional and national customers. Sourcing and offshoring decisions must consider risks like possible trade-offs of inexpensive components possibly associated with poor quality. Overall profitability of the supply chain may suffer if perceived quality should deteriorate or necessitate costly rework and delays. Another risk is involved in standardization of products to COO versions. A product that is largely standardized across multiple markets may present the lowest cost production alternative. But failure to adapt products to local customers may reduce the attractiveness of the product and hence reduce overall profitability as well.

Dornier, Ernst, Fender, and Kouvelis (1998) point out four areas that present difficulties for the globalization of supply chains. These are: (1) substantial geographic distances, (2) added forecasting difficulties, (3) foreign exchange rates and other matters of national economic policies, and (4) infrastructure inadequacies. The role of the supply chain in globalization has been discussed by several authors. Perlmutter (1969) stressed the value of each player making its unique contribution with its unique competence. Lussier, Baeder, and Corman (1998) discuss the myth of globalization as involving acquisitions without integration. Yip (1989) emphasizes the importance of location of value-adding activities. Zou and Cavusgil (1995) note the role of concentration of value-adding activities.

Copyright © 2008, IGI Global. Copying or distributing in print or electronic forms without written permission of IGI Global is prohibited.

Delfmann (2000) proposed a list of globalization drivers for supply chains based on an analysis of the literature, especially the work of Bhattacharya, Coleman, and Brace (1996); Cooper and Ellram (1993); and Ellram (1991). He proposes the five major drivers of: (1) reduction of inventory investment in the chain, (2) general reduction in costs, (3) improved customer service, (4) higher customer satisfaction, and (5) building a competitive advantage. In addition, he proposes drivers that enable a firm to gain cost or differentiation advantages based on Porter (1985). These include: linkages among various value-adding activities, economies of scale, learning from past experience and gaining knowledge, interrelationships among business units that involve shared systems, integration of value-adding activities, timing, and location factors.

Polin et al. (2005) suggest that while the COO concept was once simple, the modern supply chain has made its role less clear. What might be said of leather goods designed in Italy, but assembled in Mexico? Does this differ from a Japanese COO car manufactured in the U.S.? Is there a perceived difference between such a car assembled in the U.S. of Japanese components and one assembled in the U.S. of U.S. components? Several empirical studies, conducted in different locations indicated that COO does indeed influence buyer perceptions (Bilkey & Nes, 1982; Schooler, 1965). In the context of SCM and globalization, a U.S. COO might assemble athletic shoes in Asia at lower labor costs. While this may generate a substantial cost savings, it may not result in enhanced profitability. Label-conscious U.S. consumers might prefer the "made in USA" label. Whether such a preference is based on patriotic sentiments or real or perceived qualitative differences, supply chains must consider the impact of globalization on net profits rather than just cost minimization.

Defining COO goes well beyond actual country of manufacture or the traditional national home of a particular brand. Jaffe and Nebenzahl (2001) developed a taxonomy reflecting all six aspects of a product's origin(s): (1) home country, or consumer's country of residence, (2) designed-in country, as in Japanese-designed electronics, (3) made-in country, usually the location of the final stages of production, (4) parts country signifies the origin of the components, (5) assembly country, and (6) origin country, a symbolic or traditional country of origin, as in the case of USA for Zenith and Japan for Sony.

In the context of SCM and globalization, a traditionally North American firm might opt to assemble its athletic shoes in Asia due to the lower labor costs.

Copyright © 2008, IGI Global. Copying or distributing in print or electronic forms without written permission of IGI Global is prohibited.

While this may generate a substantial cost savings, it may not result in enhanced profitability. In this age of overseas standardization, label-conscious U.S. consumers might prefer the "made in USA" label, even at the cost of a major price increase. Whether this preference is based on patriotic sentiments or real or perceived qualitative differences, firms might best consider the impact of globalization on the "bottom line," rather than pursuing maximal cost minimization.

COO consumer perceptions may be positive or negative and can evolve over time, and hence over the industry life cycle. Studies of car buyers in Germany indicated that COO alone was a leading factor in the choice of car make among 27 percent of the respondents (Loeffler, 2001). However, the weight of COO alone may be less significant in other countries and/or product categories (Cordell, 1992). Verlegh and Steenkamp (1999) distinguish between cognitive and normative aspects of COO effects. Cognitive refers to perceptions of product quality, while normative refers to actual purchase decision. Their analysis of 41 studies revealed that the impact of COO is stronger for perceived quality than for attitudes and purchase likelihood. That is, COO appears to influence buyer perceptions, but it does not significantly alter purchase likelihood.

Lampert and Jaffe (1997) identify a direct relationship between the degree of differentiation in a product class and the market price ratio of goods in that category. Thus, there may be only very narrow price differences among different brands of commodity-like homogeneous goods such as sugar and salt, while highly differentiated product classes, such as automobiles may present larger price differences on the basis of COO. Polin et al. (2005) suggest that this relationship may be useful for overcoming a negative COO effect. Between homogenous goods and differentiated goods, Lampert and Jaffe also classify goods as of low differentiation (gasoline, tires, toothpaste, etc.) and medium differentiation (vacuum cleaners, color televisions, etc.). A broad range of automobile models are available, with an even broader range of prices. Within each of the four classifications of product differentiation identified by Lampert and Jaffe, there also exists a range of degrees of differentiation. Thus, automobile prices as a whole may vary widely, but price differences among competing models of economy cars are much smaller than price differences among competing luxury models. This implies that developing countries might export products in any category, provided they are relatively undifferentiated within their product class. While the export of Yugoslavian-made cars to the U.S. was an unquestionable failure, the

Copyright © 2008, IGI Global. Copying or distributing in print or electronic forms without written permission of IGI Global is prohibited.

relationship observed by Lampert and Jaffe suggests that the likelihood of a Yugoslavian car succeeding in the U.S. market was far greater in the sub-compact category than in the sport or luxury categories. Lampert and Jaffe (1997) also observed longitudinal changes in the COO effect. They commented that during the 1950s, a "made in Japan" label signified a cheap imitation of products made in industrialized countries. However, even before the VCR industry life cycle, that COO label had become associated with high quality, excellent workmanship and innovative products.

Practical Implications for Supply Chain Strategic Planning

Chopra and Meindl (2004) discussed the importance of achieving fit of competitive strategy with supply-chain strategy. Generally, globalization in the integrated geocentric sense is consistent with competitive strategy. At the same time, use of the supply chain as an enabler of globalization is also possible. Foreign sourcing or manufacturing can acquaint regional consumers with a product and reveal potential demand or product variations that would have higher regional appeal while still being consistent with competitive strategy. An initial positioning as ethnocentric and multilocal may yield opportunities for evolution by reexamining the globalization possibilities for competitive strategy and using opportunities opened up by efficiencies in the supply chain.

Economies of scale are clearly important drivers of globalization, but should not be pursued without regard to other considerations. Fast food firms such as McDonald's usually need to procure perishable supplies locally but can still realize economies of scale from packaging materials, fixtures, equipment, and operating procedures. More generally, cost minimization may not necessarily be consistent with profit maximization.

The COO and related effects offer both opportunities and potential liabilities. Many of these are marketing-related. Sourcing or manufacturing in a region or country with marketing appeal may be more profitable than choices with lower costs. Assembly of clothing in a third-world country likely will be both cost-efficient and appealing to customers of discount stores. However, for stylish high-priced items, this image would not be attractive and, in fact, could be less than attractive if negative associations to socially undesirable

Copyright © 2008, IGI Global. Copying or distributing in print or electronic forms without written permission of IGI Global is prohibited.

practices, like child labor, say, are involved. For each activity in the supply chain and for each customer country or region, there is a potential interaction that may be favorable or unfavorable to profitability, or may offer opportunities. A careful focus on profit or total costs should help prevent shortsighted choices.

Polin et al. (2005) introduced the concept of delayed adaptation. This was supported by the research of Schlie and Yip (2000), and observations of successful instances in the automobile and fast foods industries. The point is essentially that adaptation of products to local markets should be delayed to allow for economies of scale benefits. This would allow the firm to exert leverage over suppliers and produce an adapted product at the cost, but not necessarily at the price, of a standardized one.

Conclusion

We discussed the partial dispersion of the manufacturing facilities in this chapter. Partial dispersion becomes more reasonable than maintaining close proximity. As time passes, leading companies start locating assembly facilities in the other advanced countries for sales promotion advantages and for cost reduction. The standardization of the product including the modularity of the parts makes it possible to pursue partial dispersion.

In the VCR case, partial dispersion was conducted to avoid trade friction and to penetrate foreign markets quickly. JVC and Matsushita set assembly factories in the EU in the early 1980s. Funai, a typical cost cutter, also had factories in Europe at the same time. Funai entered the market very late, but their products were so simple that they could easily transfer to new assembly plants. In the early 1980s, none of them, except Sony, had a factory in Asia. This is because of the lack of the manufacturability and brcause it was not a necessity for sales promotion in the area. In 1980s, European countries accumulated much more technology and expertise than in Asian countries. Companies decided to set assembly factories and source the standardized parts in Europe.

Copyright © 2008, IGI Global. Copying or distributing in print or electronic forms without written permission of IGI Global is prohibited.

References

Bhattacharya, A. K., Coleman, J. L., & Brace, G. (1996). The structure conundrum in supply chain management. *International Journal of Logistics Management, 7*(1), 39-48.

Bilkey, W. J. & Nes, E. (1982) Country-of-origin effects on product evaluations. *Journal of International Business Studies, 13*(1), 89-100.

Botschen, G., & Hemetsberger, A. (1998). Diagnosing means-end structures to determine the degree of potential marketing program. *Journal of Business Research, 42*, 151-159.

Bowersox, J., & Closs, D. J. (1996). *Logistical management.* New York: McGraw-Hill.

Chopra, S., & Meindl, P. (2004). *Supply chain management: Strategy, planning and operation* (2nd ed.). Upper Saddle River, NJ: Pearson Prentice Hall.

Cooper, M. C., & Ellram, L. M. (1993). Characteristics of supply chain management and the implications for purchasing and logistics strategy. *International Journal of Logistics Management, 4*(1), 1-14.

Cordell, V. V. (1992). Effects of consumer preferences for foreign sourced products. *Journal of International Business Studies 23*(2), 251-269.

Delffmann, W. (2000). *Supply chain management in the global context.* Working Paper No. 102, Department of General Management, Business Planning and Logistics, The University of Cologne, Cologne, Germany.

Dornier, P.-P., Ernst, R., Fender, M., & Kouvelis, P. (1998). *Global operations and logistics: Text and cases.* New York: Wiley.

Dunning, J. H. (1988). *Explaining international production.* London: Unwin Hyman.

Ellram, L. M. (1991). Supply chain management: The industrial organisation perspective. *International Journal of Physical Distribution and Logistics Management, 21*(1), 13-22.

Higuchi, T. (2006). Industrial life cycle in the VCR industry. *Sakushin Policy Studies,* 6, 19-34.

Higuchi, T, Troutt, M. D., & Polin, B. A. (2004). Life cycle considerations for supply chain strategy. In C.K. Chan & H.W.J. Lee (Eds.), *Successful strategies in supply chain management* (pp. 67-89). Hershey, PA: Idea Group.

Copyright © 2008, IGI Global. Copying or distributing in print or electronic forms without written permission of IGI Global is prohibited.

Hout, T., Porter, M. A., & Rudden, E. (1982). How global companies win out. *Harvard Business Review, 60*(5), 98-108.

Itami, H. (1989). *Nihon no VTR sanngyou naze sekai wo seiha dekitanoka.* Tokyo: NTT Publishing.

Jaffe, E. D., & Nebenzahl, I. D. (2001). *National image and competitive advantage: The theory and practice of country-of-origin effect.* Herndon, VA: Copenhagen Business School Press/Books International.

JEITA (Japan Electronics and Information Technology Industries Association). (2005). *Minnseiyou dennsi kiki deta shyu.* Tokyo: JEITA.

Lampert, S. I., & Jaffe, E. D. (1998). A dynamic approach to country-of-origin effect. *European Journal of Marketing, 32*, 61-78.

Levitt, T. (1983). The globalization of markets. *Harvard Business Review, 61*(3), 92-101.

Loeffler, M. (2002). A multinational examination of the "(non-) domestic product" effect. *International Marketing Review, 19*(5), 482-498.

Lussier, R. N., Baeder, R. W., & Corman, J. (1994). Measuring global practices: Global strategic planning through company situational analysis. *Business Horizons, 37*(5), 56-63.

NHK (Japan Broadcasting Corporation.). (2000). *Project X: Challengers 1.* Tokyo: Japan Broadcast Publishing.

Perlmutter, H. V. (1969). The tortuous evolution of the multinational corporation. *Columbia Journal of World Business, 4*(1), 9-18.

Polin, B. A., Troutt, M. D., & Acar, W. (2005). Supply Chain Globalization And The Complexities of Cost Miminization. In C.K. Chan & H.W.J. Lee (Eds.), *Successful strategies in supply chain management* (pp. 109-143). Hershey, PA: Idea Group Publishing Co.

Porter, M. E. (1980). *Competitive strategy.* New York: The Free Press.

Porter, M. E. (1985). *Competitive advantage.* New York: The Free Press.

Sato, M. (1999). *The story of a media industry.* Tokyo: Nikkei Business Publications.

Schlie, E., & Yip, G. (2000). Regional follows global: Strategy mixes in the world automotive industry. *European Management Journal, 18*, 343-356.

Schooler, R. D. (1965). Product bias in the Central American common market. *Journal of Marketing Research, 2*, 394-397.

Copyright © 2008, IGI Global. Copying or distributing in print or electronic forms without written permission of IGI Global is prohibited.

Shoham, A. (1995). Global marketing standardization. *Journal of Global Marketing, 9*(1-2), 91-119.

Verlegh, P. W. J.,& Steenkamp, J.-B. E. M. (1999). A review and meta-analysis of country-of-origin research. *Journal of Economic Psychology, 20*, 521-546.

Vernon, R. (1966). International investment and international trade in the product cycle. *Quarterly Journal of Economics, 80*(1), 190-207.

Vernon, R. (1977). *Storm over the multinationals*. Boston: Harvard University Press.

Vernon, R. (1998). *In the hurricane's eye*. Boston: Harvard University Press.

Vollman, T. E., Berry, W. L., Whybark, D. C., & Jacobs, F. R. (2005). *Manufacturing planning & control for supply chain management* (5th ed.). New York: McGraw-Hill.

Yip, G. S. (1989). Global strategy…in a world of nations. *Sloan Management Review, 31*(1), 29-42.

Zou, S., & Cavusgil, S. T. (1996). Global strategy: A review and integrated conceptual framework. *European Journal of Marketing, 30*, 52-69.

Copyright © 2008, IGI Global. Copying or distributing in print or electronic forms without written permission of IGI Global is prohibited.

Chapter XIV

Total Dispersion

In the middle of the standardized stage, the severe competition in the market promotes the total dispersion of the manufacturing facilities. Various types of companies, such as leading companies, followers, and cost cutters, can exist in the same market because of the advancement of the product design and the production process invites newcomers and enables them to compete with others equally. It is natural that an appropriate strategy differs based on the company type. All types of consumers, the extreme innovator, the innovator, the early adopter, the early majority, the late majority, and the laggard, also exist in the market in this stage. Although their preferences are different, they, in particular the late majority and the laggard, want the price to fall. In addition, the demand grows globally. To respond to the consumer's request and the global demand, the total dispersion of the manufacturing facility advances globally.

Copyright © 2008, IGI Global. Copying or distributing in print or electronic forms without written permission of IGI Global is prohibited.

Divergence in the Middle of the Standardized Stage

As standardization progresses, the leading companies lose their advantages over others because little space for differentiation is left in product performance and efficiency improvements to production. The leading companies try to launch new product series with enhancements or new functions. However, the impact of such new products becomes smaller and smaller as the product design matures. Although understandably the first mover has a big advantage, others play as tough rivals in the market. The followers catch up with the leaders in the technologies, except for some of the latest ones. Instead, they add new functions to the product by way of their original technologies. In spite of the experience curve, the cost cutters are very competitive from the beginning because they can utilize standardized parts through the market while reducing the functions. They focus on the middle and low-end products, which are the largest segments of the total market in sales units. Various types of companies exist in the market and offer a wide variety of products. In addition, the productivities of the companies are almost equal because the process innovations have advanced greatly.

With regard to manufacturability, companies have ample options for location at this stage. Product design has become so standardized and stable that offshore facilities can manufacture the product easily and cope with minor changes in the product design. As the accumulated production count increases, the parts market grows worldwide, and many companies can conveniently and inexpensively obtain most standardized parts. Thus companies have fewer constraints in manufacturability. In addition, the price of the products becomes so reasonable that the demand in the low cost operation areas grows rapidly. The center of the consumers has shifted to the low cost operation areas in which the consumers are very sensitive to the price. The facilities in the low cost areas therefore become attractive from the production cost cutting and market growth perspectives.

Companies decide to open manufacturing facilities at different times and locations. Each company has a different set of core competencies and strategies. For example, a company whose main competence is high-end products tends to keep facilities in the mother country and the highly advanced countries. They transfer production of the mid-range products to the low cost operation areas. On the other hand, a company whose main forte is the mid-range and low-end products tends to source cheap standardized parts globally and establish efficient global logistics systems. They tend to be earlier than others

Copyright © 2008, IGI Global. Copying or distributing in print or electronic forms without written permission of IGI Global is prohibited.

in shifting and concentrating their manufacturing facilities to the low cost operation areas.

Regulations also have a substantial bearing on location decisions. Local governments may either induce or discourage companies to locate facilities in a country or area. While highly advanced countries may create protective tariffs, governments in the low cost operation areas may, to the contrary, give companies very favorable treatment such as tax abatements and other accommodations.

The advancements of the standardization of product design and of the production process increase manufacturability worldwide and therefore give companies a wider array of location options. They may choose locations and establish their global logistics systems according to their strengths and selected strategy. As a result, it can be said that the manufacturing facilities diverge globally from the individual company and the industry level.

Logistics decisions and issues are those related to transportation, inventory, and coordination, and are intimately related to location decisions (Chopra & Meindl, 2004). A few general principles are known. Inventory and facility costs increase with the number of facilities while transportation costs decrease. Clearly, placing facilities close to customers can decrease transportation costs and improve response times or reduce lead times of orders. In general, the number and locations of facilities should minimize total logistics costs, other conditions being equal. Little has been written about the effects of the industry life cycle on location and logistics and this chapter addresses this issue.

Severe Competition in the Middle of the Standardized Stage

In this stage, various types of companies compete in the market. The advancement of the product design and process invites newcomers and enables them to compete equally, provided they choose the right target customers and approach them effectively in the market. An appropriate strategy differs based on the company type, namely, leading companies, followers, and cost cutters. As production efficiency and capacity meet their peak, more severe competition is induced.

Copyright © 2008, IGI Global. Copying or distributing in print or electronic forms without written permission of IGI Global is prohibited.

Also at this stage, the leading companies will have exploited the domestic and foreign markets. They may then try to differentiate their products from others through their technology and brand name. They also may continue to place an emphasis on location in the highly advanced countries for sales promotion. However, they should reinforce the location in the low cost operation areas because they are also involved in a price war. They must start transferring the manufacturing lines of the standard parts and the low-end products to the low cost operation areas. In addition, the protective tariffs begin to lose their effectiveness because the products are imported from various countries at that time. Thus, leading companies must balance the allocation of facilities between the highly advanced countries and the low cost operation areas.

Follower companies also set up assembly plants first and other plants for certain parts in the highly advanced countries behind the leading companies. However, some of them locate manufacturing facilities in the low cost operation areas earlier than leading companies. They compete with the leaders by adding minor functions and reducing costs. Hence, they shift the importance of the manufacturing facilities from the highly advanced countries to the low cost operation areas quickly.

Cost cutters also locate assembly plants in the highly advanced countries behind the leading companies and followers. But their usage rates of the local content and components are much lower than those of leading companies and followers. Their manufacturing facilities in the low cost operation areas are much more important than those of others because they have simplified the product design and product functions for maximal cost reduction.

Niche companies concentrate on small and widely scattered markets. They also source low cost and quality parts globally. However, they do not have to utilize overseas production aggressively because their market is small and specialized. Thus, location in a specific region for sales generation and stimulation makes little sense for them.

In the middle of the standardized stage, various types of companies compete aggressively in the market. The location of facilities becomes a very important factor because it becomes very difficult to differentiate products solely through technology. Companies decide their balance of the location near the market or cheap resources according to their unique strategies or strengths.

In the middle of the standardized stage, in order to achieve further cost-cutting, production facilities in the low cost operation areas are reinforced, while some facilities in highly developed countries are closed. The products are then exported from the low cost operation areas to the whole world. There-

Copyright © 2008, IGI Global. Copying or distributing in print or electronic forms without written permission of IGI Global is prohibited.

fore, the distance between the manufacturing facilities and the markets is again increased.

Logistics Management in the Middle of the Standardized Stage

Multinationals that have developed a global network commonly see the world as a chessboard on which they are conducting a wide-range of campaigns. The chessboard's squares are nation-states, and an enterprise can consider entering any one of them by a number of different means—by trading with dependent firms in the country, by developing alliances with enterprises already operating in the country, or establishing a subsidiary of its own in the country. (Vernon, 1998, p. 22)

Logistics becomes the most vital activity because the importance of technology decreases. Companies provide much more variety of products then, due to the severe competition and consumer demands. In the market, there exist not only various types of companies, but also various types of customers, such as the extreme innovators, the innovators, the early adopters, the early majority, the late majority, and laggards. Usually their behaviors are quite different from others even in repeat purchase behavior. In the case of repeat purchases, the most extreme innovators and innovators want something new. The early and late majorities tend to consider the newness and the price as concerns. Most laggards still focus on the price. The market requires the companies to produce various types of the products. Companies should determine the target customer types and the range of the product line. This decision should take into account the global allocation of the facilities and inventory policies.

The global allocation of facilities is at the core of logistics management because the product and processes become so standardized that few processes can create competitive advantages by purely technical improvements. There is little room left for differentiation in quality and manufacturing cost, with the exceptions of labor and rent costs. Global allocation determines the distance to the target customers, the structure of the global sourcing, the service level, and thus the logistics costs. In the case of the leading companies, their target customers are mainly the extreme innovators, the innovators, the early

Copyright © 2008, IGI Global. Copying or distributing in print or electronic forms without written permission of IGI Global is prohibited.

adopters, and the early majority. It still is effective for them to locate facilities near the highly advanced countries because most of their target customers live in these countries. However, the standardization of the product design and process has advanced and the international physical distribution has been deregulated and systemized so that parts and products travel easily and globally. Leading companies also utilize more resources in the low cost operation areas than before.

Companies allocate facilities globally with consideration for access to markets or resources. As time passes, global sourcing becomes more influenced by market accessibility because of worldwide manufacturability and superior experience in the highly advanced countries. The efficient and effective flow of parts and products from raw materials to the customer, the logistics, are therefore the most vital considerations.

SCM in the Middle of the Standardized Stage

During this stage, the structures of supply chains change dramatically. Extreme competition puts great pressure on all supply chains to reduce costs and widen the product variety. Some supply chains shift most of their facilities to foreign countries, especially the low cost operation areas, while others make strategic alliances with foreign partners. Supply chains become more diversified than before with respect to geographic distribution and trading partners.

First, as we have seen, the supply chain becomes global. As mentioned above, most companies deploy their facilities globally for sales promotion and cost cutting. Some suppliers decide to place their manufacturing facilities near foreign assembly plants for further cost cutting and reduced lead times for deliveries. Then, overseas production exceeds the domestic production in most supply chains. In addition, the ratio of the production in the low cost operation areas becomes overwhelming within the worldwide production.

Second, some supply chain partners will have exited and chains become partners. The advancement of the standardization of the product design and parts provides an opportunity for foreign and outside companies to become partners. Those who have strength in sales promotion or cost cutting are very attractive to any supply chain. In addition, the supply chains plan to increase the flexibility of the production quantities, as well as to increase the variety of products. They want to respond to market demand quickly and,

Copyright © 2008, IGI Global. Copying or distributing in print or electronic forms without written permission of IGI Global is prohibited.

on the other hand, reduce exposure to future risks. In that way, not only the geographic distribution, but also the nationality of the supply chain partners becomes diversified.

Finally, most supply chains capitalize on strategic alliances to the fullest extent possible. Leading companies plan to keep a wide selection of products at reasonable costs. It is not practical for the leading companies to manufacture all types of products from the low-end to the high-end. However, their brand name is obviously very valuable even for the low-end products. Hence, they tend to abandon the production of low-end products to save resources and because their structure is less appropriate for low-cost operations. On the other hand, the OEMs (original equipment manufacturers) have a good opportunity to utilize economies of scale. In the case of standardized parts and service, local companies can perform sufficiently well at a reasonable cost. Thus, the transaction mode also becomes diversified.

The VCR Case in the Middle of the Standardized Stage

VCR products had essentially been perfected by the middle of the 1980s. Ever since the HR-3300, the first VHS that was launched in 1976, VCR products have continued to be improved. Major improvements were the double speed playback mode with monaural sound in 1977, the six-hour model in 1978, and the HiFi model in 1983, which were introduced by JVC. After the HiFi model, JVC developed the eight-hour model in 1989 and launched new formats, the S-VHS in 1987, the W-VHS in 1994, and the D-VHS in 1997, which were widely adopted in the market. Therefore, it can be concluded that HiFi was the popular edition until the DVD and HDD replaced the VCR in the early 2000s.

The domestic production cost of the VCR continued to decrease. However, during the period 1987-1992, the reduction rate was very small. The global demand for the VCR was about 40 million units around 1990 and about 80 percent of them were produced in Japan. After that period, the reduction rate increased temporarily from 1993 to 1998 because of global sourcing. In other words, Japanese companies imported many more parts than before. It is reasonable to infer that the manufacturing process was also perfected in the middle of the 1980s.

Copyright © 2008, IGI Global. Copying or distributing in print or electronic forms without written permission of IGI Global is prohibited.

Companies allocated their facilities in various ways in the late 1980s. There were two major movements related to allocation in the highly advanced countries and the low cost operation areas. Many companies located their facilities in the highly advanced countries in the late 1980s. Matsushita set up facilities in the US, Spain, and France where they bolstered the production of some parts such as HiFi tuners. Matsushita also hoped to make a point of the market share through sales promotions in those countries. Sanyo also reinforced their facilities in West Germany, following a strategy of reducing logistic costs. They assembled products near the market and, as a result, their overseas production rate was very high. Funai, a cost cutter, placed facilities in the UK and West Germany in the late 1980s. However, their rate of utilization of local content and components in Europe was very low, namely about 20-30 percent. On the other hand, the importance of the facilities in the highly advanced countries decreased because the low cost operation areas demonstrated manufacturability along with substantial cost advantages. When JVC produced a million units in Europe in 1988, they closed a facility in the UK and transferred the print board production to West Germany.

Most companies reinforced their VCR production in the low cost operation areas in the early 1990s. Overseas production exceeded the Japanese domestic production in the early 1990s, as Figure 1 shows. The followers placed much higher importance on the facilities in these areas than the leading companies. They reinforced the production of parts and assembly processes in these areas. They raised the ratio of local product content in these areas for further cost cutting. They assembled large numbers of products there and exported them to other countries, the highly advanced countries and other developing countries and areas. In the middle of the 1980s, Sanyo set up facilities in Taiwan, India, and Indonesia. They manufactured cylinders and magnetic heads in Indonesia and exported them to other facilities in Asia. The leading companies transferred some facilities for the standard or low-end VCRs to the low cost operation areas and kept others for the high-end VCRs in Japan and the highly advanced countries. In 1989, JVC started manufacturing some parts including print boards in Indonesia and exported them to the facilities in Japan and West Germany. Afterwards, JVC transferred the assembly process of the monaural VCRs for the US to the facility in Indonesia. In the early 1990s, Matsushita also opened facilities in Taiwan, Indonesia, Malaysia, and China. But they did not close the facilities in the highly advanced countries. Funai, a cost cutter, took a remarkable action in the early 1990s. They transferred the facilities for the low-end products to Malaysia in 1990. They started manufacturing the print boards there and exported to facilities in

Copyright © 2008, IGI Global. Copying or distributing in print or electronic forms without written permission of IGI Global is prohibited.

Japan and Europe. In 1992, they started print boards and unfinished products at a facility in China that took over the function from the facility in Malaysia. They also reinforced the procurement center in Hong Kong and, on the other hand, they closed the facilities in Europe by the early 1990s. This effectively established a division of labor between Japan and China that promoted their survival. In Japan, they assembled the parts that were produced in China and procured through the market, mainly in Hong Kong.

During the years 1985-1991, Japan exported more than 20 million units of VCRs per year. However, as the overseas production increased, the number of the domestic production decreased after the early 1990s. The overseas production in this period replaced the exports from Japan. On the other hand, Asia became the biggest importer in the 1990s. This is because the life cycle advanced much more slowly in Asia than in North America and Europe. In addition, the global division of labor was planned based on the products. The production of the low-end products shifted to Asia in the late 1980s and the early 1990s. The low-end products were shipped all over the world from Asia. Some companies manufactured the middle-range products domestically and others in Asia. The high-end products were manufactured exclusively in Japan in the early 1990s.

Figure 1. Domestic and offshore production (Source: http://www.jeita.or.jp/english/index.htm)

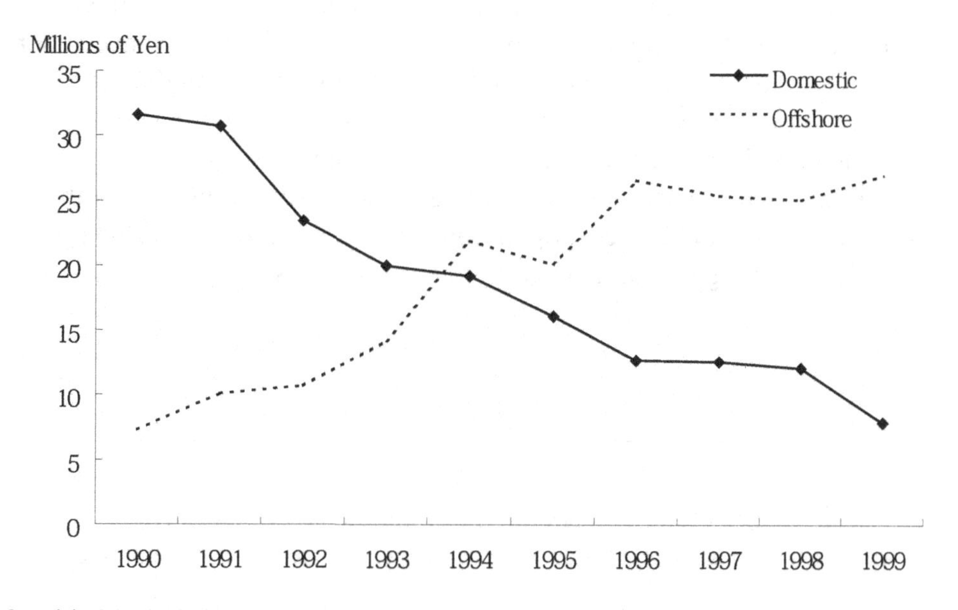

Copyright © 2008, IGI Global. Copying or distributing in print or electronic forms without written permission of IGI Global is prohibited.

Table 1. Domestic shipment, exports, and overseas production in thousands of units since 1988 (Source: EAIJ, JEITA and the Ministry of Finance)

	Domestic Shipment	Exportation from Japan to				Overseas Production
		Total	North America	Europe	Asia	
1988	7,155	21,863	10,211	5,416	4,452	5,297
1989	6,617	23,130	10,686	4,880	5,597	5,259
1990	5,712	25,839	8,955	6,776	8,092	7,298
1991	5,219	21,991	7,150	4,900	8,277	10,128
1992	4,597	17,732	6,202	2,767	7,474	10,688
1993	4,486	14,814	4,275	2,138	7,220	14,227
1994	4,854	15,235	3,776	1,840	8,691	21,929
1995	6,041	9,775	3,568	1,051	4,337	20,096
1996	6,667	6,783	3,121	902	2,410	26,647
1997	6,925	6,485	3,759	751	1,666	25,447
1998	7,156	7,159	4,569	490	1,735	25,074
1999	6,834	5,318	3,403	268	1,420	27,048
2000	6,412	3,902	2,082	261	1,308	23,376
2001	6,132	1,190	203	106	827	17,841
2002	4,729	542	133	53	346	16,944
2003	2,952	254	105	39	103	12,763
2004	1,848	131	63	27	39	8,247

In late 1980, the companies undertook various strategies for the allocation of facilities. The demand was growing so steadily in the advanced countries that companies increased the number of facilities and manufacturing capacities. Some companies started a global division of labor according to their strategies. There was some indication that the importance of the facilities in the highly advanced countries was gradually weakening due to the rise in production in the low cost operation areas. The structure of global sourcing had thus changed dramatically by the early 1990s.

Conclusion

We discussed the total dispersion of manufacturing facilities. In the middle of the standardized stage, various types of companies, such as leading

Copyright © 2008, IGI Global. Copying or distributing in print or electronic forms without written permission of IGI Global is prohibited.

companies, followers, and cost cutters, can exist in the market simultaneously. This is because the advancement of the product design and process improvements invites newcomers and enables them to compete with others equally. The late majority and the laggards are the main force in the market in this stage and their preferences are cheap standard products. The more severe competition in the market accelerates the use of offshore production. In addition, the demand grows in the developing countries. To respond to the global demand and reduce costs, companies advance the total dispersion of the manufacturing facility globally.

In the VCR case, companies allocated their facilities in the highly advanced countries and the low cost operation areas in the late 1980s and the early 1990s because of the large global demand of about 40 million units per year. However, the importance of the facilities in the highly advanced countries decreased, while the low cost operation areas demonstrated manufacturability along with substantial cost advantages. In the early 1990s, most companies reinforced their VCR production in the low cost operation areas and the offshore production exceeded the domestic production in units. In the late 1990s, even Japan imported many more parts and products than before. As a result, the price went down dramatically during 1993-1998 because of the global sourcing.

References

Bowersox, J. & Closs, D. J. (1996). *Logistical management*. New York: McGraw-Hill.

Cao D., Instlle M., Zepeda A., & Maino M. (1994). A global optimization approach to resource allocation problems with consideration of economies of scale. *Information and Decision Technologies, 19*, 257-266.

Chopra, S., & Meindl, P. (2004). *Supply chain management: Strategy, planning and operation* (2nd ed.). Upper Saddle River, NJ: Pearson Prentice Hall.

Cooper, J. C. (1993). Logistics strategies for global business. *International Journal of Physical Distribution & Logistics Management, 23*(4), 12-23.

Dunning, J. H. (1988). *Explaining international production*. London: Unwin Hyman.

Copyright © 2008, IGI Global. Copying or distributing in print or electronic forms without written permission of IGI Global is prohibited.

Higuchi, T. (2006). Industrial life cycle in the VCR industry. *Sakushin Policy Studies,* 6, 19-34.

Higuchi, T, Troutt, M. D., & Polin, B. A. (2004). Life cycle considerations for supply chain strategy. In C.K. Chan & H.W.J. Lee (Eds.), *Successful strategies in supply chain management* (pp. 67-89). Hershey, PA: Idea Group.

Itami, H. (1989). *Nihon no VTR sanngyou naze sekai wo seiha dekitanoka.* Tokyo: NTT Publishing.

JEITA (Japan Electronics and Information Technology Industries Association). (2005). *Minnseiyou dennsi kiki deta shyu.* Tokyo: JEITA.

NHK (Japan Broadcasting Corporation.). (2000). *Project X: Challengers 1.* Tokyo: Japan Broadcast Publishing.

Sato, M. (1999). *The story of a media industry.* Tokyo: Nikkei Business Publications.

Vernon, R. (1966). International investment and international trade in the product cycle. *Quarterly Journal of Economics, 80*(1), 190-207.

Vernon, R. (1977). *Storm over the multinationals.* Boston: Harvard University Press.

Vernon, R. (1998). *In the hurricane's eye.* Boston: Harvard University Press.

Copyright © 2008, IGI Global. Copying or distributing in print or electronic forms without written permission of IGI Global is prohibited.

Chapter XV

Convergence of Facilities in Low Cost Operation Areas

In this chapter, the convergence of manufacturing facilities is discussed. Very little room is left for the differentiation of products in the late standardized stage. Although companies source globally to reduce the cost, they should cut down their cost even further. In addition, the demand for a product begins to decline sharply at the end of the life cycle because of the saturation of the market or the emergence of alternative products. As a result, companies should make the most of economies of scale in a low cost operation area. Companies converge their manufacturing facilities into low cost operation areas or withdraw completely from the market.

Copyright © 2008, IGI Global. Copying or distributing in print or electronic forms without written permission of IGI Global is prohibited.

Late Standardized Stage

In the late standardized stage, almost no innovation occurs except for combined products that act as a bridge to alternative products. All companies, including the leading companies, have been involved in price wars and have encountered difficulties in differentiating their products from others. The manufacturing process has become so standardized that additionally companies require global sourcing, especially in the low cost operation areas. For further cost cutting, they should make the most of the economies of scale. Otherwise, companies can not survive under the cut throat competition in this stage.

The product has been diffused globally very extensively by the late standardized stage. The product has become spread not only in the highly advanced countries but also the low cost operation areas. As a result, the center of the demand is the repeat purchasing. The product has a much greater risk of being replaced by another product because consumers are tired of products that are not very different from the old ones, and alternative products that have competitive advantages in performance and/or price, may appear. Although some products maintain their demand at a higher level for a long time, all products face ultimate decline.

As a last resort, companies should start concentrating their facilities in a low cost area. The economies of scale are their final effort. In cases where demand for the products is declining, companies unable to utilize economies of scale would quickly withdraw from the market. The leading companies have the brand name, the niches have their domain, and the cost cutters have competitive advantages in the price war. The followers might be the first ones to withdraw from the market and seek alternative products in earnest. Surviving companies would squeeze their facilities into a low cost operations area while also considering alternative products.

Companies establish a global division of labor by product types to achieve economies of scale. Companies manufacture the high-end products, which have the biggest drop in global demand, in their domestic facilities and then export them all over the world. Other products are produced in a low-cost operation area and are also exported worldwide. Therefore, even the country of origin becomes an importer of the products because the demand for the high-end products is much smaller than that of others. Eventually companies find it difficult to secure the critical mass demand. When this occurs, companies would manufacture more than the critical mass and maintain

Copyright © 2008, IGI Global. Copying or distributing in print or electronic forms without written permission of IGI Global is prohibited.

a reserve inventory so they can respond to market demand after they stop manufacturing for a while.

Alternative Products in the Late Standardized Stage

Products in the late standardized stage face a significant risk of being replaced by alternative products. Some consumers, such as the laggards and the late majority, make much of the price while others, such as the innovators and the early majority, place high value on the innovativeness. Therefore, any successful product has a risk of being replaced because there is a great business opportunity for all companies when a product becomes highly standardized. The capriciousness of consumer needs is one of the major causes of new product creation.

All companies prepare for alternative products in the late standardized stage. However, the companies have different attitudes toward the alternative products. Leading companies are moderate. They want to protect the established order of the market conditions because they can skim off the rest of the cream. On the other hand, they also are seeking new business opportunities. Some followers who have an advance in technology or marketing are very aggressive as regards alternative products to gain the first mover advantages. Some of them who previously lost against the leading companies became followers and now want revenge. Others are somewhat aggressive because they also need the new business in situations where their sales have decreased. They prepare for the next generation product by forming strategic alliances and making a big investment. Although the cost cutters are also seeking new business opportunities, they are moderate toward the alternative products. They do not want to obtain the first mover advantage very much. On the other hand, they have sufficient ability to catch up from behind quickly and to ultimately break into the market by way of low-cost operations.

In any case, all companies prepare for alternative products in the late standardized stage. It means that they should set up facilities near the innovators and the latest technologies for the alternative products. In addition to that, the decline of the market demand and the preparation for the next generation products promotes the convergence of the facilities into the low cost operation areas and withdrawal from the market. The alternative products are good

Copyright © 2008, IGI Global. Copying or distributing in print or electronic forms without written permission of IGI Global is prohibited.

drivers for the reorganization of the allocation and the business in the late standardized stage, as well as the market size and its possibility.

SCM in the Late Standardized Stage

Supply chains should also make the most of global sourcing and the economies of scale simultaneously. It prompts them to change off their partners in a low cost operation area. By the time the standardization of the parts and process has highly advanced, companies in the low cost operation areas have acceptable manufacturability of the high-tech products at relatively lower prices. In addition, the companies do not have to keep the industrial secrets of the high-tech parts because the products have little room for differentiation and the demand is decreasing. Therefore, they should be cost oriented. Most value was added to the VCR products in a low cost operation area.

Some parts continue to be manufactured exclusively in another area adjacent to the area in which the assembly plant is allocated. This may also be a result of the utilization of global sourcing and the economies of scale at the parts level. It is very practical because they should exploit the existing facilities in the low cost operation areas. In particular, companies can transport small parts cheaply and quickly. The center of the supply chain has shifted to a low cost operation area.

It is impossible for all supply chains to remain in the market in a situation where the demand is decreasing. Headquarters of the supply chains decide to change off some partners for further cost reduction and they increase the rate of outsourcing of the parts for strategic withdrawal. The strategic withdrawal is the key word in this stage in the supply chain. They should rearrange their structure for the next business chance. Headquarters, as well as most partners, seriously seek new business opportunities. Some supply chains shift smoothly and others badly.

VCR Case in the Late Standardized Stage

Almost all VCRs had been produced in Japan. Figure 1 shows the shifts in the import and export numbers in Japan and the shifts of domestic and offshore

Copyright © 2008, IGI Global. Copying or distributing in print or electronic forms without written permission of IGI Global is prohibited.

Figure 1. Shift of exports and imports (Source: http://www.jeita.or.jp/english/index.htm)

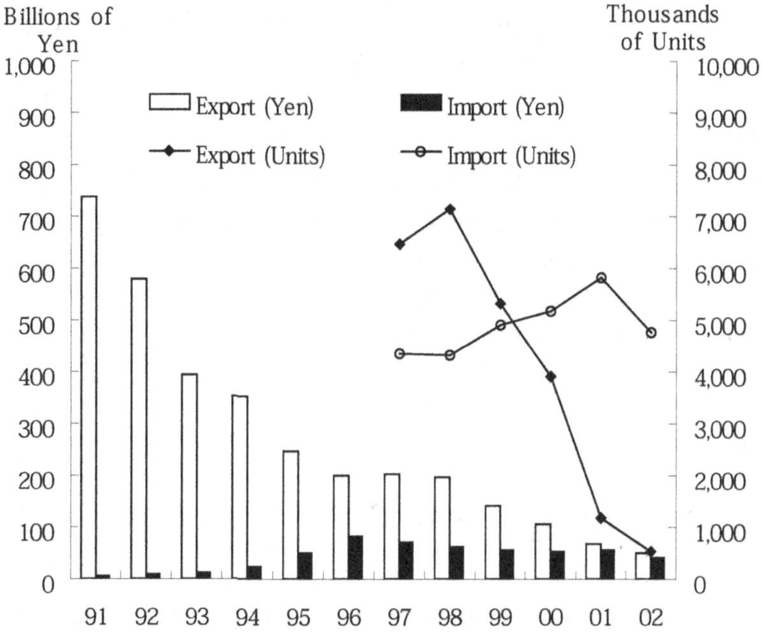

Billions of Yen
Thousands of Units
Export (Yen)
Import (Yen)
Export (Units)
Import (Units)
1,000 900 800 700 600 500 400 300 200 100 0
10,000 9,000 8,000 7,000 6,000 5,000 4,000 3,000 2,000 1,000 0
91 92 93 94 95 96 97 98 99 00 01 02

production by Japanese companies. In the early 1990s, some Japanese companies started producing VCRs in foreign countries. By 1994, the offshore production surpassed the domestic production. Finally, overseas production tripled and by 1999, imports to Japan exceeded exports.

Next to the cost cutters, the followers established low cost operation manufacturing systems that were inflexible to further major changes. By the middle of the 1990s, Toshiba manufactured 3.5 million units in Singapore, an amount that equaled 90 percent of the total of Toshiba's production. They moved the headquarters of the VCR division to Singapore in 1996. However, they moved all manufacturing facilities to Thailand in 1997. Hitachi also closed a facility in the UK and transferred its production process to other facilities in Asia in 1996. In 1997, Mitsubishi closed their domestic facilities and manufactured the S-VHS in Malaysia. They could not compete with the cost cutters in the price war but could compete with the leading companies until those companies shifted their facilities for the high-end products to a low cost operation area.

Copyright © 2008, IGI Global. Copying or distributing in print or electronic forms without written permission of IGI Global is prohibited.

DVD recorders and players replaced the VCR in the early 2000s. The possession rate of those machines per household increased from 19 percent to 49 percent during 2001-2004. The sales of DVD software reached 43 million units, an amount that far surpassed that of the VHS, namely 28 million units in 2001. The sales of DVD software in 2004 were about 0.1 billion units per year, 10 times as large as that of the VHS. It was obvious that the VCR industry was in decline by the mid-2000s.

In the early 2000s, leading companies also converged their facilities to the low cost operation areas. Matsushita stopped manufacturing VCRs and instead started the production of DVD-RAM drivers in Japan. JVC manufactured the VCR frugally by reducing the number of employees and closing the subsidiaries that produced key parts. By the middle of the 2000s, only one or two companies continued to manufacture VCRs in Japan.

Supplement to Vernon's Product Life Cycle Theory

It is important to note that we have obtained some results that are inconsistent with Vernon's Product Cycle Theory from the VCR case. Although the VCR and its manufacturing processes became highly standardized, Japan exported more VCR products than it imported in monetary terms by the early 2000s. This was because some leading Japanese companies delayed the maturity stage through product innovations such as adding new functions. They also established a global division of labor based on differentiating product types and receiving countries.

The leading companies, such as JVC and Matsushita, did anti-ageing through product innovations. Figure 2 illustrates the shift of the lowest prices adjusted by the GNP deflator in each category, monaural and HiFi sound, by the companies JVC and Funai. While JVC is a leading company, Funai is a cost cutter. JVC tried to avoid price competition by sophisticating their products. From Figure 3, it is clear that JVC had a much wider range of products than Funai. Thus, the difference in the degree of maturity of the product has a great effect on the allocation of facilities and hence the structure of the supply chain.

The structure of the global division of labor depends on the company's strategy. The cost cutters and the followers finished concentrating their facilities in the low cost operation areas and establishing low cost operation systems by 2000. On the other hand, leading companies retained their domestic fa-

Copyright © 2008, IGI Global. Copying or distributing in print or electronic forms without written permission of IGI Global is prohibited.

Figure 2. Lowest prices in each category by company (Note: Prices have been adjusted by the GNP deflator [1994:100%])

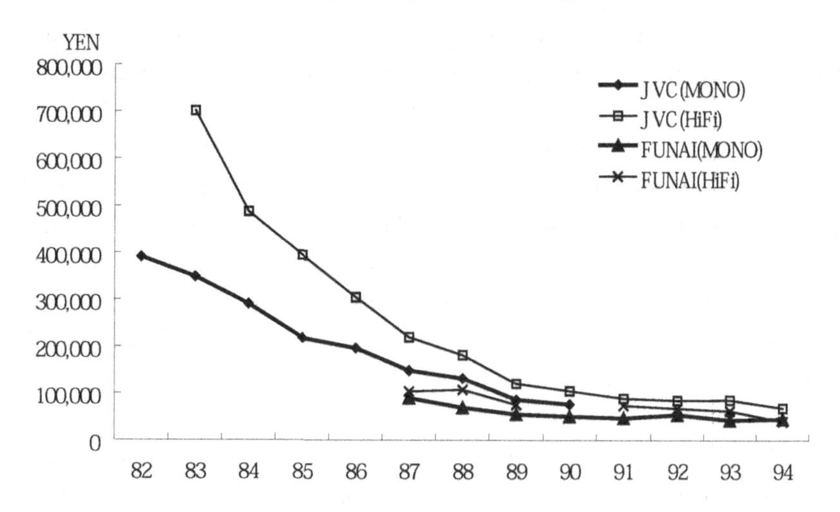

Figure 3. Price zones in 1993-1994

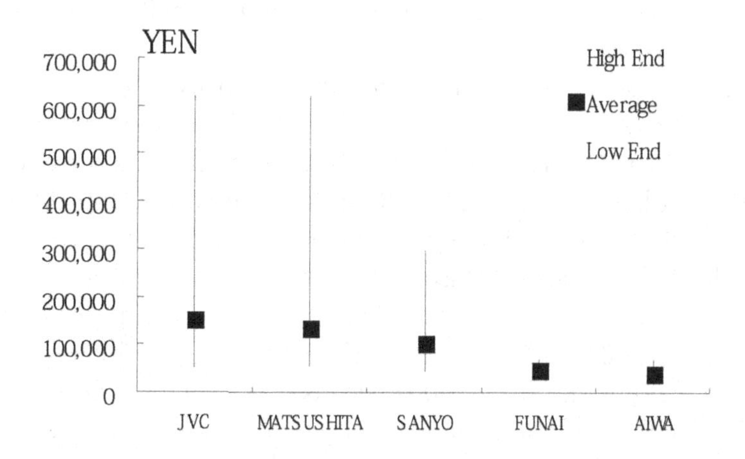

cilities for the domestic market and the high-end market. Their brand name slowed the price decline and made them able to keep their domestic facilities longer than others. As a result, Japan exported more VCR products than they imported in monetary terms in the early 2000s.

Copyright © 2008, IGI Global. Copying or distributing in print or electronic forms without written permission of IGI Global is prohibited.

In Vernon's product cycle theory, the location of factories has a close relationship to the degree of maturity of the product. Companies have their own mission and sources of competitive advantages at each stage. When the product is new (the innovative stage), companies tend to locate factories in the most advanced countries. They build a strong linkage to the customers in order to grow the new market. After the product matures, they tend to spread production facilities to other advanced countries. They should respond to the expansion of the market and the intensification of competition. Overseas investment and international strategic alliances become effective. Once the product becomes standardized, it becomes efficient to locate manufacturing facilities in the low cost operation areas. They need to pursue efficiency through global logistics. To achieve this, it becomes critical to allocate supply chain functions globally and to improve coordination among them.

The case study of the VCR illustrates many of the general principles of various life cycle theories mentioned. The facilities moved around the world due to various reasons. At first, product innovation has a great influence on the location. This innovation stage limits the spread of manufacturability globally in order to safeguard industrial secrets and adjust to major changes required for the product. After the dominant design has emerged in the market and global demand for the product has grown steadily, the distribution of customers became important. Many facilities were set up in the highly advanced countries in which the most affluent customers exist. Thus, access to the market was vital for a while. Over time, the product and its manufacturability diffused widely. This means that the importance of the highly developed countries decreased and the competition became more severe. The allocation of facilities became cost oriented. At that time global sourcing was the strongest driver for cost reduction. The demand for the VCR decreased dramatically in the early 2000s because DVD and HDD recorders replaced VCRs. In that situation, companies should concentrate their facilities to the low cost operation areas for economies of scale benefits. Companies that did not have enough demand to make the most of economies of scale withdrew from the market. The economies of scale are the last key factor.

Table 1 summarizes location aspects and considerations for supply chain location by stages and factors. In the beginning, there is little freedom in location decisions because imperfections of manufacturability still need to be solved and the demand is limited. As the demand grows, the freedom of

Copyright © 2008, IGI Global. Copying or distributing in print or electronic forms without written permission of IGI Global is prohibited.

the location increases technically and also becomes required for marketing reasons. Customers are unevenly distributed around the world. The timings of purchases, including repeat purchases, and the purchasing capabilities of customers differ greatly by country or region. It is a major and constant task for supply chain management to establish efficient and effective supply chains for the unevenly growing global demand. The main target customer and the environment change in this stage of the life cycle. The growing demand after standardization of the production process then has attracted so many entrants that cutthroat competition occurs in the market. Supply chains should respond by reallocating manufacturing facilities for efficiency. When the global demand has reached the decline phase, supply chains should integrate their facilities for further cost cutting and economies of scale benefits or withdraw from the market entirely.

The type of integrator also has a great influence on location decisions. The leading companies were the first to set up foreign facilities because they wanted to promote sales internationally and partly because they standardized manufacturing processes to some extent earlier. The followers necessarily trailed leading companies. However, they made much of the innovativeness and the cost equality. They could often be the first to utilize the advantages of global sourcing. The cost cutters entered the market after the product was diffused and the parts market was established. They reduced costs by reducing parts and process times. They simplified the product design and utilized the parts market from the beginning. They established complete low cost manufacturing systems that were also very flexible against the demand variance. Thus different company types varied in their timing for shifting to and structuring of their global sourcing decisions.

Companies also needed to be mindful of the global distribution of customer types, such as the extreme innovators, the innovators, the early adopters, the early majority, and the late majority. However, the last type, the laggards, is less influential. Most of the extreme innovators and innovators were in the highly advanced countries. After the product became popular and the price declined, the products diffused to the early and late majority in the highly advanced countries and the innovators in the less developed countries. Then the majority in these countries became the great purchasing power block. On the other hand, the repeat purchases of the innovators in the highly advanced counties gave the companies the chance to upgrade the product. Leading companies kept their domestic facilities while they still had a chance to upgrade the product. The extreme innovators and the innovators in the highly

Copyright © 2008, IGI Global. Copying or distributing in print or electronic forms without written permission of IGI Global is prohibited.

advanced countries were the first adopters of alternative products. The movement spread to other customers and went around the world gradually.

Every industry has different problems and constraints at each stage. The supply chain is a means of coping with them and launching products. The headquarters companies control the number of members and the structure of their supply chain. At the beginning, the number of members is limited in order to speed up product development and to avoid breaches of confidence. After the product specifications are confirmed, the headquarters companies start systemizing the supply chain for mass production and consistent quality of the products. When they face rapid demand growth in the early half of the life cycle, they order and support their members to reinforce their manufacturing capacity instead of outsourcing. The standardization of the production process and parts makes it possible to shift the manufacturing facilities to other countries and to outsource some parts. The outsourcing becomes popular as the price war advances. When the demand decreases at the end of the life cycle, many parts suppliers also have difficulties achieving economies of scale. As a result, some suppliers survive and others are weeded out.

We have reviewed the location aspects of the VCR industry. The VCR is representative of a durable good that was based on analog technology and

Table 1. Summary of supply chain location factors

	Physical Location	Partial Dispersion	Total Dispersion	Convergence
Purpose	Development of the product and the product process	Increase of global sales	Cost cutting	Further cost cutting (strategic withdrawal)
Main Activity	R&D Prototyping Establishment of the mass production system	Standardization of the product process Increase of the rates of the local contents	Reallocation of the manufacturing facilities	Integration of the manufacturing facilities (restructuring)
Target Customer	Extreme innovators Innovators	Early adopters Early majority	Late majority	Laggards
Driver	Close communication for the speed up of the product development and production process	Global marketing (uneven distribution of the early adopters and the early majority)	Global sourcing	Economies of scale at the global level
Problems and Constraints	Imperfection of the product and production process	Excessive growing demand	Cutthroat competition	Shrinking global demand
Logistics	Exportation from a advanced country	Partial local production in advanced countries	Global logistics	Exportation from a low cost operation area

Copyright © 2008, IGI Global. Copying or distributing in print or electronic forms without written permission of IGI Global is prohibited.

mechatronics. However, in the digital product world, the impact of physical distances becomes much weaker. The advancement of simulation technology omits some parts and the coordination required with analog technology. Furthermore, management technology guarantees smoother collaboration with new partners at the R&D phase. We should consider these factors for future challenges.

Conclusion

We discussed the convergence of facilities. In the late standardized stage, very little room is left for the differentiation of products. In addition, companies are forced to reduce the cost even further under the decreasing global demand at the end of the life cycle, by the saturation in the market, or the emergence of the alternative products. As a result, companies make the most of economies of scale in a low cost operation area for the final effort. Companies should concentrate their manufacturing facilities in a low cost operation area or withdraw from the market.

In the VCR case, in the latter part of the 1990s, the cost cutters and followers moved to low cost operation manufacturing systems, which were inflexible to further major changes. Leading companies also concentrated their facilities to the low cost operation areas by the early 2000s. Matsushita stopped manufacturing VCRs in domestic factories and started the production of DVD-RAM drivers there. JVC frugally reduced the number of employees and closed the subsidiaries that produced key parts. By the middle of the 2000s, very few companies continued to manufacture VCRs in Japan.

References

Bowersox, J. & Closs, D. J. (1996). *Logistical management*. New York: McGraw-Hill.

Cao D., Instlle M., Zepeda A., & Maino M. (1994). A global optimization approach to resource allocation problems with consideration of economies of scale. *Information and Decision Technologies*, *19*, 257-266.

Copyright © 2008, IGI Global. Copying or distributing in print or electronic forms without written permission of IGI Global is prohibited.

Cooper, J. C. (1993). Logistics strategies for global business. *International Journal of Physical Distribution & Logistics Management*, *23*(4), 12-23.

Dunning, J. H. (1988). *Explaining international production*. London: Unwin Hyman.

Higuchi, T. (2006). Industrial life cycle in the VCR industry. *Sakushin Policy Studies,* 6, 19-34.

Higuchi, T, Troutt, M. D., & Polin, B. A. (2004). Life cycle considerations for supply chain strategy. In C.K. Chan & H.W.J. Lee (Eds.), *Successful strategies in supply chain management* (pp. 67-89). Hershey, PA: Idea Group.

Itami, H. (1989). *Nihon no VTR sanngyou naze sekai wo seiha dekitanoka*. Tokyo: NTT Publishing.

JEITA (Japan Electronics and Information Technology Industries Association). (2005). *Minnseiyou dennsi kiki deta shyu*. Tokyo: JEITA.

NHK (Japan Broadcasting Corporation.). (2000). *Project X: Challengers 1*. Tokyo: Japan Broadcast Publishing.

Sato, M. (1999). *The story of a media industry*. Tokyo: Nikkei Business Publications.

Vernon, R. (1966). International investment and international trade in the product cycle. *Quarterly Journal of Economics*, *80*(1), 190-207.

Vernon, R. (1977). *Storm over the multinationals*. Boston: Harvard University Press.

Vernon, R. (1998). *In the hurricane's eye*. Boston: Harvard University Press.

Copyright © 2008, IGI Global. Copying or distributing in print or electronic forms without written permission of IGI Global is prohibited.

Section V

Conclusion

Chapter XVI

Application of Industrial Life Cycle Concept

The life cycle of products is a major concern in business all over the world. This Appendix proposes the *Industrial Life Cycle Concept* integrating the innovation, location, and marketing aspects. It is a very holistic concept in two regards. First, it deals with a whole category of products including multi-generation products. Second, it considers the whole supply chain from the consumers to suppliers. We demonstrated the industrial life cycle concept through a home-use VCR case study. The time frame is divided into five stages: *introduction stage* (before 1980); *early growth stage* (1980-1984); *late growth stage* (1985-1989); *maturity stage* (1990-2001); and *decline stage* (after 2002).

Copyright © 2008, IGI Global. Copying or distributing in print or electronic forms without written permission of IGI Global is prohibited.

Introduction

This chapter discusses an *industrial life cycle*, a life cycle of a category of products, which is related closely to business practice. The goals and appropriate strategies of companies differ at each stage of the life cycle. It often is said that the life cycles of products become shorter and shorter because new products emerge more frequently and obsolete existing products earlier than before (Bayus, 1994). Although this is true for some series of products and for individual products (Higuchi & Troutt, 2004; Loch, Stein, & Terwiesch, 1996; Millson, Raj, & Wilemon, 1992), it is questionable as a whole, in particular for categories of products. A category of products is composed of some series of products, which are incorporated some individual products. In the home-use VCR, a category of products includes all home-use VCR products and the series of products are the Monaural (standard) products, HiFi (High-Fidelity) products, S-VHS (Super Home Video System) products, and so on. An individual product is a certain product, such as HR-3300, HR-F6, and so on. Three different life cycle theories are integrated into a single *industrial life cycle* concept and focus on the life cycle of a particular category of product, the home-use VCR. This study hopefully will contribute to further research on the life cycle of products by proposing an ideal and latest case study for the life cycle by taking wide view including R&D, manufacturing, and marketing aspects. The home-use VCR industry is particularly appropriate for research on the life cycle of products for the following reasons:

1. The industry progresses with perfect timing. By the middle of 2000s, almost all companies stopped manufacturing home-use VCR products. And the majority of consumers started purchasing DVD or HDD recorders instead of home-use VCR. This industry is coming to an end.
2. The life of home-use VCR industry was about 30-40 years. This is long enough to divide into stages (in this book, five stages), yet not too long to be able to compare the earliest and final stages.
3. The home-use VCR is very lightweight. The constraint of its weight is so negligible that its logistics system can change easily according to the situation or need. Most VCR parts are so light and small that the manufacturers easily could change the location of factories and establish the effective and efficient global division of labor, for reducing the total costs including tariffs.

Copyright © 2008, IGI Global. Copying or distributing in print or electronic forms without written permission of IGI Global is prohibited.

4. Japanese companies have monopolized the industry. It is very helpful for the collection of data to be monopolized by a limited number of companies from the same country.

The goal of this chapter is to propose a framework for business strategy based on the industrial life cycle from the VCR case study. This book reinforces the product life cycle theory and the BCG Matrix (PPM: Product Portfolio Management). Product Life Cycle, a well-known life cycle theory around the world, focuses on the marketing and is divided into four stages based on the transition of sales. BCG Matrix is based on the life cycle theory and the experienced curve. BCG Matrix manages the investment of resources among the existing businesses and guides companies to focus on the competitive business by withdrawing from other businesses. Both use the transition of sales as an all-inclusive indicator to measure the stage of the life cycle of a category of products. However, the art of innovation and the shift of production facilities should also be taken into account equally. In the first half of the life cycle of products, constraints are within the industry, the design, performance, the manufacturing cost, and the quality of products. In general, the levels of early products are far behind desired products and consumers should wait for the emergence of proper products (Christensen, 1997). In the later half, the companies that enter into the industry, especially consumer electronics, at a later stage sell the product at a considerably lower price. To analyze from start to finish of a category of products, this book focuses attention equally on the product and process innovation, the location of factories, and marketing for an analysis of the whole life cycle of a category of the particular product, the home-use VCR.

Industrial Life Cycle Concept

The Industrial Life Cycle concept is developed in this chapter. There are many versions of life cycles applied to many industries (Debresson & Lampel, 1985). Three major life cycle theories are mainly considered in this book. They are concerned with the following topics: (1) the innovation (Abernathy, 1978; Abernathy, Clark, & Kantrow, 1983; Christensen, 1997; Cooper, 1998; Norman, 1998; Shintaku 1990; Utterback, 1994); (2) the location (Dunning, 1988; Vernon, 1966, 1977, 1998); and (3) the marketing (Bass, 1969; Kotler,

Copyright © 2008, IGI Global. Copying or distributing in print or electronic forms without written permission of IGI Global is prohibited.

1999: Lilien & Kotler, 1992: Sheth, 1971). Although these theories appear to be closely connected, they have not yet sufficiently been correlated. In this chapter, it is attempted to integrate them through creation of the Industrial Life Cycle.

The first life cycle, from the viewpoint of innovation, explains industry-wide product development (Abernathy, 1978; Abernathy et al., 1983). Based on their characteristics, innovations are classified into two parts, that is, product and process innovation. Usually, product innovations take place frequently in the early stages of the life cycle. They contribute to a considerable enhancement of performance of a product. They are sometimes also destructive because they obsolete old technologies and existing products by changing the basic design, components, production process, and usage of a product, such as cellular phones and portable games. In the middle stage, while product innovations gradually abate, process innovations frequently occur. Process innovations make the production system more efficient by sophisticating the product designs, components, and production processes. In the later half of the life cycle, a product innovation sometimes creates a brand-new category of products that makes an existing category of products obsolete. This phenomenon is called *de-maturity*.

The second life cycle, from the standpoint of location, is the Product Cycle Theory developed by Vernon (1966, 1977, 1998). He described the geographical location of manufacturing facilities according to the state of technology (the product and process innovations), international trade, and economies of scale. Vernon (1966) classified the time periods into three stages: new, maturing, and standardized.

The third life cycle from the viewpoint of the marketing is the well-known Product Life Cycle (Buzzell, 1966; Dhalla & Yuspeh, 1967; Kotler, 1999; Levitt, 1965; McIntyre, 1988). It proposed market strategies and explained the transition of profit based on the stages of the Product Life Cycle. The Product Life Cycle consists of four stages: introduction, growth, maturity, and decline. Its framework is applicable to the analysis of categories of products, series of products, and brands of products. This cycle is divided into four stages according to the transition in the amount of sales. The introduction stage starts with the launching of a new category of products in the market. In this stage, the amount of sales increases very slowly. In the second stage, the growth stage, the amount of sales increases rapidly. The third stage is the maturity stage. In this stage, the growth of sales declines because the market of latent customers has become saturated with the product. The final stage

Copyright © 2008, IGI Global. Copying or distributing in print or electronic forms without written permission of IGI Global is prohibited.

is the decline stage. In this stage, the amount of sales declines even further. However, the Product Life Cycle theory differs from other theories because it places much more emphasis on the individual and series rather than on the category of products. In addition, it does not pay an enough attention on product advancement.

To be consistent with other life cycles, the Product Life Cycle should be adjusted by focusing on the analysis of a category and by splitting the maturity stage. The time frame of the category of products is much longer than those of series and brands of products, and the significant impact of the price decline of products is very conspicuous for a category of products. The transition of the diffusion rate and sales volume of a product should be given much more emphasis than the transition in the amount of sales as they are much better cause indicators than the amount of sales in the case of the sharp price decline. Otherwise, in some cases, a period of steady growth of the diffusion rate might be classified the maturity stage or the decline stage.

In addition, the maturity stage is divided into three parts: growth maturity, stable maturity, and decline maturity (Kotler 1999). In growth maturity, the increase in the amount of sales begins to decrease. In stable maturity, the diffusion rate remains at almost the same level. In decline maturity, a continuing decline in the amount of sales can be noted. In this book, growth maturity is upgraded to the late growth stage. Otherwise, the growth stage is too short to explain the transition of a category of products. Figure 1 summarizes the adjusted Product Life Cycle.

Figure 1. Adjusted product life cycle

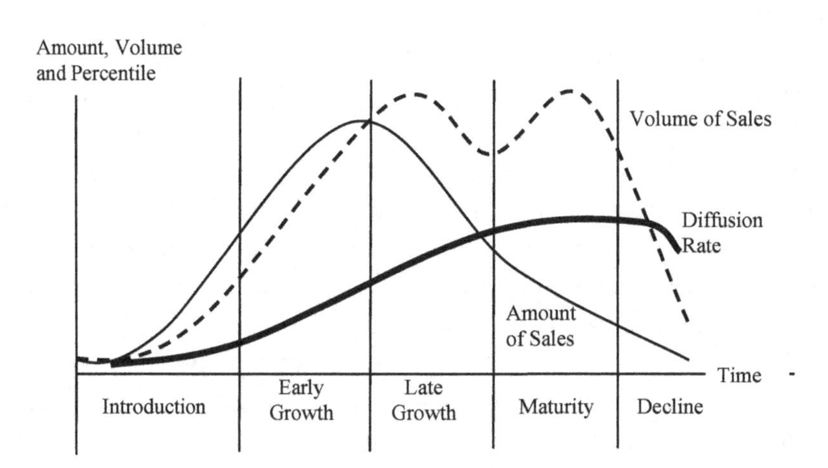

Copyright © 2008, IGI Global. Copying or distributing in print or electronic forms without written permission of IGI Global is prohibited.

Table 1. Integrated time frame of life cycles

Stage	I	II	III	IV	V
Innovation	Era of Product Innovation	Era of Process Innovation	Era of Productivity Dilemma		Decline Withdrawal
			Diversification of Products	Deadlock	
Location	New	Maturing	Standardized		
	Most Advanced Countries		Local Production Other Advanced Countries	Convergence Developing Countries	
Market	Introduction	Early Growth	Late growth	Maturity	

The previously mentioned three life cycles are coaxial and complementary to each other. All of them use a time frame as a horizontal axis and their research object is a category of products. The time frames of these life cycles can be integrated and divided into five stages as shown in Table 1. Separately, they explain the nature of innovation, rational location, and market trend. This type of integrated model is very helpful in understanding the movement of an industry. Hereafter, in this chapter, the time frame of Adjusted Product Life Cycle is used to indicate the stages.

Summary of Home-use VCR Case

VCRs use mechatronics, which combines mechanics and electronics. Technically, VCRs can be positioned as an extension of tape recorders. However, VCRs deal with an enormity of much more complex electrical signals in per second than do tape recorders. VCRs for broadcasting were developed by US companies in the 1950s. It took almost two decades to succeed in commercializing home-use VCRs by the Japanese consumer electronics companies JVC (Japan Victor), Sony, and Matsushita (Chan, 2004; Higuchi, 2006; Itami, 1989; Sato, 1999). In the mid-1970s, JVC who developed VHS (Video Home System) competed with Sony who developed Betamax for the de-facto standard and ultimately JVC prevailed. Afterward, Japanese companies exclusively developed and manufactured VCRs (Rosenbloom

Copyright © 2008, IGI Global. Copying or distributing in print or electronic forms without written permission of IGI Global is prohibited.

& Abernathy, 1982) and profited from the sale of these VCRs for a few decades (Dummer, 1983; Kohli, Lehman, & Pae, 1999). In this chapter, for the further analysis, the time frame is divided into five stages: introduction (before 1980), early growth (1980-1984), late growth (1985-1989), maturity (1990-2001), decline (after 2002).

Introduction Stage (Prior to 1980)

VCRs for broadcasting were developed by Ampex and RCA in the 1950s. The demand for VCRs for broadcasting was born in the US because broadcasting stations needed to take into account the time differences within the country. In the 1960s, VCRs became cheaper and smaller than before. As a result, they were available for utilization in business and education.

In the early 1970s, the essential functional and minimum requirements for the home-use VCR became clear. These were the long recording time (more than 120 minutes), the price (approximately less than US $1,000), and the appropriate level of size and weights for a person to move alone (approximately less than 50 pounds). Some companies, such as Ampex, Philips, and Matsushita, tried to enter and commercialize the home-use VCR market, but all of them failed; most of these products were discontinued before reaching the market.

In the mid-1970s, two prominent home-use VCRs appeared on the market. In 1975, Sony introduced the Betamax at 229,800 yen (about US $771 at US $1= 298 yen). It was the first product that satisfied the essential functional requirements of the home-use VCR market. However, in 1976 JVC introduced the HR-3300 model with the VHS format at 256,000 yen (about US $892 at US $1 = 287 yen) and thereby competed with Sony for the de facto standard. In the late 1970s, Japanese consumer electronics companies were divided into two groups, the Betamax group and the VHS group (NHK, 2000). The struggle for the de facto standard expanded around the world. The United States was considered to be the crucial battleground for the de facto standard. Finally, VHS overtook Betamax domestically and globally and established itself as the de-facto standard—the dominant design—in the home-use VCR market.

The other notable movement in the late 1970s was the advancement of the VHS format by JVC. To the VHS format, JVC added innovative new functions

Copyright © 2008, IGI Global. Copying or distributing in print or electronic forms without written permission of IGI Global is prohibited.

such as the double-speed playback mode in 1977 and the triple-recording-time mode in 1979, with keeping the compatibility among other versions. These functions were essential to the development of VHS and different from other complementary functions other JVC group contributed. JVC had been the leader of VHS group and they introduced new products with new functions. In March 1979, VCR ownership by percentage of Japanese households, reached 2.0 percent. This number is equal to 2.4 percent of the maximum 82.6 percent reached in 2004. The first 2.5 percent of users are called the innovators (Rogers, 1995) or innovative adopters (Kotler, 2005). Their evaluation is regarded as a critical factor in determining the future diffusion of the product (Moore, 1991; Rogers, 1995). The introduction stage of the home-use VCR is considered to have finished by 1980.

Early Growth Stage (1980-1984)

This stage is characterized by progress in process innovation and rapid market growth. After a specific product design was determined, process innovation progressed. It led to mass-production and promoted the price markdown and competition. Subsequent to VHS becoming the de facto standard in the home-use VCR, VHS group members began severe competition among themselves.

1980 was the first year that production volume exceeded the deci-peak (10 percent of its peak in 1986). The production volume increased rapidly because of the advancement in mass-production and the arrival of new competitors. By the early 1980s, 20 companies manufactured VCRs. As a result, the top three, JVC, Matsushita, and Sony, who collectively maintained 90 percent of the market share in 1976, lost about 30 percent of their market share.

Production costs decreased rapidly due to economies of scale, learning curve effects, and intense competition. The production cost per unit continued to show a period of tremendous decrease from the prior introduction stage. By the early 1980s, production cost had been decreased by half. On the other hand, a new series of VCRs with HiFi sound function appeared in 1983. This feature has since been widely adopted by other manufacturers. However, it had little impact on the production system because the process innovation was so advanced that the manufacturing system could localize the impact of change.

Copyright © 2008, IGI Global. Copying or distributing in print or electronic forms without written permission of IGI Global is prohibited.

Late Growth Stage (1985-1989)

During this stage, the VCR industry continued to grow. Domestic shipment met its first and second largest peak of 7,155,000 units in 1988. On the other hand, many negative signs, such as the productivity dilemma, the yen's appreciation, intense competition, and offshore production, came to appear during this stage.

In the VCR industry, the productivity dilemma wherein the opportunities for product and process innovations of the product diminish, took place during this stage. Although JVC succeeded in 1987 in developing the S-VHS, a new series of product with high quality images, it was not widely adopted because most consumers were satisfied with the former HiFi series. Product innovations then began to have less of an impact on performance and consumption. Furthermore, the reduction rate of domestic production cost per unit began to be decreased. During the period from1977 to 1989, the reduction rate kept more than 10 percent except in 1979 and 1988 (both 6 percent).

Local production of VCRs started in Europe because of the Common External Tariffs of the European Communities. All types of companies, leaders, followers, and cost-cutters, already manufactured overseas and most of them had started in Europe. Leading companies such as JVC and Matsushita were the first in the VCR industry to establish manufacturing facilities in Europe. JVC and Matsushita had already set up joint companies with European companies to assemble VCRs in West Germany prior to the Plaza Accord (in 1982 and 1983).

During the period from 1984 to 1989, the adoption rate of VCRs, that is, the average number of houses which posses a VCR or VCRs per 100 houses, in Japan increased from 18.7 percent to 63.7 percent and the growth met its peak of 45.0 percent, as shown in Figure 7. During the next five years, 1989-1994, the adoption rate increased from 63.7 percent to 72.5 percent, and its growth was only 8.8 percent. Taking into account the uneven growth of the adoption rate, it is reasonable to divide 1984-1989 and the period after 1990 into separate stages.

Maturity Stage (1990-2001)

The slowed growth of the adoption rate changed the consumer behavior. Most consumers purchased VCRs to replace old units. This meant consumers had

Copyright © 2008, IGI Global. Copying or distributing in print or electronic forms without written permission of IGI Global is prohibited.

enough experience and knowledge of the VCR to do purchase next one and their tastes had become more discriminating regarding, for example, a lower price, brand name, high quality, small size, light weight, design, convenient controls, and special features. Companies responded to customer needs, and as a result, the number of domestic shipments peaked for the second time in 1998.

By this time, most latent competitors had entered the market and had geared up for mass-production. As it is very costly for companies to deal with a variety of products, companies lost their merit from the economy of scale. There was little space left for differentiation in the market. In the maturity stage, the technological excellence lost its big advantage because others caught up and consumer tended to want the relatively cheap products rather than high quality and expensive products.

During this stage, Japan was an exporter and importer of the VCR. Japanese companies developed a global production system in which high-end products were manufactured in both Japan and in other Asian locations. Japan imported low-end and standard products from Asia. The volume of domestic shipments of VCR has steadily decreased since its peak in 1998 although for a few years it had maintained a higher level than that of the trough between the first and second peaks. However, the volume fell below the trough between the first and second peak in 2002 and decreased sharply after that. This indicates that plenty of customers stopped replacing VCRs and started purchasing alternative products such as DVD and HDD recorders. Because of this phenomenon, it is reasonable to represent this period of time as two distinct stages: 1990-2001 (maturity stage) and after 2002 (decline stage).

Decline Stage (After 2002)

The volume of domestic shipments of VCR decreased steadily beginning in 1998. During the period from 1998-2004, it diminished by a 25 percent. The yield of domestic production of 1,563,000 units in 2002, fell below 5 percent of its peak of 32,015,000 units in 1989. Finally, the domestic VCR production statistics nearly came to an end in 2004 when fewer than three companies still produces VCRs in Japan. The overseas production that accounted for 99 percent of the worldwide yield of VCRs in 2003 had also been decreasing sharply since 1996.

Copyright © 2008, IGI Global. Copying or distributing in print or electronic forms without written permission of IGI Global is prohibited.

According to the decline of the VCR industry, the number of companies that sold VCRs decreased by half to just ten. As a result, the share of the top three, JVC, Matsushita, and Sony, increased to about 60 percent in 2004. Most surviving manufacturers began to sell VCR products in combination with DVDs or HDDs or both. At this stage, these types of products bridged the gap between these machines for the alternation of generations.

The volume of the domestic shipment of DVD players and recorders caught up with that of the VCR in 2002. It reached 7,240,000 units, thereby exceeding the peak of the VCR of 7,156,000 in 1998. Although the DVD and HDD recorders are presently much more expensive than the VCR, they are capable of greater performance. The DVD is much smaller and more durable than the VHS cassette and it can record much longer and at higher quality than the VCR. In Japan, the sales volume of DVD software exceeded that of the VHS in 2001. In 2004, 100 million DVD units were sold; VHS software sold only 10 million units. The alternation of players and recorders became obvious after 2002.

Implications from the VCR Case Study

In the introduction stage of the life cycle, manufacturers need a huge amount of investment in R&D for the creation of a new category of products. Although it may be very profitable in the future, it involves great risks in the beginning. The main risk is whether or not a technology can become a standard in a new category of products. At first, it is necessary for companies to develop certain key technologies by themselves and to integrate them with others essential for the development of new products. The number and strength of the competitors, the degree of difficulty of the technology, and the range of essential technologies collectively determine the level of risk. The first step in reducing the risk is to decide whether or not to become involved in developing a new category of products. Those who are not involved in the development of the new category of products wait for the time when a standard technology or design becomes stable to avoid the fatal damages in case key technologies and the mainstream change. The second step is to decide whether or not to utilize a strategic alliance. A strategic alliance saves the cost and time of developing a new category of products (Lambe & Spekman, 1997). However, risk-sharing implies profit-sharing. The third step is

Copyright © 2008, IGI Global. Copying or distributing in print or electronic forms without written permission of IGI Global is prohibited.

deciding whether or not to go public with the confidential technologies. The publication of critical technologies is a way to establish a de facto standard within a relatively short time. On the other hand, it gives rivals a chance to catch up immediately.

In the early growth stage, the market grows rapidly. Companies still need a huge amount of investment, but their sales increase sharply. In this stage, the product and process innovations progress simultaneously. Leading companies advance the product and launch the new product series while production costs decline dramatically. At this point, they need to make two major decisions. The first is related to investment. They must decide how much to invest and which areas to emphasize, for example, R&D, the production system, advertising and marketing activities, or others. The second decision concerns the range of product offerings. The designated range of products offered correlates with available resources, such as the latest technologies, distribution channels and brand names. At the beginning of this stage, other companies also recognize the expansive possibilities of the category of products, how big the latent market is. New competitors can be divided into two groups: followers and cost-cutters. The differences between the former and the latter are in the timing of their entry, the scope of the products offered, and pricing. The followers try to maintain a wide selection of products with relatively lower prices than the leaders. In order to accomplish this, they sometimes acquire new series of products by means of OEM contracts with the leading companies. On the other hand, cost-cutters lag far behind the leaders and concentrate on a narrow selection with fairly low prices. They sometimes supply low-end and standard products to shoppers and other competitors. Different types of companies then start competing in the market. By the end of this stage, most companies with strategies appropriate to their resources begin to realize profits through growth in sales.

In the late growth stage, an industry as a whole becomes beneficial as many companies coexist because of market expansion. For a company, the goal of this stage is the maximization of throughput and a good balance between high volumes of yield and sales. The productivity dilemma evolves in a rapidly growing market as existing series of products become cheaper and of higher quality. The production system is then able to localize changes in design of a new series of the product. However, in this stage, a dramatic improvement in the performance of a new series is unlikely. Local production becomes feasible as the process innovation progresses and greatly influences global sales. In general, leading companies locate their factories near foreign markets

Copyright © 2008, IGI Global. Copying or distributing in print or electronic forms without written permission of IGI Global is prohibited.

prior to other companies doing so and thus achieve considerably higher local content ratios. On the other hand, followers and cost-cutters tend to locate their assembly plants near foreign markets and those for manufacturing parts in cheap labor zones.

In the maturity stage, the industry may enjoy profits because it has finished with huge investments in R&D and the production system and because the demand for the products maintains itself at a high level. By this stage, a good number of competitors will also have entered the market and will have advanced their products and production systems. As a result, supply far outpaces demand and the technological lead becomes meaningless. Furthermore, the decline of the market expansion makes competition much more severe to the point of becoming cutthroat. Marketing and foreign production are the keys to success in this stage. Marketing activities play an important role in developing a good brand image, understanding customer preferences, and inspiring them to purchase the product. Price also has become very important in this stage. Production systems have become so sophisticated that there is little room for improvements. For further price reductions, companies manufacture products with limited functions, predominantly in foreign countries with cheap labor zones. Some companies then manufacture and assemble almost all parts of their products in the cheap labor zones. Consolidation of factories starts in these areas of cheap labor. In the later part of this stage, some companies may decide to withdraw from the market because of the cutthroat competition and in expectation of the decline stage.

In the decline stage, surviving companies have choices concerning partial or complete withdrawal from production. In general, very few companies maintain their manufacturing facilities in highly developed countries, whereas some continue to operate only in developing countries, and others withdraw altogether from worldwide production and sales. These surviving companies are divided into two groups. One group has a well-known brand image and the other group is able to manufacture the products at considerably lower cost than their competitors or than before. However, sales eventually falls below the break-even point due to the obsolescence of the products and the emergence of alternative products; at this point a single company cannot achieve most of the economies of scale. For these reasons, the critical plan of action in this stage is strategic withdrawal.

Copyright © 2008, IGI Global. Copying or distributing in print or electronic forms without written permission of IGI Global is prohibited.

Conclusion

This book proposes the conceptual framework of the Industrial Life Cycle (Table 2). It is validated by means of a case study of the home-use VCR industry. The Industrial Life Cycle covers three aspects: product and process innovation, the location of factories, and marketing.

A company's strategies change as constraints differ over time. The constraint in the introduction stage is in product innovation. Product design and performance determine the level of potential adopters. In the Early and late growth stages, the constraints are in process innovation. The price decline expands the demand for the product, and mass-production ensures its quality and the quantity. After the maturity stage, consumers become a subject of intense interest. As they have enough experience with and knowledge of the products their tastes have become more discriminating regarding, for example, lower price, brand name, high quality, small size, light weight, design, convenient controls, and special features. In the decline stage, the volume of shipments decreases rapidly due to the obsolescence of the products and the emergence

Table 2. Summary of industrial life cycle

Stage	Introduction	Early Growth	Late Growth	Maturity	Decline
Product Innovation Type	Basic Design	Innovative Functions	Supplemental or Grand Functions		Compound Products
Process Innovation Type		Mass-Production	Flexible Manufacturing	Flexible Supply Chain	
Mission of Production Facility	Trial Manufacturing	Correspondence to Rapid Demand Growth	Correspondence to Wide Variety of Products and Global Markets	Low Cost Operation	Further Cost-cutting
No. of Manufacturers	Limited (Leaders)	Growing (Followers)	Growing (Cost-Cutters)	Peak	Decreasing
OEM Direction	From Leaders to Followers		From Cost-cutters to Leaders and Followers		
Consumer Type	Innovators	Early Adopters	Early Majority	Late Majority	Laggards
Repeat Purchase	Scarcely	Occasionally	Sometimes	Frequently	Mostly
Key Activity	R&D	Manufacturing (Domestic)	Manufacturing (Foreign)	Marketing	Marketing

Copyright © 2008, IGI Global. Copying or distributing in print or electronic forms without written permission of IGI Global is prohibited.

of alternative products that cause consumers to stop replacing the spent products and to purchase extra products for personal use.

In the interest of further research, this book deals with internal and external issues. The internal issues contain analyses at the individual company level and at the global sales level. Additional research on individual companies would help to corroborate the validity of the findings of this book, such as the classification of companies into three types, and the typal strategies, leader, follower and cost cutter type, according to life cycle stages For example, global sales had an enormous effect on the development of the VCR industry from its beginning. Sales in developed countries impacted competition for the de facto standard and growth in the early stages. Salcs in other countries became very important after the decline in sales in the developed countries. Therefore, an analysis of the transition of global sales might be very meaningful.

The external issues include the technological matters and the economic growth of the developing countries. The advancement of information technology and management techniques changes the supply chain structure. In addition to the growth of international transportation and the worldwide trend toward deregulation, the area of feasibility of a company or supply chain expands globally. Currently, many developing countries now have adequate technologies and techniques for the production of new products sooner than they previously did due to technology transfer and the production experience of former products. Furthermore, technical innovation, sophisticated design for fabrication and digital technologies enable overseas production at a much earlier stage than previously. As a result, some developing countries have greater purchasing powers than before and can no longer be ignored in the introduction stage of products. Therefore, we hope that this study will positively contribute to the investigation of these issues and trends.

References

Abernathy, William J. (1978). *The productivity dilemma*. Baltimore: The John Hopkins University Press.

Abernathy, W. J., Clark, K. B., & Kantrow, A. M. (1983). *Industrial renaissance*. New York: Basic Books, Inc.

Copyright © 2008, IGI Global. Copying or distributing in print or electronic forms without written permission of IGI Global is prohibited.

Barry, L. B. (1994). Are product life cycles really getting shorter? *Journal of Product Innovation Management, 11*(4), 300-308.

Bass, Frank M. (1969). A new product growth model for consumer durables. *Management Science, 15*, 215-227

Bowersox J., & Closs D., (1996). *Logistical management*. New York: Mc-Graw-Hill.

Buzzell, R. D. (1966). Competitive behavior and product life cycles. In J.S. Wright & J.L. Goldstucker (Eds.), *New ideas for successful marketing* (pp. 46-68). Chicago: American Marketing Association.

Christensen, C. M. (1997). *The innovator's dilemma*. Boston: Harvard Business School Press.

Cooper R. G. (2005). *Product leadership* (2nd ed.). New York: Basic Books.

Davis T. (1993). Effective supply chain management. *Sloan Management Review*, *34*(4), 35-46.

Debresson C., & Lampel, J. (1985). Beyond the life cycle: Organizational and technological design. I. An alternative perspective. *Journal of Product Innovation Management*, *2*(3), 170-187.

Dhalla, N. K., & Yuspeh, S. (1967). Forget the product life cycle concept! *Harvard Business Review*, *45*, 102-112.

Dummer, G.. W. W. (1983). *Electronics inventions and discoveries: Electronics from its earliest beginning to the present.* New York: Pergamon Press.

Dunning, J. H. (1988). *Explaining international production*. London: Unwin Hyman.

Fisher M. (1997). What is the right supply chain for your product? *Harvard Business Review*, *75*(2), 105-116.

Higuchi, T. (2006). Industrial life cycle in the VCR industry. *Sakushin Policy Studies,* 6, 19-34.

Higuchi, T., & Troutt, M.D. (2004). Dynamic simulation of the supply chain for a short life cycle product - Lessons from the *Tamagotchi* case. *Computers & Operations Research, 31*(7), 1097-1114.

Higuchi, T, Troutt, M. D., & Polin, B. A. (2004). Life cycle considerations for supply chain strategy. In C.K. Chan & H.W.J. Lee (Eds.), *Successful strategies in supply chain management* (pp. 67-89). Hershey, PA: Idea Group.

Copyright © 2008, IGI Global. Copying or distributing in print or electronic forms without written permission of IGI Global is prohibited.

Itami, H. (1989). *Nihon no VTR sanngyou naze sekai wo seiha dekitanoka.* Tokyo: NTT Publishing.

JEITA (Japan Electronics and Information Technology Industries Association). (2005). *Minnseiyou dennsi kiki deta shyu.* Tokyo: JEITA.

Kohli, R., Lehman, D. R, & Pae, J. (1999). Extent and impact of incubation time in new product diffusion. *Journal of Product Innovation Management, 16*(2), 134-144.

Kotler, P. (2005). *Marketing management* (12th ed.). Upper Saddle River, NJ: Prentice Hall.

Lambe C. J., & Spekman, R. E. (1997). Alliances, external technology acquisition, and discontinuous technological change. *Journal of Product Innovation Management, 14*(2), 102-116.

Levitt, T. (1965). Exploit the product life cycle. *Harvard Business Review, 43*, 81-94.

Lilien, G., Kotler, P., & Moorthy, K. (1992), *Marketing models*. Englewood Cliffs, NJ: Prentice-Hall.

Loch, C., Stein, L., &. Terwiesch, C. (1996). Measuring development performance in the electronics industry. *Journal of Product Innovation Management, 13*(1), 3-20.

McIntyre, S. H. (1988). Market adaptation as a process in the product life cycle of radical innovations and high technology products. *Journal of Product Innovation Management, 5*(2), 140-149.

Millson, M. R., Raj, S. P., & Wilemon, D. (1992). A survey of major aproaches for accelerating new product development. *Journal of Innovation Management, 9*(1), 53-69.

Moore, G.A. (1991). *Crossing the chasm*, New York: Harper Business.

NHK (Japan Broadcasting Corporation.). (2000). *Project X: Challengers 1.* Tokyo: Japan Broadcast Publishing.

Norman, D. A. (1998). *The invisible computer*. Cambridge, MA: MIT Press.

Rogers, E. M. (1995). *Diffusion of innovations* (4th ed.). New York: The Free Press.

Rosenbloom, R. S., & Abernathy, W. J. (1982). The climate for innovation in industry. *Research Policy, 11*(4), 209-225.

Copyright © 2008, IGI Global. Copying or distributing in print or electronic forms without written permission of IGI Global is prohibited.

Sato, M. (1999). *The story of a media industry.* Tokyo: Nikkei Business Publications.

Sheth, Jagdish N. (1971). Word of mouth in low-risk innovations. *Journal of Advertising Research, 11*(3), 15-18.

Shintaku, J. (1990). *Nihon kigyou no kyousou sennryaku* (*Competitive strategies of Japanese firms*). Tokyo: Yuhikaku.

Utterback, J. M. (1994). Mastering the dynamics of innovation: How companies can seize opportunities in the face of technological change. Boston: Harvard Business School Press.

Vernon, R. (1966). International investment and international trade in the product cycle. *Quarterly Journal of Economics, 80*(1), 190-207.

Vernon, R. (1977). *Storm over the multinationals.* Boston: Harvard University Press.

Vernon, R. (1998). *In the hurricane's eye.* Boston: Harvard University Press.

Copyright © 2008, IGI Global. Copying or distributing in print or electronic forms without written permission of IGI Global is prohibited.

About the Editors

Toru Higuchi is an associate professor of logistics in the Department of Policy Studies at Sakushin Gakuin University in Utsunomiya, Tochigi prefecture, Japan. He was an associate editor of *Journal of Public Utility Economics*. He received a MA from the School of Management at Gakushuin University in 1996. His publications have appeared in *Computers & Operations Research, Journal of Public Utility Economics,* and others. His current research interests include supply chain management, business logistics, global logistics, and transfer pricing.

Marvin D. Troutt is a professor in the Department of Management & Information Systems and in the Graduate School of Management at Kent State University, Kent, Ohio. He is a Fellow of the Decision Sciences Institute. He received a PhD in Mathematical Statistics from The University of Illinois at Chicago in 1975. His publications have appeared in *Management Science*, *Decision Sciences, Journal of the Operational Research Society*, *European Journal of Operational Research*, *Operations Research*, *Decision Support Systems*, *Naval Research Logistics*, *Statistics*, and others. He received the 2005 Distinguished Scholar Award at Kent State. He was formerly Director

Copyright © 2008, IGI Global. Copying or distributing in print or electronic forms without written permission of IGI Global is prohibited.

of the Center for Information Systems at Kent State and the Rehn Research Professor in Management at Southern Illinois University, Carbondale, Illinois. He served as Visiting Scholar in the Department of Applied Mathematics at the Hong Kong Polytechnic University during 1994-95. His current interests include supply chain management, strategy, and planning.

Copyright © 2008, IGI Global. Copying or distributing in print or electronic forms without written permission of IGI Global is prohibited.

Index

Copyright © 2008, IGI Global. Copying or distributing in print or electronic forms without written permission of IGI Global is prohibited.

Copyright © 2008, IGI Global. Copying or distributing in print or electronic forms without written permission of IGI Global is prohibited.

I

J

K

L

M

N

O

P

Copyright © 2008, IGI Global. Copying or distributing in print or electronic forms without written permission of IGI Global is prohibited.

R

S

T

U

Copyright © 2008, IGI Global. Copying or distributing in print or electronic forms without written permission of IGI Global is prohibited.

V

W

Copyright © 2008, IGI Global. Copying or distributing in print or electronic forms without written permission of IGI Global is prohibited.